Springer Undergraduate Mathematics Series

Springer
London
Berlin
Heidelberg
New York
Barcelona
Hong Kong
Milan
Paris
Singapore
Tokyo

Advisory Board

Other books in this series

Analytic Methods for Partial Differential Equations *G. Evans, J. Blackledge, P. Yardley*
Applied Geometry for Computer Graphics and CAD *D. Marsh*
Basic Linear Algebra *T.S. Blyth and E.F. Robertson*
Basic Stochastic Processes *Z. Brzeźniak and T. Zastawniak*
Elements of Logic via Numbers and Sets *D.L. Johnson*
Elementary Number Theory *G.A. Jones and J.M. Jones*
Groups, Rings and Fields *D.A.R. Wallace*
Hyperbolic Geometry *J.W. Anderson*
Introduction to Laplace Transforms and Fourier Series *P.P.G. Dyke*
Introduction to Ring Theory *P.M. Cohn*
Introductory Mathematics: Algebra and Analysis *G. Smith*
Introductory Mathematics: Applications and Methods *G.S. Marshall*
Measure, Integral and Probability *M. Capińksi and E. Kopp*
Multivariate Calculus and Geometry *S. Dineen*
Numerical Methods for Partial Differential Equations *G. Evans, J. Blackledge, P.Yardley*
Sets, Logic and Categories *P. Cameron*
Topologies and Uniformities *I.M. James*
Vector Calculus *P.C. Matthews*

Bryan P. Rynne and Martin A. Youngson

Linear Functional Analysis

Springer

Bryan Patrick Rynne, BSc, PhD
Martin Alexander Youngson, BSc, PhD

Department of Mathematics, Heriot-Watt University, Riccarton, Edinburgh EH14 4AS, UK

Cover illustration elements reproduced by kind permission of:

Aptech Systems, Inc., Publishers of the GAUSS Mathematical and Statistical System, 23804 S.E. Kent-Kangley Road, Maple Valley, WA 98038, USA. Tel: (206) 432 - 7855 Fax (206) 432 - 7832 email: info@aptech.com URL: www.aptech.com

American Statistical Association: Chance Vol 8 No 1, 1995 article by KS and KW Heiner 'Tree Rings of the Northern Shawangunks' page 32 fig 2

Springer-Verlag: Mathematica in Education and Research Vol 4 Issue 3 1995 article by Roman E Maeder, Beatrice Amrhein and Oliver Gloor 'Illustrated Mathematics: Visualization of Mathematical Objects' page 9 fig 11, originally published as a CD ROM 'Illustrated Mathematics' by TELOS: ISBN 0-387-14222-3, German edition by Birkhauser: ISBN 3-7643-5100-4.

Mathematica in Education and Research Vol 4 Issue 3 1995 article by Richard J Gaylord and Kazume Nishidate 'Traffic Engineering with Cellular Automata' page 35 fig 2. Mathematica in Education and Research Vol 5 Issue 2 1996 article by Michael Trott 'The Implicitization of a Trefoil Knot' page 14.

Mathematica in Education and Research Vol 5 Issue 2 1996 article by Lee de Cola 'Coins, Trees, Bars and Bells: Simulation of the Binomial Process page 19 fig 3. Mathematica in Education and Research Vol 5 Issue 2 1996 article by Richard Gaylord and Kazume Nishidate 'Contagious Spreading' page 33 fig 1. Mathematica in Education and Research Vol 5 Issue 2 1996 article by Joe Buhler and Stan Wagon 'Secrets of the Madelung Constant' page 50 fig 1.

ISBN 1-85233-257-3 Springer-Verlag London Berlin Heidelberg

British Library Cataloguing in Publication Data
Rynne, Bryan Patrick
Linear functional analysis. - (Springer undergraduate mathematics series)
1. Functional analysis
I. Title II. Youngson, Martin A.
515.7
ISBN 1852332573

Library of Congress Cataloging-in-Publication Data
Rynne, Bryan P., 1958-
Linear functional analysis / Bryan P. Rynne and Martin A. Youngson.
p. cm. -- (Springer undergraduate mathematics series)
Includes bibliographical references and index.
ISBN 1-85233-257-3 (alk. paper)
1. Functional analysis. I. Youngson, Martin A., 1953- II. Title. III. Series.
QA320.R96 2000
515'.7—dc21 99-057793

Printed in Great Britain

Typesetting: Camera ready by the authors
Printed and bound at the Athenæum Press Ltd., Gateshead, Tyne & Wear
12/3830-543210 Printed on acid-free paper SPIN 10661476

Preface

This book provides an introduction to the ideas and methods of linear functional analysis at a level appropriate to the final year of an undergraduate course at a British university. The prerequisites for reading it are a standard undergraduate knowledge of linear algebra and real analysis (including the theory of metric spaces).

Part of the development of functional analysis can be traced to attempts to find a suitable framework in which to discuss differential and integral equations. Often, the appropriate setting turned out to be a vector space of real or complex-valued functions defined on some set. In general, such a vector space is infinite-dimensional. This leads to difficulties in that, although many of the elementary properties of finite-dimensional vector spaces hold in infinite-dimensional vector spaces, many others do not. For example, in general infinite-dimensional vector spaces there is no framework in which to make sense of analytic concepts such as convergence and continuity. Nevertheless, on the spaces of most interest to us there is often a *norm* (which extends the idea of the length of a vector to a somewhat more abstract setting). Since a norm on a vector space gives rise to a metric on the space, it is now possible to do analysis in the space. As real or complex-valued functions are often called *functionals*, the term *functional analysis* came to be used for this topic.

We now briefly outline the contents of the book. In Chapter 1 we present (for reference and to establish our notation) various basic ideas that will be required throughout the book. Specifically, we discuss the results from elementary linear algebra and the basic theory of metric spaces which will be required in later chapters. We also give a brief summary of the elements of the theory of Lebesgue measure and integration. Of the three topics discussed in this introductory chapter, Lebesgue integration is undoubtedly the most technically difficult and the one which the prospective reader is least likely to have encountered before. Unfortunately, many of the most important spaces which arise in functional analysis are spaces of integrable functions, and it is necessary to use the Lebesgue integral to overcome various drawbacks of the elementary Riemann integral, commonly taught in real analysis courses. The reader who has not met Lebesgue integration before can still read this book by accepting that an integration process exists which coincides with the Riemann integral when this is defined, but extends to a larger class of functions, and which has the properties described in Section 1.3.

In Chapter 2 we discuss the fundamental concept of functional analysis, the *normed vector space*. As mentioned above, a norm on a vector space is simply an extension of the idea of the length of a vector to a rather more abstract setting. Via an associated metric, the norm is behind all the discussion of convergence and continuity in vector spaces in this book. The basic properties of normed vector spaces are described in this chapter. In particular we begin the study of *Banach spaces* which are complete normed vector spaces.

In finite dimensions, in addition to the length of a vector, the angle between two vectors is also used. To extend this to more abstract spaces the idea of an *inner product* on a vector space is introduced. This generalizes the well-known "dot product" used in $\mathbb{R}^3$. *Inner product spaces*, which are vector spaces possessing an inner product, are discussed in Chapter 3. Every inner product space is a normed space and, as in Chapter 2, we find that the most important inner product spaces are those which are complete. These are called *Hilbert spaces*.

Having discussed various properties of infinite-dimensional vector spaces the next step is to look at linear transformations between these spaces. The most important linear transformations are the continuous ones, and these will be called linear *operators*. In Chapter 4 we describe general properties of linear operators between normed vector spaces. Any linear transformation between finite-dimensional vector spaces is automatically continuous so questions relating to the continuity of the transformation can safely be ignored (and usually are). However, when the spaces are infinite-dimensional this is certainly not the case and the continuity, or otherwise, of individual linear transformations must be studied much more carefully. In addition, we investigate the algebraic properties of the entire set of all linear operators between given normed vector spaces. Finally, for some linear operators it is possible to define an inverse operator, and we conclude the chapter with a characterization of the invertibility of an operator.

In Chapter 5 we specialize the discussion of linear operators to those acting between Hilbert spaces. The additional structure of these spaces means that we can define the *adjoint* of a linear operator and hence the particular classes of *self-adjoint* and *unitary* operators which have especially nice properties. We also introduce the *spectrum* of linear operators acting on a Hilbert space. The spectrum of a linear operator is a generalization of the set of eigenvalues of a matrix, which is a well-known concept in finite-dimensional linear algebra.

As we have already remarked, there are many significant differences between the theory of linear transformations in finite and infinite dimensions. However, for the class of *compact* operators a great deal of the theory carries over from finite to infinite dimensions. The properties of these particular operators are discussed in detail in Chapter 6. In particular, we study compact, self-adjoint operators on Hilbert spaces, and their spectral properties.

Finally, in Chapter 7, we use the results of the preceding chapters to discuss two extremely important areas of application of functional analysis, namely integral and differential equations. As we remarked above, the study of these equations was one of the main early influences and driving forces in the growth and development of functional analysis, so it forms a fitting conclusion to this book. Nowadays, functional analysis has applications to a vast range of areas of mathematics, but limitations of space preclude us from studying further applications.

A large number of exercises are included, together with complete solutions. Many of these exercises are relatively simple, while some are considerably less so. It is strongly recommended that the student should at least attempt most of these questions before looking at the solution. This is the only way to really learn any branch of mathematics.

There is a World Wide Web site associated with this book, at the URL

`http://www.ma.hw.ac.uk/~bryan/lfa_book.html`

This site contains links to sites on the Web which give some historical background to the subject, and also contains a list of any significant misprints which have been found in the book.

Contents

1
Preliminaries

To a certain extent, functional analysis can be described as infinite-dimensional linear algebra combined with analysis, in order to make sense of ideas such as convergence and continuity. It follows that we will make extensive use of these topics, so in this chapter we briefly recall and summarize the various ideas and results which are fundamental to the study of functional analysis. We must stress, however, that this chapter only attempts to review the material and establish the notation that we will use. We do not attempt to motivate or explain this material, and any reader who has not met this material before should consult an appropriate textbook for more information.

Section 1.1 discusses the basic results from linear algebra that will be required. The material here is quite standard although, in general, we do not make any assumptions about finite-dimensionality except where absolutely necessary. Section 1.2 discusses the basic ideas of metric spaces. Metric spaces are the appropriate setting in which to discuss basic analytical concepts such as convergence of sequences and continuity of functions. The ideas are a natural extension of the usual concepts encountered in elementary courses in real analysis. In general metric spaces no other structure is imposed beyond a metric, which is used to discuss convergence and continuity. However, the essence of functional analysis is to consider vector spaces (usually infinite-dimensional) which are metric spaces and to study the interplay between the algebraic and metric structures of the spaces, especially when the spaces are complete metric spaces.

An important technical tool in the theory is Lebesgue integration. This is because many important vector spaces consist of sets of integrable functions.

In order for desirable metric space properties, such as completeness, to hold in these spaces it is necessary to use Lebesgue integration rather than the simpler Riemann integration usually discussed in elementary analysis courses. Of the three topics discussed in this introductory chapter, Lebesgue integration is undoubtedly the most technically difficult and the one which the prospective student is most likely to have not encountered before. In this book we will avoid arcane details of Lebesgue integration theory. The basic results which will be needed are described in Section 1.3, without any proofs. For the reader who is unfamiliar with Lebesgue integration and who does not wish to embark on a prolonged study of the theory, it will be sufficient to accept that an integration process exists which applies to a broad class of "Lebesgue integrable" functions and has the properties described in Section 1.3, most of which are obvious extensions of corresponding properties of the Riemann integral.

1.1 Linear Algebra

Throughout the book we have attempted to use standard mathematical notation wherever possible. Basic to the discussion is standard set theoretic notation and terminology. Details are given in, for example, [7]. Sets will usually be denoted by upper case letters, X, $Y, \ldots$, while elements of sets will be denoted by lower case letters, x, $y, \ldots$. The usual set theoretic operations will be used: $\in$, $\subset$, $\cup$, $\cap$, $\varnothing$ (the empty set), $\times$ (Cartesian product), $X \setminus Y = \{x \in X : x \notin Y\}$.

The following standard sets will be used,

$$\begin{aligned}\mathbb{R} &= \text{the set of real numbers,}\\ \mathbb{C} &= \text{the set of complex numbers,}\\ \mathbb{N} &= \text{the set of positive integers } \{1, 2, \ldots\}.\end{aligned}$$

The sets $\mathbb{R}$ and $\mathbb{C}$ are algebraic *fields*. These fields will occur throughout the discussion, associated with vector spaces. Sometimes it will be crucial to be specific about which of these fields we are using, but when the discussion applies equally well to both we will simply use the notation $\mathbb{F}$ to denote either set. The real and imaginary parts of a complex number z will be denoted by $\Re e\, z$ and $\Im m\, z$ respectively, while the complex conjugate will be denoted $\overline{z}$.

For any $k \in \mathbb{N}$ we let $\mathbb{F}^k = \mathbb{F} \times \ldots \times \mathbb{F}$ (the Cartesian product of k copies of $\mathbb{F}$). Elements of $\mathbb{F}^k$ will written in the form $x = (x_1, \ldots, x_k)$, $x_j \in \mathbb{F}$, $j = 1, \ldots, k$.

For any two sets X and Y, the notation $f : X \to Y$ will denote a function or mapping from X into Y. The set X is the *domain* of f and Y is the *codomain*. If $A \subset X$ and $B \subset Y$, we use the notation

$$f(A) = \{f(x) : x \in A\}, \quad f^{-1}(B) = \{x \in X : f(x) \in B\}.$$

If Z is a third set and $g : Y \to Z$ is another function, we define the *composition* of g and f, written $g \circ f : X \to Z$, by

$$(g \circ f)(x) = g(f(x)),$$

for all $x \in X$.

We now discuss the essential concepts from linear algebra that will be required in later chapters. Most of this section should be familiar, at least in the finite-dimensional setting, see for example [1] or [5], or any other book on linear algebra. However, we do not assume here that any spaces are finite-dimensional unless explicitly stated.

Definition 1.1

A *vector space over* $\mathbb{F}$ is a non-empty set V together with two functions, one from $V \times V$ to V and the other from $\mathbb{F} \times V$ to V, denoted by $x + y$ and αx respectively, for all x, $y \in V$ and $\alpha \in \mathbb{F}$, such that, for any α, $\beta \in \mathbb{F}$ and any x, y, $z \in V$,

(a) $x + y = y + x, \quad x + (y + z) = (x + y) + z;$

(b) there exists a unique $0 \in V$ (independent of x) such that $x + 0 = x$;

(c) there exists a unique $-x \in V$ such that $x + (-x) = 0$;

(d) $1x = x, \quad \alpha(\beta x) = (\alpha\beta)x;$

(e) $\alpha(x + y) = \alpha x + \alpha y, \quad (\alpha + \beta)x = \alpha x + \beta x.$

If $\mathbb{F} = \mathbb{R}$ (respectively, $\mathbb{F} = \mathbb{C}$) then V is a *real* (respectively, *complex*) vector space. Elements of $\mathbb{F}$ are called *scalars*, while elements of V are called *vectors*. The operation $x + y$ is called *vector addition*, while the operation αx is called *scalar multiplication*.

Many results about vector spaces apply equally well to both real or complex vector spaces so if the type of a space is not stated explicitly then the space may be of either type, and we will simply use the term "vector space".

If V is a vector space with $x \in V$ and A, $B \subset V$, we use the notation,

$$\begin{aligned} x + A &= \{x + a : a \in A\}, \\ A + B &= \{a + b : a \in A,\, b \in B\}. \end{aligned}$$

Definition 1.2

Let V be a vector space. A non-empty set $U \subset V$ is a *linear subspace* of V if U is itself a vector space (with the same vector addition and scalar multiplication as in V). This is equivalent to the condition that

$$\alpha x + \beta y \in U, \quad \text{for all } \alpha,\, \beta \in \mathbb{F} \text{ and } \ x,\, y \in U$$

(which is called the *subspace test*).

Note that, by definition, vector spaces and linear subspaces are always non-empty, while general subsets of vector spaces which are not subspaces may be empty. In particular, it is a consequence of the vector space definitions that $0x = \mathbf{0}$, for all $x \in V$ (here, 0 is the scalar zero and $\mathbf{0}$ is the vector zero; except where it is important to distinguish between the two, both will be denoted by 0). Hence, any linear subspace $U \subset V$ must contain at least the vector $\mathbf{0}$, and the set $\{\mathbf{0}\} \subset V$ is a linear subspace.

Definition 1.3

Let V be a vector space, let $\mathbf{v} = \{v_1, \ldots, v_k\} \subset V$, $k \geq 1$, be a finite set and let $A \subset V$ be an arbitrary non-empty set.

(a) A *linear combination* of the elements of $\mathbf{v}$ is any vector of the form

$$x = \alpha_1 v_1 + \ldots + \alpha_k v_k \in V, \tag{1.1}$$

for any set of scalars $\alpha_1, \ldots, \alpha_k$.

(b) $\mathbf{v}$ is *linearly independent* if the following implication holds:

$$\alpha_1 v_1 + \ldots + \alpha_k v_k = 0 \quad \Rightarrow \quad \alpha_1 = \ldots = \alpha_k = 0.$$

(c) A is *linearly independent* if every finite subset of A is linearly independent. If A is not linearly independent then it is *linearly dependent.*

(d) The *span* of A (denoted $\operatorname{Sp} A$) is the set of all linear combinations of all finite subsets of A. This set is a linear subspace of V. Equivalently, $\operatorname{Sp} A$ is the intersection of the set of all linear subspace of V which contain A. Thus, $\operatorname{Sp} A$ is the smallest linear subspace of V containing A (in the sense that if $A \subset B \subset V$ and B is a linear subspace of V then $\operatorname{Sp} A \subset B$).

(e) If $\mathbf{v}$ is linearly independent and $\operatorname{Sp} \mathbf{v} = V$, then $\mathbf{v}$ is called a *basis* for V. It can be shown that if V has such a (finite) basis then all bases of V have the same number of elements. If this number is k then V is said to be *k-dimensional* (or, more generally, *finite-dimensional*), and we write $\dim V = k$. If V does not have such a finite basis it is said to be *infinite-dimensional.*

(f) If $\mathbf{v}$ is a basis for V then any $x \in V$ can be written as a linear combination of the form (1.1), with a unique set of scalars α_j, $j = 1, \ldots, k$. These scalars (which clearly depend on x) are called the *components* of x with respect to the basis $\mathbf{v}$.

(g) The set $\mathbb{F}^k$ is a vector space over $\mathbb{F}$ and the set of vectors

$$\widehat{e}_1 = (1, 0, 0, \ldots, 0),\ \widehat{e}_2 = (0, 1, 0, \ldots, 0), \ldots,\ \widehat{e}_k = (0, 0, 0, \ldots, 1),$$

is a basis for $\mathbb{F}^k$. This notation will be used throughout the book, and this basis will be called the *standard basis* for $\mathbb{F}^k$.

We will sometimes write $\dim V = \infty$ when V is infinite-dimensional. However, this is simply a notational convenience, and should not be interpreted in the sense of ordinal or cardinal numbers (see [7]). In a sense, infinite-dimensional spaces can vary greatly in their "size"; see Section 3.4 for some further discussion of this.

Definition 1.4

Let V, W be vector spaces over $\mathbb{F}$. The Cartesian product $V \times W$ is a vector space with the following vector space operations. For any $\alpha \in \mathbb{F}$ and any $(x_j, y_j) \in V \times W$, $j = 1, 2$, let

$$(x_1, y_1) + (x_2, y_2) = (x_1 + x_2, y_1 + x_2), \quad \alpha(x_1, y_1) = (\alpha x_1, \alpha y_1)$$

(using the corresponding vector space operations in V and W).

We next describe a typical construction of vector spaces consisting of functions defined on some underlying set.

Definition 1.5

Let S be a set and let V be a vector space over $\mathbb{F}$. We denote the set of functions $f : S \to V$ by $F(S, V)$. For any $\alpha \in \mathbb{F}$ and any $f, g \in F(S, V)$, we define functions $f + g$ and αf in $F(S, V)$ by

$$(f + g)(x) = f(x) + g(x), \quad (\alpha f)(x) = \alpha f(x),$$

for all $x \in S$ (using the vector space operations in V). With these definitions the set $F(S, V)$ is a vector space over $\mathbb{F}$.

Many of the vector spaces used in functional analysis are of the above form. From now on, whenever functions are added or multiplied by a scalar the process will be as in Definition 1.5. We note that the zero element in $F(S,V)$ is the function which is identically equal to the zero element of V. Also, if S contains infinitely many elements and $V \neq \{0\}$ then $F(S,V)$ is infinite-dimensional.

Example 1.6

If S is the set of integers $\{1,\ldots,k\}$ then the set $F(S,\mathbb{F})$ can be identified with the space $\mathbb{F}^k$ (by identifying an element $x \in \mathbb{F}^k$ with the function $f \in F(S,\mathbb{F})$ defined by $f(j) = x_j$, $1 \leq j \leq k$).

Often, in the construction in Definition 1.5, the set S is a vector space and only a subset of the set of all functions $f : S \to V$ is considered. In particular, in this case the most important functions to consider are those which preserve the linear structure of the vector spaces in the sense of the following definition.

Definition 1.7

Let V, W be vector spaces over the same scalar field $\mathbb{F}$. A function $T : V \to W$ is called a *linear transformation* (or *mapping*) if, for all α, $\beta \in \mathbb{F}$ and x, $y \in V$,

$$T(\alpha x + \beta y) = \alpha T(x) + \beta T(y).$$

The set of all linear transformations $T : V \to W$ will be denoted by $L(V,W)$. With the scalar multiplication and vector addition defined in Definition 1.5 the set $L(V,W)$ is a vector space (it is a subspace of $F(V,W)$). When $V = W$ we abbreviate $L(V,V)$ to $L(V)$.

A particularly simple linear transformation in $L(V)$ is defined by $I_V(x) = x$, for $x \in V$. This is called the *identity* transformation on V (usually we use the notation I if it is clear what space the transformation is acting on).

Whenever we discuss linear transformations $T : V \to W$ it will be taken for granted, without being explicitly stated, that V and W are vector spaces over the same scalar field.

Since linear transformations are functions they can be composed (when they act on appropriate spaces). The following lemmas are immediate consequences of the definition of a linear transformation.

Lemma 1.8

Let V, W, X be vector spaces and $T \in L(V,W)$, $S \in L(W,X)$. Then the composition $S \circ T \in L(V,X)$.

Lemma 1.9

Let V be a vector space, R, S, $T \in L(V)$, and $\alpha \in \mathbb{F}$. Then:

(a) $R \circ (S \circ T) = (R \circ S) \circ T$;

(b) $R \circ (S + T) = R \circ S + R \circ T$;

(c) $(S + T) \circ R = S \circ R + T \circ R$;

(d) $I_V \circ T = T \circ I_V = T$;

(e) $(\alpha S) \circ T = \alpha(S \circ T) = S \circ (\alpha T)$.

These properties also hold for linear transformations between different spaces when the relevant operations make sense (for instance, (a) holds when $T \in L(V, W)$, $S \in L(W, X)$ and $R \in L(X, Y)$, for vector spaces V, W, X, Y).

The five properties listed in Lemma 1.9 are exactly the extra axioms which a vector space must satisfy in order to be an *algebra*. Since this is the only example of an algebra which we will meet in this book we will not discuss this further, but we note that an algebra is both a vector space and a *ring*.

When dealing with the composition of linear transformations S, T it is conventional to omit the symbol $\circ$ and simply write ST. Eventually we will do this, but for now we retain the symbol $\circ$.

The following lemma gives some further elementary properties of linear transformations.

Lemma 1.10

Let V, W be vector spaces and $T \in L(V, W)$.

(a) $T(0) = 0$.

(b) If U is a linear subspace of V then the set $T(U)$ is a linear subspace of W and $\dim T(U) \leq \dim U$ (as either finite numbers or ∞).

(c) If U is a linear subspace of W then the set $\{x \in V : T(x) \in U\}$ is a linear subspace of V.

We can now state some standard terminology.

Definition 1.11

Let V, W be vector spaces and $T \in L(V, W)$.

(a) The *image* of T (often known as the *range* of T) is the subspace $\operatorname{Im} T = T(V)$; the *rank* of T is the number $r(T) = \dim(\operatorname{Im} T)$.

(b) The *kernel* of T (often known as the *null-space* of T) is the subspace $\operatorname{Ker} T = \{x \in V : T(x) = 0\}$; the *nullity* of T is the number $n(T) = \dim(\operatorname{Ker} T)$.

The rank and nullity, $r(T)$, $n(T)$, may have the value ∞.

(c) T has *finite rank* if $r(T)$ is finite.

(d) T is *one-to-one* if, for any $y \in W$, the equation $T(x) = y$ has at most one solution x.

(e) T is *onto* if, for any $y \in W$, the equation $T(x) = y$ has at least one solution x.

(f) T is *bijective* if, for any $y \in W$, the equation $T(x) = y$ has exactly one solution x (that is, T is both one-to-one and onto).

Lemma 1.12

Let V, W be vector spaces and $T \in L(V, W)$.

(a) T is one-to-one if and only if the equation $T(x) = 0$ has only the solution $x = 0$. This is equivalent to $\operatorname{Ker} T = \{0\}$ or $n(T) = 0$.

(b) T is onto if and only if $\operatorname{Im} T = W$. If $\dim W$ is finite this is equivalent to $r(T) = \dim W$.

(c) $T \in L(V, W)$ is bijective if and only if there exists a transformation $S \in L(W, V)$ which is bijective and $S \circ T = I_V$ and $T \circ S = I_W$.

If V is k-dimensional then

$$n(T) + r(T) = k$$

(in particular, $r(T)$ is necessarily finite, irrespective of whether W is finite-dimensional). Hence, if W is also k-dimensional then T is bijective if and only if $n(T) = 0$.

Related to the bijectivity, or otherwise, of a transformation T from a space to itself we have the following definition, which will be extremely important later.

Definition 1.13

Let V be a vector space and $T \in L(V)$. A scalar $\lambda \in \mathbb{F}$ is an *eigenvalue* of T if the equation $T(x) = \lambda x$ has a non-zero solution $x \in V$, and any such non-zero solution is an *eigenvector*. The subspace $\operatorname{Ker}(T - \lambda I) \subset V$ is called the *eigenspace* (corresponding to λ) and the *multiplicity* of λ is the number $m_\lambda = n(T - \lambda I)$.

Lemma 1.14

Let V be a vector space and let $T \in L(V)$. Let $\{\lambda_1, \ldots, \lambda_k\}$ be a set of distinct eigenvalues of T, and for each $1 \le j \le k$ let x_j be an eigenvector corresponding to λ_j. Then the set $\{x_1, \ldots, x_k\}$ is linearly independent.

Linear transformations between finite-dimensional vector spaces are closely related to matrices. For any integers m, $n \ge 1$, let $M_{mn}(\mathbb{F})$ denote the set of all $m \times n$ matrices with entries in $\mathbb{F}$. A typical element of $M_{mn}(\mathbb{F})$ will be written as $[a_{ij}]$ (or $[a_{ij}]_{mn}$ if it is necessary to emphasize the size of the matrix). Any matrix $C = [c_{ij}] \in M_{mn}(\mathbb{F})$ induces a linear transformation $T_C \in L(\mathbb{F}^n, \mathbb{F}^m)$ as follows: for any $x \in \mathbb{F}^n$, let $T_C x = y$, where $y \in \mathbb{F}^m$ is defined by

$$y_i = \sum_{j=1}^{n} c_{ij} x_j, \quad 1 \le i \le m.$$

Note that, if we were to regard x and y as column vectors then this transformation corresponds to standard matrix multiplication. However, mainly for notational purposes, it is generally convenient to regard elements of $\mathbb{F}^k$ as row vectors. This convention will always be used below, except when we specifically wish to perform computations of matrices acting on vectors, and then it will be convenient to use column vector notation.

On the other hand, if U and V are finite-dimensional vector spaces then a linear transformation $T \in L(U, V)$ can be represented in terms of a matrix. To fix our notation we briefly review this representation (see Chapter 7 of [1] for further details). Suppose that U is n-dimensional and V is m-dimensional, with bases $\mathbf{u} = \{u_1, \ldots, u_n\}$ and $\mathbf{v} = \{v_1, \ldots, v_m\}$ respectively. Any vector $a \in U$ can be represented in the form

$$a = \sum_{j=1}^{n} \alpha_j u_j,$$

for a unique collection of scalars $\alpha_1, \ldots, \alpha_n$. We define the column matrix

$$A = \begin{bmatrix} \alpha_1 \\ \vdots \\ \alpha_n \end{bmatrix} \in M_{n1}(\mathbb{F}).$$

The mapping $a \to A$ is a bijective linear transformation from U to $M_{n1}(\mathbb{F})$, that is, there is a one-to-one correspondence between vectors $a \in U$ and column matrices $A \in M_{n1}(\mathbb{F})$. There is a similar correspondence between vectors $b \in V$ and column matrices $B \in M_{m1}(\mathbb{F})$. Now, for any $1 \le j \le n$, the vector Tu_j has the representation

$$Tu_j = \sum_{i=1}^{m} \tau_{ij} v_i,$$

for appropriate (unique) scalars τ_{ij}, $i = 1, \ldots, m$, $j = 1, \ldots, n$. It follows from this, by linearity, that for any $a \in U$,

$$Ta = \sum_{j=1}^{n} \alpha_j T u_j = \sum_{i=1}^{m} \Big(\sum_{j=1}^{n} \tau_{ij} \alpha_j \Big) v_i,$$

and hence, letting M_T denote the matrix $[\tau_{ij}]$, the matrix representation B of the vector $b = Ta$ has the form

$$B = M_T A$$

(using standard matrix multiplication here). We will write $M_{\mathbf{v}}^{\mathbf{u}}(T) = M_T$ for the above matrix representation of T with respect to the bases $\mathbf{u}$, $\mathbf{v}$ (the notation emphasizes that the representation $M_{\mathbf{v}}^{\mathbf{u}}(T)$ depends on $\mathbf{u}$ and $\mathbf{v}$ as well as on T). This matrix representation has the following properties.

Theorem 1.15

(a) The mapping $T \to M_{\mathbf{v}}^{\mathbf{u}}(T)$ is a bijective linear transformation from $L(U, V)$ to $M_{mn}(\mathbb{F})$, that is, if S, $T \in L(U, V)$ and $\alpha \in \mathbb{F}$, then

$$M_{\mathbf{v}}^{\mathbf{u}}(\alpha T) = \alpha M_{\mathbf{v}}^{\mathbf{u}}(T), \quad M_{\mathbf{v}}^{\mathbf{u}}(S + T) = M_{\mathbf{v}}^{\mathbf{u}}(S) + M_{\mathbf{v}}^{\mathbf{u}}(T).$$

(b) If $T \in L(U, V)$, $S \in L(V, W)$ (where W is l-dimensional, with basis $\mathbf{w}$) then (again using standard matrix multiplication here).

$$M_{\mathbf{w}}^{\mathbf{u}}(ST) = M_{\mathbf{w}}^{\mathbf{v}}(S) M_{\mathbf{v}}^{\mathbf{u}}(T)$$

When $U = \mathbb{F}^n$ and $V = \mathbb{F}^m$, the above constructions of an operator from a matrix and a matrix from an operator are consistent, in the following sense.

Lemma 1.16

Let $\mathbf{u}$ be the standard basis of $\mathbb{F}^n$ and let $\mathbf{v}$ be the standard basis of $\mathbb{F}^m$. Let $C \in M_{mn}(\mathbb{F})$ and $T \in L(\mathbb{F}^n, \mathbb{F}^m)$. Then,

(a) $M_{\mathbf{v}}^{\mathbf{u}}(T_C) = C$;

(b) $T_B = T$ (where $B = M_{\mathbf{v}}^{\mathbf{u}}(T)$).

The above results show that although matrices and linear transformations between finite-dimensional vector spaces are logically distinct concepts, there is a close connection between them, and much of their theory is, in essence, identical. This will be particularly apparent in Chapter 5.

1.2 Metric Spaces

Metric spaces are an abstract setting in which to discuss basic analytical concepts such as convergence of sequences and continuity of functions. The fundamental tool required for this is a distance function or "metric". The following definition lists the crucial properties of a distance function.

Definition 1.17

A *metric* on a set M is a function $d : M \times M \to \mathbb{R}$ with the following properties. For all x, y, $z \in M$,

(a) $d(x, y) \geq 0$;

(b) $d(x, y) = 0 \iff x = y$;

(c) $d(x, y) = d(y, x)$;

(d) $d(x, z) = d(x, y) + d(y, z)$ (the *triangle inequality*).

If d is a metric on M, then the pair (M, d) is called a *metric space.*

Any given set M can have more than one metric (unless it consists of a single point). If it is clear what the metric is we often simply write "the metric space M", rather than "the metric space (M, d)".

Example 1.18

For any integer $k \geq 1$, the function $d : \mathbb{F}^k \times \mathbb{F}^k \to \mathbb{R}$ defined by

$$d(x, y) = \Big(\sum_{j=1}^{k} |x_j - y_j|^2 \Big)^{1/2}, \tag{1.2}$$

is a metric on the set $\mathbb{F}^k$. This metric will be called the *standard metric* on $\mathbb{F}^k$ and, unless otherwise stated, $\mathbb{F}^k$ will be regarded as a metric space with this metric. An example of an alternative metric on $\mathbb{F}^k$ is the function $d_1 : \mathbb{F}^k \times \mathbb{F}^k \to \mathbb{R}$ defined by

$$d_1(x, y) = \sum_{j=1}^{k} |x_j - y_j|.$$

Definition 1.19

Let (M, d) be a metric space and let N be a subset of M. Define $d_N : N \times N \to \mathbb{R}$ by $d_N(x, y) = d(x, y)$ for all x, $y \in N$ (that is, d_N is the restriction of d to the subset N). Then d_N is a metric on N, called the metric *induced* on N by d.

Whenever we consider subsets of metric spaces we will regard them as metric spaces with the induced metric unless otherwise stated. Furthermore, we will normally retain the original notation for the metric (that is, in the notation of Definition 1.19, we will simply write d rather than d_N for the induced metric).

The idea of a sequence should be familiar from elementary analysis courses. Formally, a *sequence* in a set X is often defined to be a function $s : \mathbb{N} \to X$, see [7]. Alternatively, a sequence in X can be regarded as an ordered list of elements of X, written in the form $(x_1, x_2, \ldots)$, with $x_n = s(n)$ for each $n \in \mathbb{N}$. The function definition is logically precise, but the idea of an ordered list of elements is often more helpful intuitively. For brevity, we will usually use the notation $\{x_n\}$ for a sequence (or $\{x_n\}_{n=1}^{\infty}$ if it is necessary to emphasize which variable is indexing the sequence). Strictly speaking, this notation could lead to confusion between a sequence $\{x_n\}$ (which has an ordering) and the corresponding set $\{x_n : n \in \mathbb{N}\}$ (which has no ordering) or the set consisting of the single element x_n, but this is rarely a problem in practice. The notation (x_n) is also sometimes used for sequences, but when we come to look at functions of sequences this notation can make it seem that an individual element x_n is the argument of the function, so this notation will not be used in this book. A *subsequence* of $\{x_n\}$ is a sequence of the form $\{x_{n(r)}\}_{r=1}^{\infty}$, where $n(r) \in \mathbb{N}$ is a strictly increasing function of $r \in \mathbb{N}$.

Example 1.20

Using the definition of a sequence as a function from $\mathbb{N}$ to $\mathbb{F}$ we see that the space $F(\mathbb{N}, \mathbb{F})$ (see Definition 1.5) can be identified with the space consisting of all sequences in $\mathbb{F}$ (compare this with Example 1.6).

A fundamental concept in analysis is the convergence of sequences. Convergence of sequences in metric spaces will now be defined (we say that $\{x_n\}$ is a sequence in a metric space (M, d) if it is a sequence in the set M).

Definition 1.21

A sequence $\{x_n\}$ in a metric space (M, d) *converges to* $x \in M$ (or the sequence $\{x_n\}$ is *convergent*) if, for every $\epsilon > 0$, there exists $N \in \mathbb{N}$ such that

$$d(x, x_n) < \epsilon, \quad \text{for all } n \geq N.$$

As usual, we write $\lim_{n\to\infty} x_n = x$ or $x_n \to x$. A sequence $\{x_n\}$ in (M, d) is a *Cauchy sequence* if, for every $\epsilon > 0$, there exists $N \in \mathbb{N}$ such that

$$d(x_m, x_n) < \epsilon, \quad \text{for all } m,\, n \geq N.$$

Note that, using the idea of convergence of a sequence of real numbers, the above definitions are equivalent to

$$d(x, x_n) \to 0, \quad \text{as } n \to \infty; \qquad d(x_m, x_n) \to 0, \quad \text{as } m,\, n \to \infty,$$

respectively.

Theorem 1.22

Suppose that $\{x_n\}$ is a convergent sequence in a metric space (M, d). Then:

(a) the limit $x = \lim_{n \to \infty} x_n$ is unique;

(b) any subsequence of $\{x_n\}$ also converges to x;

(c) $\{x_n\}$ is a Cauchy sequence.

Various properties and classes of subsets of metric spaces can now be defined in terms of the metric.

Definition 1.23

Let (M, d) be a metric space. For any point $x \in M$ and any number $r > 0$, the set

$$B_x(r) = \{y \in M : d(x, y) < r\}$$

is said to be the *open ball* with centre x and radius r. A ball $B_x(1)$ with radius 1 is said to be an open *unit ball.*

Definition 1.24

Let (M, d) be a metric space and let $A \subset M$.

(a) A is *bounded* if there is a number $b > 0$ such that $d(x, y) < b$ for all x, $y \in A$.

(b) A is *open* if, for each point $x \in A$, there is an $\epsilon > 0$ such that $B_x(\epsilon) \subset A$.

(c) A is *closed* if the set $M \setminus A$ is open.

(d) A point $x \in A$ is a *closure point* of A if, for every $\epsilon > 0$, there is a point $y \in A$ with $d(x, y) < \epsilon$ (equivalently, if there exists a sequence $\{y_n\} \subset A$ such that $y_n \to x$).

(e) The *closure* of A, denoted by $\overline{A}$ or A^-, is the set of all closure points of A.

(f) A is *dense* (in M) if $\overline{A} = M$.

We will use both notations $\overline{A}$ or A^- for the closure of A; the notation $\overline{A}$ is very common, but the notation A^- can be useful to avoid possible confusion with complex conjugate or for denoting the closure of a set given by a complicated formula.

Note that if $x \in A$ then x is, by definition, a closure point of A (in the definition, put $y = x$ for every $\epsilon > 0$), so $A \subset \overline{A}$. It need not be true that $A = \overline{A}$, as we see in the following theorem.

Theorem 1.25

Let (M, d) be a metric space and let $A \subset M$.

(a) $\overline{A}$ is closed and is equal to the intersection of the collection of all closed subsets of M which contain A (thus, $\overline{A}$ is the smallest closed set containing A).

(b) A is closed if and only if $A = \overline{A}$.

(c) A is closed if and only if, whenever $\{x_n\}$ is a sequence in A which converges to an element $x \in M$, then $x \in A$.

(d) $x \in \overline{A}$ if and only if $\inf\{d(x, y) : y \in A\} = 0$.

(e) For any $x \in M$ and $r > 0$, the "open ball" $B_x(r)$ in Definition 1.23 is actually open according to Definition 1.24, and its closure is the set

$$\overline{B_x(r)} = \{y \in M : d(x, y) \leq r\}$$

(this set will be called the *closed ball* with centre x and radius r; the set $\overline{B_x(r)}$ will be called a *closed unit ball*).

(f) A is dense if and only if, for any element $x \in M$ and any number $\epsilon > 0$, there exists a point $y \in A$ with $d(x, y) < \epsilon$ (equivalently, for any element $x \in M$ there exists a sequence $\{y_n\} \subset A$ such that $y_n \to x$).

Heuristically, part (f) of Theorem 1.25 says that a set A is dense in M if any element $x \in M$ can be "approximated arbitrarily closely by elements of A", in the sense of the metric on M.

Recall that if (M, d) is a metric space and $N \subset M$, then (N, d) is also a metric space (Definition 1.19). Thus all the above concepts also make sense in (N, d). However, it is important to be clear which of these spaces is being used in any given context as the results may be different.

Example 1.26

Let $M = \mathbb{R}$, with the standard metric, and let $N = (0,1] \subset M$. If $A = (0,1)$ then the closure of A in N is equal to N (so A is dense in N), but the closure of A in M is $[0,1]$.

In real analysis the idea of a "continuous function" can be defined in terms of the standard metric on $\mathbb{R}$, so the idea can also be extended to the general metric space setting.

Definition 1.27

Let (M, d_M) and (N, d_N) be metric spaces and let $f : M \to N$ be a function.

(a) f is *continuous at* a point $x \in M$ if, for every $\epsilon > 0$, there exists $\delta > 0$ such that, for $y \in M$,

$$d_M(x, y) < \delta \Rightarrow d_N(f(x), f(y)) < \epsilon.$$

(b) f is *continuous* (on M) if it is continuous at each point of M.

(c) f is *uniformly continuous* (on M) if, for every $\epsilon > 0$, there exists $\delta > 0$ such that, for all x, $y \in M$,

$$d_M(x, y) < \delta \Rightarrow d_N(f(x), f(y)) < \epsilon$$

(that is, the number δ can be chosen independently of x, $y \in M$).

As in real analysis the idea of continuity is closely connected with sequences, and also with open and closed sets.

Theorem 1.28

Suppose that (M, d_M), (N, d_N), are metric spaces and that $f : M \to N$. Then:

(a) f is continuous at $x \in M$ if and only if, for every sequence $\{x_n\}$ in (M, d_M) with $x_n \to x$, the sequence $\{f(x_n)\}$ in (N, d_N) satisfies $f(x_n) \to f(x)$;

(b) f is continuous on M if and only if either of the following conditions holds:

(i) for any open set $A \subset N$, the set $f^{-1}(A) \subset M$ is open;

(ii) for any closed set $A \subset N$, the set $f^{-1}(A) \subset M$ is closed.

Corollary 1.29

Suppose that (M, d_M), (N, d_N), are metric spaces, A is a dense subset of M and f, $g : M \to N$ are continuous functions with the property that $f(x) = g(x)$ for all $x \in A$. Then $f = g$ (that is, $f(x) = g(x)$ for all $x \in M$).

In many spaces the converse of part (c) of Theorem 1.22 also holds, that is, any Cauchy sequence is convergent. This property is so useful that metric spaces with this property have a special name.

Definition 1.30

A metric space (M, d) is *complete* if every Cauchy sequence in (M, d) is convergent. A set $A \subset M$ is *complete* (in (M, d)) if every Cauchy sequence lying in A converges to an element of A.

Theorem 1.31

For each $k \geq 1$, the space $\mathbb{F}^k$ with the standard metric is complete.

Unfortunately, not all metric spaces are complete. However, most of the spaces that we consider in this book are complete (partly because most of the important spaces are complete, partly because we choose to avoid incomplete spaces). The following theorem is a deep result in metric space theory which is used frequently in functional analysis. It is one of the reasons why complete metric spaces are so important in functional analysis.

Theorem 1.32 (Baire's category theorem)

If (M, d) is a complete metric space and $M = \bigcup_{j=1}^{\infty} A_j$, where each $A_j \subset M$, $j = 1, 2, \ldots$, is closed, then at least one of the sets A_j contains an open ball.

The concept of compactness should be familiar from real analysis, and it can also be extended to the metric space setting.

Definition 1.33

Let (M, d) be a metric space. A set $A \subset M$ is *compact* if every sequence $\{x_n\}$ in A contains a subsequence that converges to an element of A. A set $A \subset M$ is *relatively compact* if the closure $\overline{A}$ is compact. If the set M itself is compact then we say that (M, d) is a *compact metric space*.

Theorem 1.34

Suppose that (M, d) is a metric space and $A \subset M$. Then:

(a) if A is complete then it is closed;

(b) if M is complete then A is complete if and only if it is closed;

(c) if A is compact then it is closed and bounded;

(d) (Bolzano–Weierstrass theorem) every closed, bounded subset of $\mathbb{F}^k$ is compact.

Compactness is a very powerful and useful property but it is often difficult to prove. Thus the above Bolzano–Weierstrass theorem, which gives a very simple criterion for the compactness of a set in $\mathbb{F}^k$, is very useful.

Metric spaces consisting of sets of functions defined on other spaces are extremely common in the study of functional analysis (in fact, such spaces could almost be regarded as the defining characteristic of the subject, and the inspiration for the term "functional analysis"). We now describe one of the most important such constructions, involving continuous $\mathbb{F}$-valued functions.

Theorem 1.35

Suppose that (M, d) is a compact metric space and $f : M \to \mathbb{F}$ is continuous. Then there exists a constant $b > 0$ such that $|f(x)| \leq b$ for all $x \in M$ (we say that f is *bounded*). Hence the numbers $\sup\{f(x) : x \in M\}$ and $\inf\{f(x) : x \in M\}$, exist and are finite. Furthermore, there exist points x_s, $x_i \in M$ such that $f(x_s) = \sup\{f(x) : x \in M\}$, $f(x_i) = \inf\{f(x) : x \in M\}$.

Definition 1.36

Let (M, d) be a compact metric space. The set of continuous functions $f : M \to \mathbb{F}$ will be denoted by $C_{\mathbb{F}}(M)$. We define a metric on $C_{\mathbb{F}}(M)$ by

$$d(f, g) = \sup\{|f(x) - g(x)| : x \in M\}$$

(it can easily be verified that for any f, $g \in C_{\mathbb{F}}(M)$, the function $f - g$ is continuous so $d(f, g)$ is well-defined, by Theorem 1.35, and that d is a metric on $C_{\mathbb{F}}(M)$). This metric will be called the *uniform metric* and, unless otherwise stated, $C_{\mathbb{F}}(M)$ will always be assumed to have this metric.

Notation

Most properties of the space $C_{\mathbb{F}}(M)$ hold equally well in both the real and complex cases so, except where it is important to distinguish between these cases, we will omit the subscript and simply write $C(M)$. A similar convention will be adopted below for other spaces with both real and complex versions. Also, when M is a bounded interval $[a,b] \subset \mathbb{R}$ we write $C[a,b]$.

Definition 1.37

Suppose that (M,d) is a compact metric space and $\{f_n\}$ is a sequence in $C(M)$, and let $f : M \to \mathbb{F}$ be a function.

(a) $\{f_n\}$ converges *pointwise* to f if $|f_n(x) - f(x)| \to 0$ for all $x \in M$.

(b) $\{f_n\}$ converges *uniformly* to f if $\sup\{|f_n(x) - f(x)| : x \in M\} \to 0$.

Clearly, uniform convergence implies pointwise convergence, but not conversely. Also, uniform convergence implies that $f \in C(M)$, but this is not true for pointwise convergence, see [3]. Thus uniform convergence provides a more useful definition of convergence in $C(M)$ than pointwise convergence, and in fact uniform convergence corresponds to convergence with respect to the uniform metric on $C(M)$ in Definition 1.36.

We now have the following crucial theorem concerning $C(M)$.

Theorem 1.38 (Theorem 3.45 in [3])

The metric space $C(M)$ is complete.

Now suppose that M is a compact subset of $\mathbb{R}$. We denote the set of real-valued polynomials by $\mathcal{P}_{\mathbb{R}}$. Any polynomial $p \in \mathcal{P}_{\mathbb{R}}$ can be regarded as a function $p : M \to \mathbb{R}$, simply by restricting the domain of p to M, so in this sense $\mathcal{P}_{\mathbb{R}} \subset C_{\mathbb{R}}(M)$. The following theorem is a special case of Theorem 5.62 in [3]. It will enable us to approximate general continuous functions on M by "simple" functions (that is, polynomials).

Theorem 1.39 (The Stone–Weierstrass theorem)

For any compact set $M \subset \mathbb{R}$, the set $\mathcal{P}_{\mathbb{R}}$ is dense in $C_{\mathbb{R}}(M)$.

Another way of stating this result is that, if f is a real-valued, continuous function on M, then there exists a sequence of polynomials $\{p_n\}$ which converge uniformly to f on the set M. The polynomials $\{p_n\}$ are, of course, defined on the whole of $\mathbb{R}$, but their behaviour outside the set M is irrelevant here.

General metric spaces can, in a sense, be very large and pathological. Our final definitions in this section describe a concept that will be used below to restrict the "size" of some of the spaces we encounter and avoid certain types of "bad" behaviour (see, in particular, Section 3.4).

Definition 1.40

A set X is *countable* if it contains either a finite number of elements or infinitely many elements and can be written in the form $X = \{x_n : n \in \mathbb{N}\}$; in the latter case X is said to be *countably infinite.*

A metric space (M, d) is *separable* if it contains a countable, dense subset. The empty set is regarded as separable.

Heuristically, a countably infinite set is the same size as the set of positive integers $\mathbb{N}$, and so it can be regarded as a "small" infinite set (see [7], or any book on set theory, for a further discussion of this point). Also, the set $X = \{x_n : n \in \mathbb{N}\}$ in the above definition can be regarded as a sequence, in the obvious manner, and in fact we will often construct countable sets in the form of a sequence (usually by induction). Thus, in the above definition of separability, "countable subset" may be replaced by "sequence".

Example 1.41

The space $\mathbb{R}$ is separable since the set of rational numbers is countably infinite (see [7]) and dense in $\mathbb{R}$.

A separable space is one for which all its elements can be approximated arbitrary closely by elements of a "small" (countably infinite) set. In practice, one would also hope that the elements of this approximating set have "nicer" properties than general elements of the space. For instance, Theorem 1.39 shows that general elements of the space $C_{\mathbb{R}}(M)$ can be approximated by polynomials. Although the set $\mathcal{P}_{\mathbb{R}}$ is not countable, we will deduce from this, in Section 3.5, that the space $C_{\mathbb{R}}(M)$ is separable.

Theorem 1.42

Suppose that (M, d) is a metric space and $A \subset M$.

(a) If A is compact then it is separable.

(b) If A is separable and $B \subset A$ then B is separable.

1.3 Lebesgue Integration

In Definition 1.36 we introduced the uniform metric d on the space $C[a,b]$, and noted that the metric space $(C[a,b],d)$ is complete. However, there are other useful metrics on $C[a,b]$ which are defined by integrals. Let $\int_a^b f(x)\,dx$ denote the usual Riemann integral of a function $f \in C[a,b]$, as defined in elementary real analysis courses (see, for instance, [2]). Then, for $1 \le p < \infty$, the function $d_p : C[a,b] \times C[a,b] \to \mathbb{R}$ defined by

$$d_p(f,g) = \Big(\int_a^b |f(x) - g(x)|^p\,dx \Big)^{1/p}$$

is a metric on $C[a,b]$. Unfortunately, the metric space $(C[a,b],d_p)$ is not complete. Since complete metric spaces are much more important than incomplete ones (for instance, Baire's category theorem holds in complete metric spaces), this situation is undesirable. Heuristically, the problem is that Cauchy sequences in $(C[a,b],d_p)$ "converge" to functions which do not belong to $C[a,b]$, and may not be Riemann integrable. This is a weakness of the Riemann integral and the remedy is to replace the Riemann integral with the so-called "Lebesgue integral" (the Riemann integral also suffers from other mathematical drawbacks, which are described in Section 1.2 of [4]). The Lebesgue integral is a more powerful theory of integration which enables a wider class of functions to be integrated. A very readable summary (without proofs) of the main elements of this theory, as required for the study of functional analysis, is given in Chapter 2 of [8], while the theory (with proofs) is described in detail in [4], at a similar mathematical level to that of this book. There are of course many other more advanced books on the subject.

Here we will give a very short summary of the results that will be required in this book. For the reader who does not wish to embark on a prolonged study of the theory of Lebesgue integration it will be sufficient to accept that an integration process exists which applies to a broad class of "Lebesgue integrable" functions and has the properties described below, most of which are obvious extensions of corresponding properties of the Riemann integral (see, in particular, Theorem 1.51 and the results following this theorem). The problem of the lack of completeness of the space $(C[a,b],d_p)$ (with $1 \le p < \infty$) is resolved by introducing the metric space $L^p[a,b]$ in Definition 1.53. This space contains $C[a,b]$ and has the same metric d_p. Furthermore, Theorems 1.60 and 1.61 show that $C[a,b]$ is dense in $L^p[a,b]$ and that $L^p[a,b]$ is complete (in terms of abstract metric space theory, $L^p[a,b]$ is said to be the "completion" of the space $(C[a,b],d_p)$, but we will not use the concept of completion any further here). In addition, the spaces ℓ^p, $1 \le p \le \infty$, introduced in Example 1.56 will be very useful, but these spaces can be understood without any knowledge of Lebesgue integration.

Fundamental to the theory of Lebesgue integration is the idea of the size or "measure" of a set. For instance, for any bounded interval $I = [a, b]$, $a \leq b$, we say that I has *length* $\ell(I) = b - a$. To define the Lebesgue integral on $\mathbb{R}$ it is necessary to be able to assign a "length" or "measure" to a much broader class of sets than simply intervals. Unfortunately, in general it is not possible to construct a useful definition of "measure" which applies to every subset of $\mathbb{R}$ (there is a rather subtle set-theoretic point here which we will ignore, see [4] for further discussion). Because of this it is necessary to restrict attention to certain classes of subsets of $\mathbb{R}$ with useful properties. An obvious first step is to define the "measure" of a finite union of disjoint intervals to be simply the sum of their lengths. However, to define an integral which behaves well with regard to taking limits, it is also desirable to be able to deal with countable unions of sets and to be able to calculate their measure in terms of the measures of the individual sets. Furthermore, in the general theory it is just as easy to replace $\mathbb{R}$ with an abstract set X, and consider abstract measures on classes of subsets of X. These considerations inspire Definitions 1.44 and 1.43 below.

Before giving these definitions we note that many sets must clearly be regarded as having infinite measure (e.g., $\mathbb{R}$ has infinite length). To deal with this it is convenient to introduce the following extended sets of real numbers, $\overline{\mathbb{R}} = \mathbb{R} \cup \{-\infty, \infty\}$ and $\overline{\mathbb{R}}^+ = [0, \infty) \cup \{\infty\}$. The standard algebraic operations are defined in the obvious manner (e.g., $\infty + \infty = \infty$), except that the products $0.\infty$, $0.(-\infty)$ are defined to be zero, while the operations $\infty - \infty$ and ∞/∞ are forbidden.

Definition 1.43

A *σ-algebra* (also known as a *σ-field*) is a class Σ of subsets of a set X with the properties:

(a) $\emptyset, X \in \Sigma$;

(b) $S \in \Sigma \Rightarrow X \setminus S \in \Sigma$;

(c) $S_n \in \Sigma,\ n = 1, 2, \ldots, \Rightarrow \displaystyle\bigcup_{n=1}^{\infty} S_n \in \Sigma$.

A set $S \in \Sigma$ is said to be *measurable*.

Definition 1.44

Let X be a set and let Σ be a σ-algebra of subsets of X. A function $\mu : \Sigma \to \overline{\mathbb{R}}^+$ is a *measure* if it has the properties:

(a) $\mu(\emptyset) = 0$;

(b) μ is *countably additive*, that is, if $S_j \in \Sigma$, $j = 1, 2, \ldots$, are pairwise disjoint sets then

$$\mu\Big(\bigcup_{j=1}^{\infty} S_j\Big) = \sum_{j=1}^{\infty} \mu(S_j).$$

The triple (X, Σ, μ) is called a *measure space.*

In many applications of measure theory, sets whose measure is zero are regarded as "negligible" and it is useful to have some terminology for such sets.

Definition 1.45

Let (X, Σ, μ) be a measure space. A set $S \in \Sigma$ with $\mu(S) = 0$ is said to have *measure zero* (or is a *null set*). A given property $P(x)$ of points $x \in X$ is said to hold *almost everywhere* if the set $\{x : P(x) \text{ is false}\}$ has measure zero; alternatively, the property P is said to hold for *almost every* $x \in X$. The abbreviation a.e. will denote either of these terms.

Example 1.46 (Counting measure)

Let $X = \mathbb{N}$, let Σ_c be the class of all subsets of $\mathbb{N}$ and, for any $S \subset \mathbb{N}$, define $\mu_c(S)$ to be the number of elements of S. Then Σ_c is a σ-algebra and μ_c is a measure on Σ_c. This measure is called *counting measure* on $\mathbb{N}$. The only set of measure zero in this measure space is the empty set.

Example 1.47 (Lebesgue measure)

There is a σ-algebra Σ_L in $\mathbb{R}$ and a measure μ_L on Σ_L such that any finite interval $I = [a, b] \in \Sigma_L$ and $\mu_L(I) = \ell(I)$. The sets of measure zero in this space are exactly those sets A with the following property: for any $\epsilon > 0$ there exists a sequence of intervals $I_j \subset \mathbb{R}$, $j = 1, 2, \ldots$, such that

$$A \subset \bigcup_{j=1}^{\infty} I_j \quad \text{and} \quad \sum_{j=1}^{\infty} \ell(I_j) < \epsilon.$$

This measure is called *Lebesgue measure* and the sets in Σ_L are said to be *Lebesgue measurable.*

The above two properties of Lebesgue measure uniquely characterize this measure space. There are other measures, for instance *Borel measure*, which coincide with Lebesgue measure on intervals, but for which some sets with the above "covering" property are not actually measurable. This distinction is a rather technical feature of measure theory which will be unimportant here. It is discussed in more detail in [4].

For any integer $k > 1$ there is a σ-algebra Σ_L in $\mathbb{R}^k$ and Lebesgue measure μ_L on Σ_L, which extends the idea of the area of a set when $k = 2$, the volume when $k = 3$, and the "generalized volume" when $k \geq 4$.

Now suppose that we have a fixed measure space (X, Σ, μ). In the following sequence of definitions we describe the construction of the integral of appropriate functions $f : X \to \overline{\mathbb{R}}$. Proofs and further details are given in Chapter 4 of [4].

For any subset $A \subset X$ the *characteristic function* $\chi_A : X \to \mathbb{R}$ of A is defined by

$$\chi_A(x) = \begin{cases} 1, & \text{if } x \in A, \\ 0, & \text{if } x \notin A. \end{cases}$$

A function $\phi : X \to \mathbb{R}$ is *simple* if it has the form

$$\phi = \sum_{j=1}^{k} \alpha_j \chi_{S_j},$$

for some $k \in \mathbb{N}$, where $\alpha_j \in \mathbb{R}$ and $S_j \in \Sigma$, $j = 1, \ldots, k$. If ϕ is non-negative and simple then the *integral* of ϕ (over X, with respect to μ) is defined to be

$$\int_X \phi \, d\mu = \sum_{j=1}^{k} \alpha_j \mu(S_j)$$

(we allow $\mu(S_j) = \infty$ here, and we use the algebraic rules in $\overline{\mathbb{R}}^+$ mentioned above to evaluate the right-hand side – since ϕ is non-negative we do not encounter any differences of the form $\infty - \infty$). The value of the integral may be ∞. A function $f : X \to \overline{\mathbb{R}}$ is said to be *measurable* if, for every $\alpha \in \mathbb{R}$,

$$\{x \in X : f(x) > \alpha\} \in \Sigma.$$

If f is measurable then the functions $|f| : X \to \overline{\mathbb{R}}$ and $f^{\pm} : X \to \overline{\mathbb{R}}$, defined by

$$|f|(x) = |f(x)|, \quad f^{\pm}(x) = \max\{\pm f(x), 0\},$$

are measurable. If f is non-negative and measurable then the *integral* of f is defined to be

$$\int_X f \, d\mu = \sup \left\{ \int_X \phi \, d\mu : \phi \text{ is simple and } 0 \leq \phi \leq f \right\}.$$

If f is measurable and $\int_X |f|\, d\mu < \infty$ then f is said to be *integrable* and the *integral* of f is defined to be

$$\int_X f\, d\mu = \int_X f^+\, d\mu - \int_X f^-\, d\mu$$

(it can be shown that if f is integrable then each of the terms on the right of this definition are finite, so there is no problem with a difference such as $\infty - \infty$ arising in this definition). A complex-valued function f is said to be *integrable* if the real and imaginary parts $\Re\mathrm{e}\, f$ and $\Im\mathrm{m}\, f$ (these functions are defined in the obvious manner) are integrable and the *integral* of f is defined to be

$$\int_X f\, d\mu = \int_X \Re\mathrm{e}\, f\, d\mu + i \int_X \Im\mathrm{m}\, f\, d\mu.$$

Finally, suppose that $S \in \Sigma$ and f is a real or complex-valued function on S. Extend f to a function $\tilde{f}$ on X by defining $\tilde{f}(x) = 0$ for $x \notin S$. Then f is said to be integrable (over S) if $\tilde{f}$ is integrable (over X), and we define

$$\int_S f\, d\mu = \int_X \tilde{f}\, d\mu.$$

The set of $\mathbb{F}$-valued integrable functions on X will be denoted by $\mathcal{L}^1_{\mathbb{F}}(X)$ (or $\mathcal{L}^1_{\mathbb{F}}(S)$ for functions on $S \in \Sigma$). The reason for the superscript 1 will be seen below. As we remarked when discussing the space $C(M)$, except where it is important to distinguish between the real and complex versions of a space, we will omit the subscript indicating the type, so in this instance we simply write $\mathcal{L}^1(X)$. Also, when M is a compact interval $[a, b]$ we write $\mathcal{L}^1[a, b]$. Similar notational conventions will be adopted for other spaces of integrable functions below.

Example 1.48 (Counting measure)

Suppose that $(X, \Sigma, \mu) = (\mathbb{N}, \Sigma_c, \mu_c)$ (see Example 1.46). Any function $f : \mathbb{N} \to \mathbb{F}$ can be regarded as an $\mathbb{F}$-valued sequence $\{a_n\}$ (with $a_n = f(n)$, $n \geq 1$), and since all subsets of $\mathbb{N}$ are measurable, every such sequence $\{a_n\}$ can be regarded as a measurable function. It follows from the above definitions that the sequence $\{a_n\}$ is integrable (with respect to μ_c) if and only if $\sum_{n=1}^{\infty} |a_n| < \infty$, and then the integral of $\{a_n\}$ is simply the sum $\sum_{n=1}^{\infty} a_n$. Instead of the general notation $\mathcal{L}^1(\mathbb{N})$, the space of such sequences will be denoted by ℓ^1 (or $\ell^1_{\mathbb{F}}$).

Definition 1.49 (Lebesgue integral)

Let $(X, \Sigma, \mu) = (\mathbb{R}^k, \Sigma_L, \mu_L)$, for some $k \geq 1$. If $f \in \mathcal{L}^1(\mathbb{R}^k)$ (or $f \in \mathcal{L}^1(S)$, with $S \in \Sigma_L$) then f is said to be *Lebesgue integrable*.

The class of Lebesgue integrable functions is much larger than the class of Riemann integrable functions. However, when both integrals are defined they agree.

Theorem 1.50

If $I = [a, b] \subset \mathbb{R}$ is a bounded interval and $f : I \to \mathbb{R}$ is bounded and Riemann integrable on I, then f is Lebesgue integrable on I, and the values of the two integrals of f coincide. In particular, continuous functions on I are Lebesgue integrable.

In view of Theorem 1.50 the Lebesgue integral of f on $I = [a, b]$ will be denoted by

$$\int_I f(x)\,dx \quad \text{or} \quad \int_a^b f(x)\,dx$$

(that is, the same notation is used for Riemann and Lebesgue integrals). It also follows from Theorem 1.50 that, for Riemann integrable functions, Lebesgue integration gives nothing new and the well-known methods for evaluating Riemann integrals (based on the fundamental theorem of calculus) still apply.

We now list some of the basic properties of the integral.

Theorem 1.51

Let (X, Σ, μ) be a measure space and let $f \in \mathcal{L}^1(X)$.

(a) If $f(x) = 0$ a.e., then $f \in \mathcal{L}^1(X)$ and $\int_X f\,d\mu = 0$.

(b) If $\alpha \in \mathbb{R}$ and $f,\, g \in \mathcal{L}^1(X)$ then the functions $f + g$ and αf (see Definition 1.5) belong to $\mathcal{L}^1(X)$ and

$$\int_I (f+g)\,d\mu = \int_X f\,d\mu + \int_X g\,d\mu, \qquad \int_I \alpha f\,d\mu = \alpha \int_X f\,d\mu.$$

In particular, $\mathcal{L}^1(X)$ is a vector space.

(c) If $f,\, g \in \mathcal{L}^1(X)$ and $f(x) \le g(x)$ for all $x \in X$, then $\int_X f\,d\mu \le \int_X g\,d\mu$. If, in addition, $f(x) < g(x)$ for all $x \in S$, with $\mu(S) > 0$, then $\int_X f\,d\mu < \int_X g\,d\mu$.

It follows from part (a) of Theorem 1.51 that the values of f on sets of measure zero do not affect the integral. In particular, bounds on f which hold almost everywhere are often more appropriate than those which hold everywhere (especially since we allow measurable functions to take the value ∞).

Definition 1.52

Suppose that f is a measurable function and there exists a number b such that $f(x) \le b$ a.e. Then we can define the *essential supremum* of f to be

$$\text{ess sup} f = \inf\{b : f(x) \le b \text{ a.e.}\}.$$

It is a simple (but not completely trivial) consequence of this definition that $f(x) \le \text{ess sup} f$ a.e. The *essential infimum* of f can be defined similarly.

A measurable function f is said to be *essentially bounded* if there exists a number b such that $|f(x)| \le b$ a.e.

We would now like to define a metric on the space $\mathcal{L}^1(X)$, and an obvious candidate for this is the function

$$d_1(f,g) = \int_X |f - g|\, d\mu,$$

for all $f,\ g \in \mathcal{L}^1(X)$. It follows from the properties of the integral in Theorem 1.51 that this function satisfies all the requirements for a metric except part (b) of Definition 1.17. Unfortunately, there are functions $f,\ g \in \mathcal{L}^1(X)$ with $f = g$ a.e. but $f \ne g$ (for any $f \in \mathcal{L}^1(X)$ we can construct such a g simply by changing the values of f on a set of measure zero), and so, by part (a) of Theorem 1.51, $d_1(f,g) = 0$. Thus the function d_1 is not a metric on $\mathcal{L}^1(X)$. To circumvent this problem we will agree to "identify", or regard as "equivalent", any two functions $f,\ g$ which are a.e. equal. More precisely, we define an equivalence relation $\equiv$ on $\mathcal{L}^1(X)$ by

$$f \equiv g \iff f(x) = g(x) \text{ for a.e. } x \in X$$

(precise details of this construction are given in Section 5.1 of [4]). This equivalence relation partitions the set $\mathcal{L}^1(X)$ into a space of equivalence classes, which we will denote by $L^1(X)$. By defining addition and scalar multiplication appropriately on these equivalence classes the space $L^1(X)$ becomes a vector space, and it follows from Theorem 1.51 that $f \equiv g$ if and only if $d_1(f,g) = 0$, and consequently the function d_1 yields a metric on the set $L^1(X)$. Thus from now on we will use the space $L^1(X)$ rather than the space $\mathcal{L}^1(X)$.

Strictly speaking, when using the space $L^1(X)$ one should distinguish between a function $f \in \mathcal{L}^1(X)$ and the corresponding equivalence class in $L^1(X)$ consisting of all functions a.e. equal to f. However, this is cumbersome and, in practice, is rarely done, so we will consistently talk about the "function" $f \in L^1(X)$, meaning some representative of the appropriate equivalence class. We note, in particular, that if $X = I$ is an interval of positive length in $\mathbb{R}$, then an equivalence class can contain at most one continuous function on I (it need not contain any), and if it does then we will always take this as the representative of the class. In particular, if I is compact then continuous functions on I belong to the space $L^1(I)$ in this sense (since they are certainly Riemann integrable on I).

We can now define some other spaces of integrable functions.

Definition 1.53

Define the spaces

$$\begin{aligned} \mathcal{L}^p(X) &= \{f : f \text{ is measurable and } \left(\int_X |f|^p \, d\mu\right)^{1/p} < \infty\}, \quad 1 \le p < \infty; \\ \mathcal{L}^\infty(X) &= \{f : f \text{ is measurable and } \operatorname{ess\,sup}|f| < \infty\}. \end{aligned}$$

We also define the corresponding sets $L^p(X)$ by identifying functions in $\mathcal{L}^p(X)$ which are a.e. equal and considering the corresponding spaces of equivalence classes (in practice, we again simply refer to representative functions of these equivalence classes rather than the classes themselves).

When X is a bounded interval $[a,b] \subset \mathbb{R}$ and $1 \le p \le \infty$, we write $L^p[a,b]$.

The case $p = 1$ in Definition 1.53 coincides with the previous definitions.

Theorem 1.54 (Theorem 2.5.3 in [8])

Suppose that f and g are measurable functions. Then the following inequalities hold (infinite values are allowed).
Minkowski's inequality (for $1 \le p < \infty$):

$$\begin{aligned} \left(\int_X |f+g|^p \, d\mu\right)^{1/p} &\le \left(\int_X |f|^p \, d\mu\right)^{1/p} + \left(\int_X |g|^p \, d\mu\right)^{1/p}, \\ \operatorname{ess\,sup}|f+g| &\le \operatorname{ess\,sup}|f| + \operatorname{ess\,sup}|g|. \end{aligned}$$

Hölder's inequality (for $1 < p < \infty$ and $p^{-1} + q^{-1} = 1$):

$$\begin{aligned} \int_X |fg| \, d\mu &\le \left(\int_X |f|^p \, d\mu\right)^{1/p} \left(\int_X |g|^q \, d\mu\right)^{1/q}, \\ \int_X |fg| \, d\mu &\le \operatorname{ess\,sup}|f| \int_X |g| \, d\mu. \end{aligned}$$

Corollary 1.55

Suppose that $1 \le p \le \infty$.

(a) $L^p(X)$ is a vector space (essentially, this follows from Minkowski's inequality together with simple properties of the integral).

(b) The function

$$d_p(f,g) = \begin{cases} \left(\int_X |f-g|^p \, d\mu\right)^{1/p}, & 1 \le p < \infty, \\ \text{ess sup} \, |f-g|, & p = \infty, \end{cases}$$

is a metric on $L^p(X)$ (condition (b) in Definition 1.17 follows from properties (a) and (c) in Theorem 1.51, together with the construction of the spaces $L^p(X)$, while Minkowski's inequality shows that d_p satisfies the triangle inequality). This metric will be called the *standard metric* on $L^p(X)$ and, unless otherwise stated, $L^p(X)$ will be assumed to have this metric.

Example 1.56 (Counting measure)

Suppose that $1 \le p \le \infty$. In the special case where $(X, \Sigma, \mu) = (\mathbb{N}, \Sigma_c, \mu_c)$, the space $L^p(\mathbb{N})$ consists of the set of sequences $\{a_n\}$ in $\mathbb{F}$ with the property that

$$\Big(\sum_{n=1}^{\infty} |a_n|^p\Big)^{1/p} < \infty, \quad \text{for } 1 \le p < \infty,$$

$$\sup\{|a_n| : n \in \mathbb{N}\} < \infty, \quad \text{for } p = \infty.$$

These spaces will be denoted by ℓ^p (or $\ell^p_{\mathbb{F}}$). Note that since there are no sets of measure zero in this measure space, there is no question of taking equivalence classes here. By Corollary 1.55, the spaces ℓ^p are both vector spaces and metric spaces. The *standard metric* on ℓ^p is defined analogously to the above expressions in the obvious manner.

By using counting measure and letting x and y be sequences in $\mathbb{F}$ (or elements of $\mathbb{F}^k$ for some $k \in \mathbb{N}$ – in this case x and y can be regarded as sequences with only a finite number of non-zero elements), we can obtain the following important special case of Theorem 1.54.

Corollary 1.57

Minkowski's inequality (for $1 \le p < \infty$):

$$\Big(\sum_{j=1}^{k} |x_j + y_j|^p\Big)^{1/p} \le \Big(\sum_{j=1}^{k} |x_j|^p\Big)^{1/p} + \Big(\sum_{j=1}^{k} |y_j|^p\Big)^{1/p}.$$

Hölder's inequality (for $1 < p < \infty$ and $p^{-1} + q^{-1} = 1$):

$$\sum_{j=1}^{k} |x_j y_j| \le \Big(\sum_{j=1}^{k} |x_j|^p\Big)^{1/p} \Big(\sum_{j=1}^{k} |y_j|^q\Big)^{1/q}.$$

Here, k and the values of the sums may be ∞.

For future reference we specifically state the most important special case of this result, namely Hölder's inequality with $p = q = 2$.

Corollary 1.58

$$\sum_{j=1}^{k} |x_j||y_j| \leq \Big(\sum_{j=1}^{k} |x_j|^2\Big)^{1/2} \Big(\sum_{j=1}^{k} |y_j|^2\Big)^{1/2}.$$

Some particular elements of ℓ^p will now be defined, which will be extremely useful below.

Definition 1.59

Let

$$\widetilde{e}_1 = (1, 0, 0, \dots),\ \widetilde{e}_2 = (0, 1, 0, \dots),\ \dots .$$

For any $n \in \mathbb{N}$ the sequence $\widetilde{e}_n \in \ell^p$ for all $1 \leq p \leq \infty$.

The vectors $\widetilde{e}_n$ in the infinite-dimensional space ℓ^p, introduced in Definition 1.59, bear some resemblance to the vectors $\widehat{e}_n$ in the finite-dimensional space $\mathbb{F}^k$, introduced in Definition 1.3. However, although the collection of vectors $\{\widehat{e}_1, \dots, \widehat{e}_k\}$ is a basis for $\mathbb{F}^k$, we must emphasize that at present we have no concept of a basis for infinite-dimensional vector spaces so we cannot make any analogous assertion regarding the collection $\{\widehat{e}_n : n \in \mathbb{N}\}$.

Finally, we can state the following two theorems. These will be crucial to much of our use of the spaces discussed in this section.

Theorem 1.60 (Theorems 5.1, 5.5 in [4])

Suppose that $1 \leq p \leq \infty$. Then the metric space $L^p(X)$ is complete. In particular, the sequence space ℓ^p is complete.

Theorem 1.61 (Theorem 2.5.6 in [8])

Suppose that $[a, b]$ is a bounded interval and $1 \leq p < \infty$. Then the set $C[a, b]$ is dense in $L^p[a, b]$.

As discussed at the beginning of this section, these results show that the space $L^p[a, b]$ is the "completion" of the metric space $(C[a, b], d_p)$ (see Section 3.5 of [3] for more details on the completion of a metric space – we will not use this concept further here, the above theorems will suffice for our needs).

Theorem 1.61 shows that the space $L^p[a,b]$ is "close to" the space $(C[a,b], d_p)$ in the sense that if $f \in L^p[a,b]$, then there exists a sequence of functions $\{f_n\}$ in $C[a,b]$ which converges to f in the $L^p[a,b]$ metric. This is a relatively weak type of convergence (but extremely useful). In particular, it does not imply pointwise convergence (not even for a.e. $x \in [a,b]$).

2
Normed Spaces

2.1 Examples of Normed Spaces

When the vector spaces $\mathbb{R}^2$ and $\mathbb{R}^3$ are pictured in the usual way, we have the idea of the length of a vector in $\mathbb{R}^2$ or $\mathbb{R}^3$ associated with each vector. This is clearly a bonus which gives us a deeper understanding of these vector spaces. When we turn to other (possibly infinite-dimensional) vector spaces, we might hope to get more insight into these spaces if there is some way of assigning something similar to the length of a vector for each vector in the space. Accordingly we look for a set of axioms which is satisfied by the length of a vector in $\mathbb{R}^2$ and $\mathbb{R}^3$. This set of axioms will define the "norm" of a vector and throughout this book we will mainly consider normed vector spaces. In this chapter we investigate the elementary properties of normed vector spaces.

Definition 2.1

(a) Let X be a vector space over $\mathbb{F}$. A *norm* on X is a function $\|\cdot\| : X \to \mathbb{R}$ such that for all x, $y, \in X$ and $\alpha \in \mathbb{F}$,

(i) $\|x\| \geq 0$;

(ii) $\|x\| = 0$ if and only if $x = 0$;

(iii) $\|\alpha x\| = |\alpha| \|x\|$;

(iv) $\|x + y\| \leq \|x\| + \|y\|$.

(b) A vector space X on which there is a norm is called a *normed vector space* or just a *normed space*.

(c) If X is a normed space, a *unit vector* in X is a vector x such that $\|x\| = 1$.

As a motivation for looking at norms we implied that the length of a vector in $\mathbb{R}^2$ and $\mathbb{R}^3$ satisfies the axioms of a norm. This will be verified in Example 2.2 but we note at this point that property (iv) of Definition 2.1 is often called the *triangle inequality* since, in $\mathbb{R}^2$, it is simply the fact that the length of one side of a triangle is less than or equal to the sum of the lengths of the other two sides.

Example 2.2

The function $\|\cdot\| : \mathbb{F}^n \to \mathbb{R}$ defined by

$$\|(x_1, \ldots, x_n)\| = \Big(\sum_{j=1}^{n} |x_j|^2 \Big)^{\frac{1}{2}}$$

is a norm on $\mathbb{F}^n$ called the *standard norm* on $\mathbb{F}^n$.

We will not give the solution to Example 2.2 as we generalize this example in Example 2.3. As $\mathbb{F}^n$ is perhaps the easiest normed space to visualize, when any new properties of normed vector spaces are introduced later, it can be useful to try to see what they mean first in the space $\mathbb{F}^n$ even though this is finite-dimensional.

Example 2.3

Let X be a finite-dimensional vector space over $\mathbb{F}$ with basis $\{e_1, e_2, \ldots, e_n\}$. Any $x \in X$ can be written as $x = \sum_{j=1}^{n} \lambda_j e_j$ for unique $\lambda_1, \lambda_2, \ldots, \lambda_n \in \mathbb{F}$. Then the function $\|\cdot\| : X \to \mathbb{R}$ defined by

$$\|x\| = \Big(\sum_{j=1}^{n} |\lambda_j|^2 \Big)^{\frac{1}{2}}$$

is a norm on X.

Solution

Let $x = \sum_{j=1}^{n} \lambda_j e_j$ and $y = \sum_{j=1}^{n} \mu_j e_j$ and let $\alpha \in \mathbb{F}$. Then $\alpha x = \sum_{j=1}^{n} \alpha \lambda_j e_j$ and the following results verify that $\|\cdot\|$ is a norm.

(i) $\|x\| = \Big(\sum_{j=1}^{n} |\lambda_j|^2 \Big)^{\frac{1}{2}} \geq 0$.

(ii) If $x = 0$ then $\|x\| = 0$. Conversely, if $\|x\| = 0$ then $\left(\sum_{j=1}^{n} |\lambda_j|^2 \right)^{\frac{1}{2}} = 0$ so that $\lambda_j = 0$ for $1 \le j \le n$. Hence $x = 0$.

(iii) $\|\alpha x\| = \left(\sum_{j=1}^{n} |\alpha\lambda_j|^2 \right)^{\frac{1}{2}} = |\alpha| \left(\sum_{j=1}^{n} |\lambda_j|^2 \right)^{\frac{1}{2}} = |\alpha|\|x\|$.

(iv) By Corollary 1.58,

$$\begin{aligned}
\|x+y\|^2 &= \sum_{j=1}^{n} |\lambda_j + \mu_j|^2 \\
&= \sum_{j=1}^{n} |\lambda_j|^2 + \sum_{j=1}^{n} \overline{\lambda_j}\mu_j + \sum_{j=1}^{n} \overline{\mu_j}\lambda_j + \sum_{j=1}^{n} |\mu_j|^2 \\
&= \sum_{j=1}^{n} |\lambda_j|^2 + 2\sum_{j=1}^{n} \Re\mathrm{e}(\overline{\lambda_j}\mu_j) + \sum_{j=1}^{n} |\mu_j|^2 \\
&\le \sum_{j=1}^{n} |\lambda_j|^2 + 2\sum_{j=1}^{n} |\lambda_j||\mu_j| + \sum_{j=1}^{n} |\mu_j|^2 \\
&\le \sum_{j=1}^{n} |\lambda_j|^2 + 2\left(\sum_{j=1}^{n} |\lambda_j|^2 \right)^{\frac{1}{2}} \left(\sum_{j=1}^{n} |\mu_j|^2 \right)^{\frac{1}{2}} + \sum_{j=1}^{n} |\mu_j|^2 \\
&= \|x\|^2 + 2\|x\|\|y\| + \|y\|^2 \\
&= (\|x\| + \|y\|)^2.
\end{aligned}$$

Hence $\|x+y\| \le \|x\| + \|y\|$. □

Many interesting normed vector spaces are not finite-dimensional.

Example 2.4

Let M be a compact metric space and let $C_{\mathbb{F}}(M)$ be the vector space of continuous, $\mathbb{F}$-valued functions defined on M. Then the function $\|\cdot\| : C_{\mathbb{F}}(M) \to \mathbb{R}$ defined by

$$\|f\| = \sup\{|f(x)| : x \in M\}$$

is a norm on $C_{\mathbb{F}}(M)$ called the *standard norm* on $C_{\mathbb{F}}(M)$.

Solution

Let $f, g \in C_{\mathbb{F}}(M)$ and let $\alpha \in \mathbb{F}$.

(i) $\|f\| = \sup\{|f(x)| : x \in M\} \ge 0$.

(ii) If f is the zero function then $f(x) = 0$ for all $x \in M$ and hence $\|f\| = \sup\{|f(x)| : x \in M\} = 0$. Conversely, if $\|f\| = 0$ then $\sup\{|f(x)| : x \in M\} = 0$ and so $f(x) = 0$ for all $x \in M$. Hence f is the zero function.

(iii) $\|\alpha f\| = \sup\{|\alpha f(x)| : x \in M\} = |\alpha| \sup\{|f(x)| : x \in M\} = |\alpha| \|f\|$.

(iv) If $y \in M$ then

$$|(f+g)(y)| \leq |f(y)| + |g(y)| \leq \|f\| + \|g\|,$$

and so $\|f+g\| = \sup\{|(f+g)(x)| : x \in M\} \leq \|f\| + \|g\|$. □

In the next example we show that some of the vector spaces of integrable functions defined in Chapter 1 have norms. We recall that if (X, Σ, μ) is a measure space and $1 \leq p \leq \infty$, then the space $L^p(X)$ was introduced in Definition 1.53.

Example 2.5

Let (X, Σ, μ) be a measure space.

(a) If $1 \leq p < \infty$ then $\|f\|_p = \left(\int_X |f|^p d\mu \right)^{1/p}$ is a norm on $L^p(X)$ called the *standard norm* on $L^p(X)$.

(b) $\|f\|_\infty = \operatorname{ess\,sup}\{|f(x)| : x \in X\}$ is a norm on $L^\infty(X)$ called the *standard norm* on $L^\infty(X)$.

Solution

(a) Let $f, g \in L^p(X)$ and let $\alpha \in \mathbb{F}$. Then $\|f\|_p \geq 0$ and $\|f\|_p = 0$ if and only if $f = 0$ a.e (which is the equivalence class of the zero function) by Theorem 1.51. Also,

$$\|\alpha f\|_p = \left(\int_X |\alpha f|^p d\mu \right)^{1/p} = |\alpha| \left(\int_X |f|^p d\mu \right)^{1/p} = |\alpha| \|f\|_p,$$

and the triangle inequality follows from Theorem 1.54.

(b) Let $f, g \in L^\infty(X)$ and let $\alpha \in \mathbb{F}$. Then $\|f\|_\infty \geq 0$ and $\|f\|_\infty = 0$ if and only if $f = 0$ a.e (which is the equivalence class of the zero function). If $\alpha = 0$ then $\|\alpha f\|_\infty = |\alpha| \|f\|_\infty$ so suppose that $\alpha \neq 0$. As $|\alpha f(x)| \leq |\alpha| \|f\|_\infty$ a.e. it follows that

$$\|\alpha f\|_\infty \leq |\alpha| \|f\|_\infty.$$

Applying the same argument to $\alpha^{-1} f$ it follows that

$$\|f\|_\infty = \|\alpha^{-1} \alpha f\|_\infty \leq |\alpha^{-1}| \|\alpha f\|_\infty \leq |\alpha^{-1}| |\alpha| \|f\|_\infty = \|f\|_\infty.$$

Therefore $\|\alpha f\|_\infty = |\alpha| \|f\|_\infty$. Finally, the triangle inequality follows from Theorem 1.54. □

Specific notation was introduced in Chapter 1 for the case of counting measure on $\mathbb{N}$. Recall that ℓ^p is the vector space of all sequences $\{x_n\}$ in $\mathbb{F}$ such that $\sum_{n=1}^{\infty} |x_n|^p < \infty$ for $1 \leq p < \infty$ and ℓ^∞ is the vector space of all bounded sequences in $\mathbb{F}$. Therefore, if we take counting measure on $\mathbb{N}$ in Example 2.5 we deduce that ℓ^p for $1 \leq p < \infty$ and ℓ^∞ are normed spaces. For completeness we define the norms on these spaces in Example 2.6.

Example 2.6

(a) If $1 \leq p < \infty$ then $\|\{x_n\}\|_p = \left(\sum_{n=1}^{\infty} |x_n|^p \right)^{1/p}$ is a norm on ℓ^p called the *standard norm* on ℓ^p.

(b) $\|\{x_n\}\|_\infty = \sup\{|x_n| : n \in \mathbb{N}\}$ is a norm on ℓ^∞ called the *standard norm* on ℓ^∞.

In future, if we write down any of the spaces in Examples 2.2, 2.4, 2.5 and 2.6 without explicitly mentioning a norm it will be assumed that the norm on the space is the standard norm.

Given one normed space it is possible to construct many more. The solution to Example 2.7 is so easy it is omitted.

Example 2.7

Let X be a vector space with a norm $\|\cdot\|$ and let S be a linear subspace of X. Let $\|\cdot\|_S$ be the restriction of $\|\cdot\|$ to S. Then $\|\cdot\|_S$ is a norm on S.

The solution of Example 2.8 is only slightly harder than that of Example 2.7 so it is left as an exercise.

Example 2.8

Let X and Y be vector spaces over $\mathbb{F}$ and let $Z = X \times Y$ be the Cartesian product of X and Y. This is a vector space by Definition 1.4. If $\|\cdot\|_1$ is a norm on X and $\|\cdot\|_2$ is a norm on Y then $\|(x, y)\| = \|x\|_1 + \|y\|_2$ defines a norm on Z.

As we see from the above examples there are many different normed spaces and this partly explains why the study of normed spaces is important. Since the norm of a vector is a generalization of the length of a vector in $\mathbb{R}^3$ it is perhaps not surprising that each normed space is a metric space in a very natural way.

Lemma 2.9

Let X be a vector space with norm $\|\cdot\|$. If $d : X \times X \to \mathbb{R}$ is defined by $d(x, y) = \|x - y\|$ then (X, d) is a metric space.

Proof

Let x, y, $z \in X$. Using the properties of the norm we see that:

(a) $d(x, y) = \|x - y\| \geq 0$;

(b) $d(x, y) = 0 \iff \|x - y\| = 0 \iff x - y = 0 \iff x = y$;

(c) $d(x, y) = \|x - y\| = \|(-1)(y - x)\| = |(-1)|\|y - x\| = \|y - x\| = d(y, x)$;

(d) $d(x, z) = \|x - z\| = \|(x - y) + (y - z)\| \leq \|(x - y)\| + \|(y - z)\|$
$= d(x, y) + d(y, z)$.

Hence d satisfies the axioms for a metric. □

Notation

If X is a vector space with norm $\|\cdot\|$ and d is the metric defined by $d(x, y) = \|x - y\|$ then d is called the metric *associated* with $\|\cdot\|$.

Whenever we use a metric or a metric space concept, for example, convergence, continuity or completeness, in a normed space then we will always use the metric associated with the norm even if this is not explicitly stated. The metrics associated with the standard norms are already familiar.

Example 2.10

The metrics associated with the standard norms on the following spaces are the standard metrics.

(a) $\mathbb{F}^n$;

(b) $C_{\mathbb{F}}(M)$ where M is a compact metric space;

(c) $L^p(X)$ for $1 \leq p < \infty$ where (X, Σ, μ) is a measure space;

(d) $L^\infty(X)$ where (X, Σ, μ) is a measure space.

Solution

(a) If $x, y \in \mathbb{F}^n$ then

$$d(x, y) = \|x - y\| = \left(\sum_{j=1}^{n} |x_j - y_j|^2 \right)^{\frac{1}{2}},$$

and so d is the standard metric on $\mathbb{F}^n$.

(b) If $f, g \in C_{\mathbb{F}}(M)$ then

$$d(f,g) = \|f-g\| = \sup\{|f(x)-g(x)| : x \in M\},$$

and so d is the standard metric on $C_{\mathbb{F}}(M)$.

(c) If $f, g \in L^p(X)$ then

$$d(f,g) = \|f-g\| = \Big(\int_X |f-g|^p \, d\mu \Big)^{1/p},$$

and so d is the standard metric on $L^p(X)$.

(d) If $f, g \in L^\infty(X)$ then

$$d(f,g) = \|f-g\| = \operatorname{ess\,sup}\{|f(x)-g(x)| : x \in X\},$$

and so d is the standard metric on $L^\infty(X)$. □

Using counting measure on $\mathbb{N}$ it follows that the metrics associated with the standard norms on the ℓ^p and ℓ^∞ are also the standard metrics on these spaces.

We conclude this section with some basic information about convergence of sequences in normed vector spaces.

Theorem 2.11

Let X be a vector space over $\mathbb{F}$ with norm $\|\cdot\|$. Let $\{x_n\}$ and $\{y_n\}$ be sequences in X which converge to x, y in X respectively and let $\{\alpha_n\}$ be a sequence in $\mathbb{F}$ which converges to α in $\mathbb{F}$. Then:

(a) $|\,\|x\| - \|y\|\,| \leq \|x-y\|$;

(b) $\lim_{n\to\infty} \|x_n\| = \|x\|$;

(c) $\lim_{n\to\infty} (x_n + y_n) = x + y$;

(d) $\lim_{n\to\infty} \alpha_n x_n = \alpha x$.

Proof

(a) By the triangle inequality, $\|x\| = \|(x-y)+y\| \leq \|x-y\| + \|y\|$ and so

$$\|x\| - \|y\| \leq \|x-y\|.$$

Interchanging x and y we obtain $\|y\|-\|x\| \leq \|y-x\|$. However, as $\|x-y\| = \|(-1)(y-x)\| = \|y-x\|$ we have

$$-\|x-y\| \leq \|x\| - \|y\| \leq \|x-y\|.$$

Hence $|\|x\| - \|y\|| \leq \|x-y\|$.

(b) Since $\lim_{n\to\infty} x_n = x$ and $|\|x\| - \|x_n\|| \leq \|x - x_n\|$ for all $n \in \mathbb{N}$, it follows that $\lim_{n\to\infty} \|x_n\| = \|x\|$.

(c) Since $\lim_{n\to\infty} x_n = x$, $\lim_{n\to\infty} y_n = y$ and

$$\|(x_n + y_n) - (x+y)\| = \|(x_n - x) + (y_n - y)\| \leq \|x_n - x\| + \|y_n - y\|$$

for all $n \in \mathbb{N}$, it follows that $\lim_{n\to\infty} (x_n + y_n) = x + y$.

(d) Since $\{\alpha_n\}$ is convergent it is bounded, so there exists $K > 0$ such that $|\alpha_n| \leq K$ for all $n \in \mathbb{N}$. Also,

$$\begin{aligned}\|\alpha_n x_n - \alpha x\| &= \|\alpha_n(x_n - x) + (\alpha_n - \alpha)x\| \\ &\leq |\alpha_n|\|x_n - x\| + |\alpha_n - \alpha|\|x\| \\ &\leq K\|x_n - x\| + |\alpha_n - \alpha|\|x\|\end{aligned}$$

for all $n \in \mathbb{N}$. Hence $\lim_{n\to\infty} \alpha_n x_n = \alpha x$. □

A different way of stating the results in Theorem 2.11 parts (b), (c) and (d) is that the norm, addition and scalar multiplication are continuous functions, as can be seen using the sequential characterization of continuity.

EXERCISES

2.1 Give the solution of Example 2.8.

2.2 Let S be any non-empty set and let X be a normed space over $\mathbb{F}$. Let $F_b(S, X)$ be the linear subspace of $F(S, X)$ of all functions $f : S \to X$ such that $\{\|f(s)\| : s \in S\}$ is bounded. Show that $F_b(S, X)$ has a norm defined by

$$\|f\|_b = \sup\{\|f(s)\| : s \in S\}.$$

2.3 For each $n \in \mathbb{N}$ let $f_n : [0,1] \to \mathbb{R}$ be defined by $f_n(x) = x^n$. Find the norm of f_n in the following cases:

(a) in the normed space $C_{\mathbb{R}}([0,1])$;

(b) in the normed space $L^1[0,1]$.

2.4 Let X be a normed linear space. If $x \in X\backslash\{0\}$ and $r > 0$, find $\alpha \in \mathbb{R}$ such that $\|\alpha x\| = r$.

2.5 Let X be a vector space with norm $\|\cdot\|_1$ and let Y be a vector space with norm $\|\cdot\|_2$. Let $Z = X \times Y$ have norm given in Example 2.8. Let $\{(x_n, y_n)\}$ be a sequence in Z.

(a) Show that $\{(x_n, y_n)\}$ converges to (x, y) in Z if and only if $\{x_n\}$ converges to x in X and $\{y_n\}$ converges in y in Y.

(b) Show that $\{(x_n, y_n)\}$ is Cauchy in Z if and only if $\{x_n\}$ is Cauchy in X and $\{y_n\}$ is Cauchy in Y.

2.2 Finite-dimensional Normed Spaces

The simplest vector spaces to study are the finite-dimensional ones, so a natural place to start our study of normed spaces is with finite-dimensional normed spaces. We have already seen in Example 2.3 that each finite-dimensional space has a norm, but this norm depends on the choice of basis. This suggests that there can be many different norms on each finite-dimensional space. Even in $\mathbb{R}^2$ we have already seen that there are at least two norms:

(a) the standard norm defined in Example 2.2;

(b) the norm $\|(x, y)\| = |x| + |y|$, defined in Example 2.8.

To show how different these two norms are it is instructive to sketch the set $\{(x, y) \in \mathbb{R}^2 : \|(x, y)\| = 1\}$ for each norm, which will be done in Exercise 2.7. However, even when we have two norms on a vector space, if these norms are not too dissimilar, it is possible that the metric space properties of the space could be the same for both norms. A more precise statement of what is meant by being "not too dissimilar" is given in Definition 2.12.

Definition 2.12

Let X be a vector space and let $\|\cdot\|_1$ and $\|\cdot\|_2$ be two norms on X. The norm $\|\cdot\|_2$ is *equivalent* to the norm $\|\cdot\|_1$ if there exists $M, m > 0$ such that for all $x \in X$

$$m\|x\|_1 \leq \|x\|_2 \leq M\|x\|_1.$$

In view of the terminology used, it should not be a surprise that this defines an equivalence relation on the set of all norms on X, as we now show.

Lemma 2.13

Let X be a vector space and let $\|\cdot\|_1$, $\|\cdot\|_2$ and $\|\cdot\|_3$ be three norms on X. Let $\|\cdot\|_2$ be equivalent to $\|\cdot\|_1$ and let $\|\cdot\|_3$ be equivalent to $\|\cdot\|_2$.

(a) $\|\cdot\|_1$ is equivalent to $\|\cdot\|_2$.

(b) $\|\cdot\|_3$ is equivalent to $\|\cdot\|_1$.

Proof

By the hypothesis, there exists $M, m > 0$ such that $m\|x\|_1 \leq \|x\|_2 \leq M\|x\|_1$ for all $x \in X$ and there exists $K, k > 0$ such that $k\|x\|_2 \leq \|x\|_3 \leq K\|x\|_2$ for all $x \in X$. Hence:

(a) $\dfrac{1}{M}\|x\|_2 \leq \|x\|_1 \leq \dfrac{1}{m}\|x\|_2$ for all $x \in X$ and so $\|\cdot\|_1$ is equivalent to $\|\cdot\|_2$;

(b) $km\|x\|_1 \leq \|x\|_3 \leq KM\|x\|_1$ for all $x \in X$ and so $\|\cdot\|_3$ is equivalent to $\|\cdot\|_1$. □

We now show that in a vector space with two equivalent norms the metric space properties are the same for both norms.

Lemma 2.14

Let X be a vector space and let $\|\cdot\|$ and $\|\cdot\|_1$ be norms on X. Let d and d_1 be the metrics defined by $d(x, y) = \|x - y\|$ and $d_1(x, y) = \|x - y\|_1$. Suppose that there exists $K > 0$ such that $\|x\| \leq K\|x\|_1$ for all $x \in X$. Let $\{x_n\}$ be a sequence in X.

(a) If $\{x_n\}$ converges to x in the metric space (X, d_1) then $\{x_n\}$ converges to x in the metric space (X, d).

(b) If $\{x_n\}$ is Cauchy in the metric space (X, d_1) then $\{x_n\}$ is Cauchy in the metric space (X, d).

Proof

(a) Let $\epsilon > 0$. There exists $N \in \mathbb{N}$ such that $\|x_n - x\|_1 < \dfrac{\epsilon}{K}$ when $n \geq \mathbb{N}$. Hence, when $n \geq \mathbb{N}$,

$$\|x_n - x\| \leq K\|x_n - x\|_1 < \epsilon.$$

Therefore $\{x_n\}$ converges to x in the metric space (X, d).

(b) As this proof is similar to part (a), it is left as an exercise. □

Corollary 2.15

Let X be a vector space and let $\|\cdot\|$ and $\|\cdot\|_1$ be equivalent norms on X. Let d and d_1 be the metrics defined by $d(x, y) = \|x - y\|$ and $d_1(x, y) = \|x - y\|_1$. Let $\{x_n\}$ be a sequence in X.

(a) $\{x_n\}$ converges to x in the metric space (X, d) if and only if $\{x_n\}$ converges to x in the metric space (X, d_1).

(b) $\{x_n\}$ is Cauchy in the metric space (X, d) if and only if $\{x_n\}$ is Cauchy in the metric space (X, d_1).

(c) (X, d) is complete if and only if (X, d_1) is complete.

Proof

As $\|\cdot\|$ and $\|\cdot\|_1$ are equivalent norms on X there exists $M, m > 0$ such that $m\|x\| \leq \|x\|_1 \leq M\|x\|$ for all $x \in X$.

(a) As $\|x\|_1 \leq M\|x\|$ for all $x \in X$, if $\{x_n\}$ converges to x in the metric space (X, d), then $\{x_n\}$ converges to x in the metric space (X, d_1) by Lemma 2.14.

Conversely, as $\|x\| \leq \frac{1}{m}\|x\|_1$ for all $x \in X$, if $\{x_n\}$ converges to x in the metric space (X, d_1), then $\{x_n\}$ converges to x in the metric space (X, d) by Lemma 2.14.

(b) As in Lemma 2.14, this proof is similar to part (a) so is again left as an exercise.

(c) Suppose that (X, d) is complete and let $\{x_n\}$ be a Cauchy sequence in the metric space (X, d_1). Then $\{x_n\}$ is Cauchy in the metric space (X, d) by part (b) and hence $\{x_n\}$ converges to some point x in the metric space (X, d), since (X, d) is complete. Thus $\{x_n\}$ converges to x in the metric space (X, d_1) by part (a) and so (X, d_1) is complete. The converse is true by symmetry. □

If X is a vector space with two equivalent norms $\|\cdot\|$ and $\|\cdot\|_1$ and $x \in X$ it is likely that $\|x\| \neq \|x\|_1$. However, by Corollary 2.15, as far as many metric space properties are concerned it does not matter whether we consider one norm or the other. This is important as sometimes one of the norms is easier to work with than the other.

If X is a finite-dimensional space then we know from Example 2.3 that X has at least one norm. We now show that any other norm on X is equivalent to this norm, and hence derive many metric space properties of finite-dimensional normed vector spaces.

Theorem 2.16

Let X be a finite-dimensional vector space with norm $\|\cdot\|$ and let $\{e_1, e_2, \ldots, e_n\}$ be a basis for X. Another norm on X was defined in Example 2.3 by

$$\|\sum_{j=1}^{n} \lambda_j e_j\|_1 = \Big(\sum_{j=1}^{n} |\lambda_j|^2\Big)^{\frac{1}{2}}. \tag{2.1}$$

The norms $\|\cdot\|$ and $\|\cdot\|_1$ are equivalent.

Proof

Let $M = \Big(\sum_{j=1}^{n} \|e_j\|^2\Big)^{\frac{1}{2}}$. Then $M > 0$ as $\{e_1, e_2, \ldots, e_n\}$ is a basis for X. Also

$$\begin{aligned}
\|\sum_{j=1}^{n} \lambda_j e_j\| &\leq \sum_{j=1}^{n} \|\lambda_j e_j\| \\
&= \sum_{j=1}^{n} |\lambda_j| \|e_j\| \\
&\leq \Big(\sum_{j=1}^{n} |\lambda_j|^2\Big)^{\frac{1}{2}} \Big(\sum_{j=1}^{n} \|e_j\|^2\Big)^{\frac{1}{2}} \\
&= M \|\sum_{j=1}^{n} \lambda_j e_j\|_1.
\end{aligned}$$

Now let $f : \mathbb{F}^n \to \mathbb{F}$ be defined by

$$f(\lambda_1, \ldots, \lambda_n) = \|\sum_{j=1}^{n} \lambda_j e_j\|.$$

The function f is continuous with respect to the standard metric on $\mathbb{F}^n$ by Theorem 2.11. Now if

$$S = \{(\lambda_1, \lambda_2, \ldots, \lambda_n) \in \mathbb{F}^n : \sum_{j=1}^{n} |\lambda_j|^2 = 1\},$$

then S is compact, so there exists $(\mu_1, \mu_2, \ldots, \mu_n) \in S$ such that $m = f(\mu_1, \mu_2, \ldots, \mu_n) \leq f(\lambda_1, \lambda_2, \ldots, \lambda_n)$ for all $(\lambda_1, \lambda_2, \ldots, \lambda_n) \in S$.

If $m = 0$ then $\|\sum_{j=1}^{n} \mu_j e_j\| = 0$ so $\sum_{j=1}^{n} \mu_j e_j = 0$ which contradicts the fact that $\{e_1, e_2, \ldots, e_n\}$ is a basis for X. Hence $m > 0$. Moreover, by the definition of $\|\cdot\|_1$, if $\|x\|_1 = 1$ then $\|x\| \geq m$. Therefore, if $y \in X \setminus \{0\}$, since $\left\|\frac{y}{\|y\|_1}\right\|_1 = 1$, we must have $\left\|\frac{y}{\|y\|_1}\right\| \geq m$ and so

$$\|y\| \geq m\|y\|_1.$$

As $\|y\| \geq m\|y\|_1$ when $y = 0$ it follows that $\|\cdot\|$ and $\|\cdot\|_1$ are equivalent. □

Corollary 2.17

If $\|\cdot\|$ and $\|\cdot\|_2$ are any two norms on a finite-dimensional vector space X then they are equivalent.

Proof

Let $\{e_1, e_2, \ldots, e_n\}$ be a basis for X and let $\|\cdot\|_1$ be the norm on X defined by (2.1). Then both $\|\cdot\|$ and $\|\cdot\|_2$ are equivalent to $\|\cdot\|_1$, by Theorem 2.16 and so $\|\cdot\|_2$ is equivalent to $\|\cdot\|$, by Lemma 2.13. □

Now that we have shown that all norms on a finite-dimensional space are equivalent we can obtain the metric space properties of the metric associated with any norm simply by considering one particular norm.

Lemma 2.18

Let X be a finite-dimensional vector space over $\mathbb{F}$ and let $\{e_1, e_2, \ldots, e_n\}$ be a basis for X. If $\|\cdot\|_1 : X \to \mathbb{R}$ is the norm on X defined by (2.1) then X is a complete metric space.

Proof

Let $\{x_m\}$ be a Cauchy sequence in X and let $\epsilon > 0$. Each element of the sequence can be written as $x_m = \sum_{j=1}^{n} \lambda_{j,m} e_j$ for some $\lambda_{j,m} \in \mathbb{F}$. As $\{x_m\}$ is Cauchy there exists $N \in \mathbb{N}$ such that when $k, m \geq N$,

$$\sum_{j=1}^{n} |\lambda_{j,k} - \lambda_{j,m}|^2 = \|x_k - x_m\|_1^2 \leq \epsilon^2.$$

Hence $|\lambda_{j,k} - \lambda_{j,m}|^2 \leq \epsilon^2$ for $k, m \geq N$ and $1 \leq j \leq n$. Thus $\{\lambda_{j,m}\}$ is a Cauchy sequence in $\mathbb{F}$ for $1 \leq j \leq n$ and since $\mathbb{F}$ is complete there exists $\lambda_j \in \mathbb{F}$ such that $\lim_{m \to \infty} \lambda_{j,m} = \lambda_j$. Therefore there exists $N_j \in \mathbb{N}$ such that when $m \geq N_j$

$$|\lambda_{j,m} - \lambda_j|^2 \leq \frac{\epsilon^2}{n}.$$

Let $N_0 = \max(N_1, N_2, \ldots N_n)$ and let $x = \sum_{j=1}^{n} \lambda_j e_j$. Then when $m \geq N_0$,

$$\|x_m - x\|_1^2 = \sum_{j=1}^{n} |\lambda_{j,m} - \lambda_j|^2 \leq \sum_{j=1}^{n} \frac{\epsilon^2}{n} = \epsilon^2.$$

Hence $\{x_m\}$ converges to x so X is complete. □

Corollary 2.19

If $\|\cdot\|$ is any norm on a finite-dimensional space X then X is a complete metric space.

Proof

Let $\{e_1, e_2, \ldots, e_n\}$ be a basis for X and let $\|\cdot\|_1 : X \to \mathbb{R}$ be a second norm on X defined by (2.1). The norms $\|\cdot\|$ and $\|\cdot\|_1$ are equivalent by Corollary 2.17 and X with norm $\|\cdot\|_1$ is complete by Lemma 2.18. Hence X with norm $\|\cdot\|$ is also complete by Corollary 2.15. □

Corollary 2.20

If Y is a finite-dimensional subspace of a normed vector space X, then Y is closed.

Proof

The space Y is itself a normed vector space and so it is a complete metric space by Corollary 2.19. Hence Y is closed since any complete subset of a metric space X is closed. □

These results show that the metric space properties of all finite-dimensional normed spaces are similar to those of $\mathbb{F}^n$. However, each norm on a finite-dimensional space will give different normed space properties. At the start of this section we gave a geometrical example to illustrate this. A different example of this is the difficulty in getting a good estimate of the smallest possible range we can take for $[m, M]$ in Corollary 2.17.

EXERCISES

2.6 Let $\mathcal{P}$ be the vector space of polynomials defined on $[0, 1]$. As $\mathcal{P}$ is a linear subspace of $C_{\mathbb{F}}([0, 1])$ it has a norm $\|p\|_1 = \sup\{|p(x)| : x \in [0, 1]\}$ and as $\mathcal{P}$ is a linear subspace of $L^1[0, 1]$ it has another norm $\|p\|_2 = \int_0^1 |p(x)|\, dx$. Show that $\|\cdot\|_1$ and $\|\cdot\|_2$ are not equivalent on $\mathcal{P}$.

2.7 Sketch the set $\{(x, y) \in \mathbb{R}^2 : \|(x, y)\| = 1\}$ when:

(a) $\|\cdot\|$ is the standard norm on $\mathbb{R}^2$ given in Example 2.2;

(b) $\|(x,y)\| = |x| + |y|$.

2.8 Prove Lemma 2.14(b).

2.9 Prove Corollary 2.15(b).

2.3 Banach Spaces

When we turn to an infinite-dimensional vector space X we saw in Exercise 2.6, for example, that there may be two norms on X which are not equivalent. Therefore many of the methods used in the previous section do not extend to infinite-dimensional vector spaces and so we might expect that many of these results are no longer true. For example, every finite-dimensional linear subspace of a normed vector space is closed, by Corollary 2.20. This is not true for all infinite-dimensional linear subspaces of normed spaces as we now see.

Example 2.21

Let

$$S = \{\{x_n\} \in \ell^\infty : \text{ there exists } N \in \mathbb{N} \text{ such that } x_n = 0 \text{ for } n \geq N\},$$

so that S is the linear subspace of ℓ^∞ consisting of sequences having only finitely many non-zero terms. Then S is not closed.

Solution

If $x = \left(1, \frac{1}{2}, \frac{1}{3}, \dots\right)$ then $x \in \ell^\infty \setminus S$. Let $x_n = \left(1, \frac{1}{2}, \frac{1}{3}, \dots, \frac{1}{n}, 0, 0, \dots\right)$. Then $x_n \in S$ and

$$\|x - x_n\| = \|\left(0, 0, \dots, 0, \frac{1}{n+1}, \frac{1}{n+2}, \dots\right)\| = \frac{1}{n+1}.$$

Hence $\lim_{n\to\infty} \|x - x_n\| = 0$ and so $\lim_{n\to\infty} x_n = x$. Therefore $x \in \overline{S} \setminus S$ and thus S is not closed. □

We will see below that closed linear subspaces are more important than linear subspaces which are not closed. Thus, if a given linear subspace S is not closed it will often be advantageous to consider its closure $\overline{S}$ instead (recall that for any subset A of a metric space, $A \subset \overline{A}$ and $A = \overline{A}$ if and only if A is closed). However, we must first show that the closure of a linear subspace of a normed vector space is also a linear subspace.

Lemma 2.22

If X is a normed vector space and S is a linear subspace of X then $\overline{S}$ is a linear subspace of X.

Proof

We use the subspace test. Let $x, y \in \overline{S}$ and let $\alpha \in \mathbb{F}$. Since $x, y \in \overline{S}$ there are sequences $\{x_n\}$ and $\{y_n\}$ in S such that

$$x = \lim_{n\to\infty} x_n, \quad y = \lim_{n\to\infty} y_n.$$

Since S is a linear subspace $x_n + y_n \in S$ for all $n \in \mathbb{N}$, so

$$x + y = \lim_{n\to\infty} (x_n + y_n) \in \overline{S}.$$

Similarly $\alpha x_n \in S$ for all $n \in \mathbb{N}$, so

$$\alpha x = \lim_{n\to\infty} \alpha x_n \in \overline{S}.$$

Hence $\overline{S}$ is a linear subspace. □

Suppose that X is a normed vector space and let E be any non-empty subset of X. We recall from Definition 1.3 that the span of E is the set of all linear combinations of elements of E or equivalently the intersection of all linear subspaces containing E. Since X is a closed linear subspace of X containing E we can form a similar intersection with closed linear subspaces.

Definition 2.23

Let X be a normed vector space and let E be any non-empty subset of X. The *closed linear span* of E, denoted by $\overline{\mathrm{Sp}}\, E$, is the intersection of all the closed linear subspaces of X which contain E.

The notation used for the closed linear span of E suggests that there is a link between $\overline{\mathrm{Sp}}\, E$ and $\mathrm{Sp}\, E$. This link is clarified in Lemma 2.24.

Lemma 2.24

Let X be a normed space and let E be any non-empty subset of X.

(a) $\overline{\mathrm{Sp}}\, E$ is a closed linear subspace of X which contains E.

(b) $\overline{\mathrm{Sp}}\, E = \overline{\mathrm{Sp}\, E}$, that is, $\overline{\mathrm{Sp}}\, E$ is the closure of $\mathrm{Sp}\, E$.

Proof

(a) As the intersection of any family of closed sets is closed, $\overline{\mathrm{Sp}}\, E$ is closed, and as the intersection of any family of linear subspaces is a linear subspace $\overline{\mathrm{Sp}}\, E$ is a linear subspace. Hence $\overline{\mathrm{Sp}}\, E$ is a closed linear subspace of X which contains E.

(b) Since $\overline{\mathrm{Sp}\, E}$ is a closed linear subspace containing E, we have $\overline{\mathrm{Sp}}\, E \subseteq \overline{\mathrm{Sp}\, E}$. On the other hand, $\overline{\mathrm{Sp}}\, E$ is a linear subspace of X containing E, so $\mathrm{Sp}\, E \subseteq \overline{\mathrm{Sp}}\, E$. Since $\overline{\mathrm{Sp}}\, E$ is closed, $\overline{\mathrm{Sp}\, E} \subseteq \overline{\mathrm{Sp}}\, E$. Therefore $\overline{\mathrm{Sp}\, E} = \overline{\mathrm{Sp}}\, E$. □

The usual way to find $\overline{\mathrm{Sp}}\, E$ is to find $\mathrm{Sp}\, E$, then $\overline{\mathrm{Sp}\, E}$ and then use Lemma 2.24.

The importance of closed linear subspaces is illustrated in Theorem 2.25. If the word "closed" is omitted from the statement of this theorem the result need not be true.

Theorem 2.25 (Riesz' lemma)

Suppose that X is a normed vector space, Y is a closed linear subspace of X such that $Y \neq X$ and α is a real number such that $0 < \alpha < 1$. Then there exists $x_\alpha \in X$ such that $\|x_\alpha\| = 1$ and $\|x_\alpha - y\| > \alpha$ for all $y \in Y$.

Proof

As $Y \neq X$ there is a point $x \in X \setminus Y$. Also, since Y is a closed set,

$$d = \inf\{\|x - z\| : z \in Y\} > 0,$$

by Theorem 1.25. Thus $d < d\alpha^{-1}$ since $0 < \alpha < 1$, so that there exists $z \in Y$ such that $\|x - z\| < d\alpha^{-1}$. Let $x_\alpha = \dfrac{x - z}{\|x - z\|}$. Then $\|x_\alpha\| = 1$ and for any $y \in Y$,

$$\begin{aligned}
\|x_\alpha - y\| &= \left\| \frac{x - z}{\|x - z\|} - y \right\| \\
&= \left\| \frac{x}{\|x - z\|} - \frac{z}{\|x - z\|} - \frac{\|x - z\| y}{\|x - z\|} \right\| \\
&= \frac{1}{\|x - z\|} \|x - (z + \|x - z\| y)\| \\
&> (\alpha d^{-1}) d \\
&= \alpha
\end{aligned}$$

as $z + \|x - z\| y \in Y$, since Y is a linear subspace. □

Theorem 2.26

If X is an infinite-dimensional normed vector space then neither $D = \{x \in X : \|x\| \leq 1\}$ nor $K = \{x \in X : \|x\| = 1\}$ is compact.

Proof

Let $x_1 \in K$. Then, as X is not finite-dimensional, $\mathrm{Sp}\{x_1\} \neq X$ and as $\mathrm{Sp}\{x_1\}$ is finite-dimensional, $\mathrm{Sp}\{x_1\}$ is closed by Corollary 2.20. Hence, by Riesz' lemma, there exists $x_2 \in K$ such that

$$\|x_2 - \alpha x_1\| \geq \frac{3}{4}$$

for all $\alpha \in \mathbb{F}$. Similarly $\mathrm{Sp}\{x_1, x_2\} \neq X$ and as $\mathrm{Sp}\{x_1, x_2\}$ is finite-dimensional $\mathrm{Sp}\{x_1, x_2\}$ is closed. Hence, by Riesz' lemma, there exists $x_3 \in K$ such that

$$\|x_3 - \alpha x_1 - \beta x_2\| \geq \frac{3}{4}$$

for all α, $\beta \in \mathbb{F}$.

Continuing this way we obtain a sequence $\{x_n\}$ in K such that $\|x_n - x_m\| \geq 3/4$ when $n \neq m$. This cannot have a convergent subsequence and so neither D nor K is compact. □

Compactness can be a very useful metric space property as we saw for example in the proof of Corollary 2.17. We recall that in any finite-dimensional normed space any closed bounded set is compact, but unfortunately there are not as many compact sets in infinite-dimensional normed spaces as there are in finite-dimensional spaces by Theorem 2.26. This is a major difference between the metric space structure of finite-dimensional and infinite-dimensional normed spaces. Therefore the deeper results in normed spaces will likely only occur in those spaces which are complete.

Definition 2.27

A *Banach space* is a normed vector space which is complete under the metric associated with the norm.

Fortunately, many of our examples of normed vector spaces are Banach spaces. It is convenient to list all of these together in one theorem even though many of the details have been given previously.

Theorem 2.28

(a) Any finite-dimensional normed vector space is a Banach space.

(b) If X is a compact metric space then $C_{\mathbb{F}}(X)$ is a Banach space.

(c) If (X, Σ, μ) is a measure space then $L^p(X)$ is a Banach space for $1 \leq p \leq \infty$.

(d) ℓ^p is a Banach space for $1 \leq p \leq \infty$.

(e) If X is a Banach space and Y is a linear subspace of X then Y is a Banach space if and only if Y is closed in X.

Proof

(a) This is given in Corollary 2.19.

(b) This is given in Theorem 1.38.

(c) This is given in Theorem 1.60.

(d) This is a special case of part (c) by taking counting measure on $\mathbb{N}$.

(e) Y is a normed linear subspace from Example 2.7 and Y is a Banach space if and only if it is complete. However, a subset of a complete metric space is complete if and only if it is closed by Theorem 1.34. Hence Y is a Banach space if and only if Y is closed in X. □

To give a small demonstration of the importance that completeness will play in the rest of this book, we conclude this chapter with an analogue of the absolute convergence test for series which is valid for Banach spaces. To state this, we first need the definition of convergence of a series in a normed space.

Definition 2.29

Let X be a normed space and let $\{x_k\}$ be a sequence in X. For each positive integer n let $s_n = \sum_{k=1}^{n} x_k$ be the nth partial sum of the sequence. The series $\sum_{k=1}^{\infty} x_k$ is said to *converge* if $\lim_{n \to \infty} s_n$ exists in X and, if so, we define

$$\sum_{k=1}^{\infty} x_k = \lim_{n \to \infty} s_n.$$

Theorem 2.30

Let X be a Banach space and let $\{x_n\}$ be a sequence in X. If the series $\sum_{k=1}^{\infty} \|x_k\|$ converges then the series $\sum_{k=1}^{\infty} x_k$ converges.

Proof

Let $\epsilon > 0$ and let $s_n = \sum_{k=1}^{n} x_k$ be the nth partial sum of the sequence. As $\sum_{k=1}^{\infty} \|x_k\|$ converges the partial sums of $\sum_{k=1}^{\infty} \|x_k\|$ form a Cauchy sequence so there exists $N \in \mathbb{N}$ such that $\sum_{k=n+1}^{m} \|x_k\| < \epsilon$ when $m > n \geq N$. Therefore, by the triangle inequality, when $m > n \geq N$

$$\|s_m - s_n\| \leq \sum_{k=n+1}^{m} \|x_k\| < \epsilon.$$

Hence $\{s_n\}$ is a Cauchy sequence and so converges as X is complete. Thus $\sum_{k=1}^{\infty} x_k$ converges. □

EXERCISES

2.10 If

$$S = \{\{x_n\} \in \ell^2 : \exists N \in \mathbb{N} \text{ such that } x_n = 0 \text{ for } n \geq N\},$$

so that S is a linear subspace of ℓ^2 consisting of sequences having only finitely many non-zero terms, show that S is not closed.

2.11 Let X be a normed space, let $x \in X \setminus \{0\}$ and let Y be a linear subspace of X.

(a) If there is $\eta > 0$ such that $\{y \in X : \|y\| < \eta\} \subseteq Y$, show that $\dfrac{\eta x}{2\|x\|} \in Y$.

(b) If Y is open, show that $Y = X$.

2.12 Let X be a normed linear space, let $T = \{x \in X : \|x\| \leq 1\}$ and let $S = \{x \in X : \|x\| < 1\}$.

(a) Show that T is closed.

(b) If $x \in T$ and $x_n = (1 - \dfrac{1}{n})x$, for $n \in \mathbb{N}$, show that $\lim_{n\to\infty} x_n = x$ and hence show that $\overline{S} = T$.

2.13 Let X be a Banach space with norm $\|\cdot\|_1$ and let Y be a Banach space with norm $\|\cdot\|_2$. If $Z = X \times Y$ with the norm given in Example 2.8, show that Z is a Banach space.

2.14 Let S be any non-empty set, let X be a Banach space over $\mathbb{F}$ and let $F_b(S, X)$ be the vector space given in Exercise 2.2 with the norm $\|f\|_b = \sup\{\|f(s)\| : s \in S\}$. Show that $F_b(S, X)$ is a Banach space.

3
Inner Product Spaces, Hilbert Spaces

3.1 Inner Products

The previous chapter introduced the concept of the norm of a vector as a generalization of the idea of the length of a vector. However, the length of a vector in $\mathbb{R}^2$ or $\mathbb{R}^3$ is not the only geometric concept which can be expressed algebraically. If $x = (x_1, x_2, x_3)$ and $y = (y_1, y_2, y_3)$ are vectors in $\mathbb{R}^3$ then the angle, θ, between them can be obtained using the scalar product $(x, y) = x_1y_1 + x_2y_2 + x_3y_3 = \|x\|\|y\| \cos\theta$, where $\|x\| = \sqrt{x_1^2 + x_2^2 + x_3^2} = \sqrt{(x, x)}$ and $\|y\| = \sqrt{(y, y)}$ are the lengths of x and y respectively. The scalar product is such a useful concept that we would like to extend it to other spaces. To do this we look for a set of axioms which are satisfied by the scalar product in $\mathbb{R}^3$ and which can be used as the basis of a definition in a more general context. It will be seen that it is necessary to distinguish between real and complex spaces.

Definition 3.1

Let X be a real vector space. An *inner product on* X is a function $(\cdot\,,\cdot) : X \times X \to \mathbb{R}$ such that for all x, y, $z \in X$ and α, $\beta \in \mathbb{R}$,

(a) $(x, x) \geq 0$;

(b) $(x, x) = 0$ if and only if $x = 0$;

(c) $(\alpha x + \beta y, z) = \alpha(x, z) + \beta(y, z)$;

(d) $(x, y) = (y, x)$.

The first example shows, in particular, that the scalar product on $\mathbb{R}^3$ is an inner product.

Example 3.2

The function $(\cdot,\cdot) : \mathbb{R}^k \times \mathbb{R}^k \to \mathbb{R}$ defined by $(x,y) = \sum_{n=1}^{k} x_n y_n$, is an inner product on $\mathbb{R}^k$. This inner product will be called the *standard inner product* on $\mathbb{R}^k$.

Solution

We show that the above formula defines an inner product by verifying that all the properties in Definition 3.1 or 3.3 hold.

(a) $(x,x) = \sum_{n=1}^{k} x_n^2 \geq 0$;

(b) $(x,x) = 0$ implies that $x_n = 0$ for all n, so $x = 0$;

(c) $$(\alpha x + \beta y, z) = \sum_{n=1}^{k} (\alpha x_n + \beta y_n) z_n = \alpha \sum_{n=1}^{k} x_n z_n + \beta \sum_{n=1}^{k} y_n z_n = \alpha(x,z) + \beta(y,z);$$

(d) $$(x,y) = \sum_{n=1}^{k} x_n y_n = \sum_{n=1}^{k} y_n x_n = (y,x).$$ □

Before we turn to other examples we consider what modifications need to be made to define a suitable inner product on complex spaces. Let us consider the space $\mathbb{C}^3$ and, by analogy with Example 3.2, let us examine what seems to be the natural analogue of the scalar product on $\mathbb{R}^3$, namely $(x,y) = \sum_{n=1}^{3} x_n y_n$, x, $y \in \mathbb{C}^3$. An immediate problem is that in the complex case the quantity $(x,x) = x_1^2 + x_2^2 + x_3^2$ need not be real and so, in particular, need not be positive. Thus, in the complex case property (a) in Definition 3.1 need not hold. Furthermore, the quantity $\sqrt{(x,x)}$, which gives the length of x in the real case, need not be a real number, and hence does not give a very good idea of the length of x. On the other hand, the quantities $|x_n|^2 = x_n \overline{x}_n$, $n = 1, 2, 3$ (the bar denotes complex conjugate), are real and positive, so we can avoid these problems by redefining the scalar product on $\mathbb{C}^3$ to be $(x,y) = \sum_{n=1}^{3} x_n \overline{y}_n$. However, the complex conjugate on the y-variables in this definition forces us to make a slight modification in the general definition of inner products on complex spaces.

Definition 3.3

Let X be a complex vector space. An *inner product on* X is a function $(\cdot\,,\cdot) : X \times X \to \mathbb{C}$ such that for all x, y, $z \in X$, α, $\beta \in \mathbb{C}$,

(a) $(x, x) \in \mathbb{R}$ and $(x, x) \geq 0$;

(b) $(x, x) = 0$ if and only if $x = 0$;

(c) $(\alpha x + \beta y, z) = \alpha(x, z) + \beta(y, z)$;

(d) $(x, y) = \overline{(y, x)}$.

Example 3.4

The function $(\cdot\,,\cdot) : \mathbb{C}^k \times \mathbb{C}^k \to \mathbb{C}$ defined by $(x, y) = \sum_{n=1}^{k} x_n \overline{y}_n$, is an inner product on $\mathbb{C}^k$. This inner product will be called the *standard inner product* on $\mathbb{C}^k$.

Solution

The proof follows that of Example 3.2, using properties of the complex conjugate where appropriate. □

Definition 3.5

A real or complex vector space X with an inner product $(\cdot\,,\cdot)$ is called an *inner product space.*

Strictly speaking, we should distinguish the vector space X with the inner product $(\cdot\,,\cdot)$ from the same space with a different inner product $(\cdot\,,\cdot)'$. However, since it will always be obvious which inner product is meant on any given vector space X we (in common with most authors) will ignore this distinction. In particular, from now on (unless otherwise stated) $\mathbb{R}^k$ and $\mathbb{C}^k$ will always denote the inner product spaces with inner products as in Examples 3.2 and 3.4.

The only difference between inner products in real and complex spaces lies in the occurrence of the complex conjugate in property (d) in Definition 3.3. Similarly, except for the occurrence of complex conjugates, most of the results and proofs that we discuss below apply equally to both real and complex spaces. Thus from now on, unless explicitly stated otherwise, vector spaces may be either real or complex, but we will include the complex conjugate in the discussion at appropriate points, on the understanding that in the real case this should be ignored.

In general, an inner product can be defined on any finite-dimensional vector space. We leave the solution of the next example, which is a generalization of Examples 3.2 and 3.4, to Exercise 3.2.

Example 3.6

Let X be a k-dimensional vector space with basis $\{e_1, \ldots, e_k\}$. Let $x, y \in X$ have the representation $x = \sum_{n=1}^k \lambda_n e_n$, $y = \sum_{n=1}^k \mu_n e_n$. The function $(\cdot\,,\cdot) : X \times X \to \mathbb{F}$ defined by $(x, y) = \sum_{n=1}^k \lambda_n \overline{\mu}_n$, is an inner product on X.

Clearly, the above inner product depends on the basis chosen, and so we only obtain a "standard" inner product when there is some natural "standard" basis for the space.

Now let (X, Σ, μ) be a measure space, and recall the vector spaces $L^p(X)$ from Definition 1.53.

Example 3.7

If $f, g \in L^2(X)$ then $f\overline{g} \in L^1(X)$ and the function $(\cdot\,,\cdot) : \mathbb{F}^k \times \mathbb{F}^k \to \mathbb{F}$ defined by $(f, g) = \int_X f\overline{g}\, d\mu$ is an inner product on $L^2(X)$. This inner product will be called the *standard inner product* on $L^2(X)$.

Solution

Let $f, g \in L^2(X)$. Then by Hölder's inequality, with $p = q = 2$ (Theorem 1.54) and the definition of $L^2(X)$,

$$\int_X |f\overline{g}|\, d\mu \leq \left(\int_X |f|^2\, d\mu\right)^{1/2} \left(\int_X |g|^2\, d\mu\right)^{1/2} < \infty,$$

so $f\overline{g} \in L^1(X)$ and the formula $(f, g) = \int_X f\overline{g}\, d\mu$ is well-defined. We now show that the above formula defines an inner product on $L^2(X)$ by verifying that all the properties in Definition 3.1 or 3.3 hold. It follows from the properties of the integral described in Section 1.3 that:

(a) $(f, f) = \int_X |f|^2\, d\mu \geq 0$;

(b) $(f, f) = 0 \iff \int_X |f|^2\, d\mu = 0 \iff f = 0$ a.e.;

(c) $(\alpha f + \beta g, h) = \int_X (\alpha f + \beta g)\overline{h}\, d\mu = \alpha \int_X f\overline{h} + \beta \int_X g\overline{h} = \alpha(f, h) + \beta(g, h)$;

(d) $(f, g) = \int_X f\overline{g}\, d\mu = \overline{\int_X g\overline{f}\, d\mu} = \overline{(g, f)}$. □

The next example shows that the sequence space ℓ^2 defined in Example 1.56 is an inner product space. This is in fact a special case of the previous example, using counting measure on $\mathbb{N}$ (see Chapter 1), but it seems useful to work through the solution from first principles, which we leave to Exercise 3.3.

Example 3.8

If $a = \{a_n\}$, $b = \{b_n\} \in \ell^2$ then the sequence $\{a_n b_n\} \in \ell^1$ and the function $(\cdot\,,\cdot) : \mathbb{F}^k \times \mathbb{F}^k \to \mathbb{F}$ defined by $(a, b) = \sum_{n=1}^{\infty} a_n \overline{b}_n$ is an inner product on ℓ^2. This inner product will be called the *standard inner product* on ℓ^2.

Our next examples show that subspaces and Cartesian products of inner product spaces are also inner product spaces, with the inner product defined in a natural manner. These result are similar to Examples 2.7 and 2.8 for normed spaces. The proofs are again easy so are omitted.

Example 3.9

Let X be an inner product space with inner product $(\cdot\,,\cdot)$ and let S be a linear subspace of X. Let $(\cdot\,,\cdot)_S$ be the restriction of $(\cdot\,,\cdot)$ to S. Then $(\cdot\,,\cdot)_S$ is an inner product on S.

Example 3.10

Let X and Y be inner product spaces with inner products $(\cdot\,,\cdot)_1$ and $(\cdot\,,\cdot)_2$ respectively, and let $Z = X \times Y$ be the Cartesian product space (see Definition 1.4). Then the function $(\cdot\,,\cdot) : Z \times Z \to \mathbb{F}$ defined by $((u, v), (x, y)) = (u, x)_1 + (v, y)_2$ is an inner product on Z.

Remark 3.11

We should note that, although the definitions in Examples 2.8 and 3.10 are natural, the norm induced on Z by the above inner product has the form $\sqrt{\|x\|_1^2 + \|y\|_2^2}$ (where $\|\cdot\|_1$, $\|\cdot\|_2$ are the norms induced by the inner products $(\cdot\,,\cdot)_1$, $(\cdot\,,\cdot)_2$), whereas the norm defined on Z in Example 2.8 has the form $\|x\|_1 + \|y\|_2$. These two norms are not equal, but they are equivalent so in discussing analytic properties it makes no difference which one is used. However, the induced norm is somewhat less convenient to manipulate due to the square root term. Thus, in dealing with Cartesian product spaces one generally uses the norm in Example 2.8 if only norms are involved, but one must use the induced norm if inner products are also involved.

We now state some elementary algebraic identities which inner products satisfy.

Lemma 3.12

Let X be an inner product space, x, y, $z \in X$ and α, $\beta \in \mathbb{F}$. Then,

(a) $(0, y) = (x, 0) = 0$;

(b) $(x, \alpha y + \beta z) = \overline{\alpha}(x, y) + \overline{\beta}(x, z)$;

(c) $(\alpha x + \beta y, \alpha x + \beta y) = |\alpha|^2(x, x) + \alpha\overline{\beta}(x, y) + \beta\overline{\alpha}(y, x) + |\beta|^2(y, y)$.

Proof

(a) $(\mathbf{0}, y) = (0.\mathbf{0}, y) = 0(\mathbf{0}, y) = 0$ and $(x, \mathbf{0}) = \overline{(\mathbf{0}, x)} = \overline{0} = 0$
(to distinguish the zero vector in X from the scalar zero we have, temporarily, denoted the zero vector by $\mathbf{0}$ here).

(b) $(x, \alpha y + \beta z) = \overline{(\alpha y + \beta z, x)} = \overline{\alpha(y, x) + \beta(z, x)} = \overline{\alpha}(x, y) + \overline{\beta}(x, z)$
(using the properties of the inner product in Definition 3.3).

(c)
$$\begin{aligned}(\alpha x + \beta y, \alpha x + \beta y) &= \overline{\alpha}(\alpha x + \beta y, x) + \overline{\beta}(\alpha x + \beta y, y)\\ &= \overline{\alpha}\alpha(x, x) + \overline{\alpha}\beta(y, x) + \overline{\beta}\alpha(x, y) + \overline{\beta}\beta(y, y),\end{aligned}$$

hence the result follows from $\alpha\overline{\alpha} = |\alpha|^2$, $\beta\overline{\beta} = |\beta|^2$. □

From part (c) of Definition 3.3 and part (b) of Lemma 3.12, we can say that an inner product $(\cdot\,,\cdot)$ on a complex space is linear with respect to its first variable and is *conjugate linear* with respect to its second variable (an inner product on a real space is linear with respect to both variables).

In the introduction to this chapter we noted that if $x \in \mathbb{R}^3$ and $(\cdot\,,\cdot)$ is the usual inner product on $\mathbb{R}^3$, then the formula $\sqrt{(x, x)}$ gives the usual Euclidean length, or norm, of x. We now want to show that for a general inner product space X the same formula defines a norm on X.

Lemma 3.13

Let X be an inner product space and let x, $y \in X$. Then:

(a) $|(x, y)|^2 \le (x, x)(y, y)$, x, $y \in X$;

(b) the function $\|\cdot\| : X \to \mathbb{R}$ defined by $\|x\| = (x, x)^{1/2}$, is a norm on X.

Proof

(a) If $x = 0$ or $y = 0$ the result is true, so we suppose that neither x nor y is zero. Putting $\alpha = -\overline{(x,y)}/(x,x)$ and $\beta = 1$ in part (c) of Lemma 3.12 we obtain

$$\begin{aligned} 0 &\le (\alpha x + y, \alpha x + y) \\ &= \left|\frac{\overline{(x,y)}}{(x,x)}\right|^2 (x,x) - \frac{\overline{(x,y)}(x,y)}{(x,x)} - \frac{(x,y)(y,x)}{(x,x)} + (y,y) \\ &= \frac{|(x,y)|^2}{(x,x)} - \frac{|(x,y)|^2}{(x,x)} - \frac{|(x,y)|^2}{(x,x)} + (y,y) \qquad (\text{since } (y,x) = \overline{(x,y)}) \end{aligned}$$

and hence $|(x,y)|^2 \le (x,x)(y,y)$.

(b) Using the above properties of inner products we now verify that the given definition of $\|x\|$ satisfies all the defining properties of a norm:

(i) $\|x\| = (x,x)^{1/2} \ge 0$;

(ii) $\|x\| = 0 \iff (x,x)^{1/2} = 0 \iff x = 0$;

(iii) $\|\alpha x\| = (\alpha x, \alpha x)^{1/2} = (\alpha\overline{\alpha})^{1/2}(x,x)^{1/2} = |\alpha|\|x\|$;

(iv) $$\begin{aligned} \|x+y\|^2 &= \|x\|^2 + 2\Re\mathrm{e}\,(x,y) + \|y\|^2 \quad &&(\text{by Lemma 3.12b) with } \alpha = \beta = 1) \\ &\le \|x\|^2 + 2\|x\|\|y\| + \|y\|^2 &&(\text{by (a)}) \\ &= (\|x\| + \|y\|)^2. \end{aligned}$$

□

The norm $\|x\| = (x,x)^{1/2}$ defined in Lemma 3.13 on the inner product space X is said to be *induced* by the inner product $(\cdot\,,\cdot)$. The lemma shows that, by using the induced norm, every inner product space can be regarded as a normed space. From now on, whenever we use a norm on an inner product space X it will be the induced norm – we will not mention this specifically each time. By examining the standard norms and inner products that we have defined so far (on $\mathbb{F}^k$, ℓ^2 and $L^2(X)$), we see that each of these norms are induced by the corresponding inner products. Also, with this convention, the inequality in part (a) of Lemma 3.13 can be rewritten as

$$|(x,y)| \le \|x\|\|y\|. \tag{3.1}$$

The inequality in Corollary 1.58 is a special case of inequality (3.1) (obtained by substituting the standard inner product and norm on $\mathbb{F}^k$ into (3.1)). Thus inequality (3.1) is also called the *Cauchy–Schwarz inequality*. This will prove to be an extremely useful inequality below. We note also that it is frequently more convenient to work with the square of the induced norm to avoid having to take square roots.

Since every inner product space has an induced norm, a natural question is whether *every* norm is induced by an inner product. The answer is no – norms induced by inner products have some special properties that norms in general do not possess. We leave the proofs of the identities in the following lemma and theorem as exercises (see Exercises 3.4 and 3.5).

Lemma 3.14

Let X be an inner product space with inner product $(\cdot\,,\cdot)$. Then for all u, v, x, $y \in X$:

(a) $(u+v, x+y) - (u-v, x-y) = 2(u,y) + 2(v,x)$;

(b) $4(u,y) = (u+v, x+y) - (u-v, x-y) + i(u+iv, x+iy)$
$\qquad -i(u-iv, x-iy)$
(for complex X).

Theorem 3.15

Let X be an inner product space with inner product $(\cdot\,,\cdot)$ and induced norm $\|\cdot\|$. Then for all x, $y \in X$:

(a)
$$\|x+y\|^2 + \|x-y\|^2 = 2\left(\|x\|^2 + \|y\|^2 \right)$$
(the *parallelogram rule*);

(b) if X is real then
$$4(x,y) = \|x+y\|^2 - \|x-y\|^2;$$

(c) if X is complex then
$$4(x,y) = \|x+y\|^2 - \|x-y\|^2 + i\|x+iy\|^2 - i\|x-iy\|^2$$
(the *polarization identity*).

One way to show that a given norm on a vector space is not induced by an inner product is to show that it does not satisfy the parallelogram rule.

Example 3.16

The standard norm on the space $C[0,1]$ is not induced by an inner product.

Solution

Consider the functions $f,\ g \in C[0,1]$ defined by $f(x) = 1,\ g(x) = x,\ x \in [0,1]$. From the definition of the standard norm on $C[0,1]$ we have

$$\begin{aligned} \|f+g\| + \|f-g\| &= 2+1 = 3, \\ 2(\|f\| + \|g\|) &= 2(1+1) = 4. \end{aligned}$$

Thus the parallelogram rule does not hold and so the norm cannot be induced by an inner product. □

Since an inner product space X is a normed space (with the induced norm), it is also a metric space with the metric associated with the norm (see Chapter 2). From now on, any metric space concepts that we use on X will be defined in terms of this metric. An important property of this metric is that the inner product $(\cdot\,,\cdot)$ on X, which was defined solely in terms of its algebraic properties, is a continuous function on X in the following sense.

Lemma 3.17

Let X be an inner product space and suppose that $\{x_n\}$ and $\{y_n\}$ are convergent sequences in X, with $\lim_{n\to\infty} x_n = x$, $\lim_{n\to\infty} y_n = y$. Then

$$\lim_{n\to\infty} (x_n, y_n) = (x, y).$$

Proof

$$\begin{aligned} |(x_n,y_n) - (x,y)| &= |(x_n,y_n) - (x_n,y) + (x_n,y) - (x,y)| \\ &\le |(x_n,y_n) - (x_n,y)| + |(x_n,y) - (x,y)| \\ &= |(x_n, y_n - y)| + |(x_n - x, y)| \\ &\le \|x_n\|\|y_n - y\| + \|x_n - x\|\|y\| \qquad \text{(by (3.1))}. \end{aligned}$$

Since the sequence $\{x_n\}$ is convergent, $\|x_n\|$ is bounded, so the right-hand side of this inequality tends to zero as $n \to \infty$. Hence $\lim_{n\to\infty}(x_n, y_n) = (x,y)$. □

EXERCISES

3.1 Let X be an inner product space and let $u,\ v \in X$. If $(x,u) = (x,v)$ for all $x \in X$, show that $u = v$.

3.2 Give the solution of Example 3.6.

3.3 Prove the following inequalities, for any a, $b \in \mathbb{C}$.

(a) $2|a||b| \leq |a|^2 + |b|^2$.

(b) $|a+b|^2 \leq 2(|a|^2 + |b|^2)$.

The second inequality is a special case of Minkowski's inequality, see Corollary 1.58, but it is worthwhile to prove this simple case directly.

Use these inequalities to show, from first principles, that the set ℓ^2 is a vector space and that the formula in Example 3.8 gives an inner product on ℓ^2.

3.4 Give the proof of Lemma 3.14.

3.5 Use the results of Exercise 3.4 to prove Theorem 3.15.

3.6 Draw a picture to illustrate the parallelogram rule and to show why it has this name.

3.7 Show that the non-standard norm $\|x\|_1 = \sum_{n=1}^{k} |x_n|$ on the space $\mathbb{R}^k$ is not induced by an inner product.

3.2 Orthogonality

The reason we introduced inner products was in the hope of extending the concept of angles between vectors. From the Cauchy–Schwarz inequality (3.1) for real inner product spaces, if x, y are non-zero vectors, then

$$-1 \leq \frac{(x,y)}{\|x\|\|y\|} \leq 1,$$

and so the angle between x and y can be defined to be

$$\theta = \cos^{-1}\left(\frac{(x,y)}{\|x\|\|y\|}\right).$$

For complex inner product spaces the position is more difficult (the inner product (x,y) may be complex, and it is not clear what a complex "angle" would mean). However, an important special case can be considered, namely when $(x,y) = 0$. In this case we can regard the vectors as being perpendicular, or orthogonal.

Definition 3.18

Let X be an inner product space. The vectors x, $y \in X$ are said to be *orthogonal* if $(x,y) = 0$.

From linear algebra we are also familiar with the concept of orthonormal sets of vectors in finite-dimensional inner product spaces. This concept can be extended to arbitrary inner product spaces.

Definition 3.19

Let X be an inner product space. The set $\{e_1, \ldots, e_k\} \subset X$ is said to be *orthonormal* if $\|e_n\| = 1$ for $1 \leq n \leq k$, and $(e_m, e_n) = 0$ for all $1 \leq m, n \leq k$ with $m \neq n$.

The results in the following lemma about orthonormal sets in finite dimensional inner product spaces may well be familiar, but we recall them here for later reference.

Lemma 3.20

(a) An orthonormal set $\{e_1, \ldots, e_k\}$ in any inner product space X is linearly independent. In particular, if X is k-dimensional then the set $\{e_1, \ldots, e_k\}$ is a basis for X and any vector $x \in X$ can be expressed in the form

$$x = \sum_{n=1}^{k} (x, e_n) e_n \tag{3.2}$$

(in this case $\{e_1, \ldots, e_k\}$ is usually called an *orthonormal basis* and the numbers (x, e_n) are the *components* of x with respect to this basis).

(b) Let $\{v_1, \ldots, v_k\}$ be a linearly independent subset of an inner product space X, and let $S = \mathrm{Sp}\,\{v_1, \ldots, v_k\}$. Then there is an orthonormal basis $\{e_1, \ldots, e_k\}$ for S.

Proof

(a) Suppose that $\sum_{n=1}^{k} \alpha_n e_n = 0$, for some $\alpha_n \in \mathbb{F}$, $n = 1, \ldots, k$. Then taking the inner product with e_m and using orthonormality we obtain

$$0 = \Big(\sum_{n=1}^{k} \alpha_n e_n, e_m \Big) = \alpha_m,$$

for $m = 1, \ldots, k$. Hence, the set $\{e_1, \ldots, e_k\}$ is linearly independent.

Next, if $\{e_1, \ldots, e_k\}$ is a basis there exists $\lambda_n \in \mathbb{F}$, $n = 1, \ldots, k$, such that $x = \sum_{n=1}^{k} \lambda_n e_n$. Then taking the inner product of this formula with e_m

and using orthonormality we obtain

$$(x, e_m) = \Big(\sum_{n=1}^{k} \lambda_n e_n, e_m \Big) = \sum_{n=1}^{k} \lambda_n (e_n, e_m) = \lambda_m, \quad m = 1, \ldots, k.$$

(b) The proof is by induction on k.

For $k = 1$, since $v_1 \neq 0$, $\|v_1\| \neq 0$ so we can take $e_1 = v_1/\|v_1\|$, and $\{e_1\}$ is the required basis.

Now suppose that the result is true for an arbitrary integer $k \geq 1$. Let $\{v_1, \ldots, v_{k+1}\}$ be a linearly independent set and let $\{e_1, \ldots, e_k\}$ be the orthonormal basis for $\mathrm{Sp}\{v_1, \ldots, v_k\}$ given by the inductive hypothesis. Since $\{v_1, \ldots, v_{k+1}\}$ is linearly independent, $v_{k+1} \notin \mathrm{Sp}\{v_1, \ldots, v_k\}$ so $v_{k+1} \notin \mathrm{Sp}\{e_1, \ldots, e_k\}$. Let $b_{k+1} = v_{k+1} - \sum_{n=1}^{k} (v_{k+1}, e_n) e_n$. Then $b_{k+1} \in \mathrm{Sp}\{v_1, \ldots, v_{k+1}\}$ and $b_{k+1} \neq 0$ (otherwise $v_{k+1} \in \mathrm{Sp}\{e_1, \ldots, e_k\}$). Also, for each $m = 1, \ldots, k$,

$$\begin{aligned}
(b_{k+1}, e_m) &= (v_{k+1}, e_m) - \sum_{n=1}^{k} (v_{k+1}, e_n)(e_n, e_m) \\
&= (v_{k+1}, e_m) - (v_{k+1}, e_m) = 0
\end{aligned}$$

(using orthonormality of the set $\{e_1, \ldots, e_k\}$). Thus, b_{k+1} is orthogonal to all the vectors $\{e_1, \ldots, e_k\}$. Let $e_{k+1} = b_{k+1}/\|b_{k+1}\|$. Then $\{e_1, \ldots, e_{k+1}\}$ is an orthonormal set with $\mathrm{Sp}\{e_1, \ldots, e_{k+1}\} \subset \mathrm{Sp}\{v_1, \ldots, v_{k+1}\}$. But both these subspaces are $(k+1)$-dimensional so they must be equal, which completes the inductive proof. □

Remark 3.21

The inductive construction of the basis in part (b) of Lemma 3.20, using the formulae

$$b_{k+1} = v_{k+1} - \sum_{n=1}^{k} (v_{k+1}, e_n) e_n, \qquad e_{k+1} = \frac{b_{k+1}}{\|b_{k+1}\|},$$

is called the *Gram–Schmidt algorithm*, and is described in more detail in [5] (see Fig. 3.1 for an illustration of this algorithm at the stage $k = 2$).

Using an orthonormal basis in a (finite-dimensional) inner product space makes it easy to work out the norm of a vector in terms of its components. The following theorem is a generalization of Pythagoras' theorem.

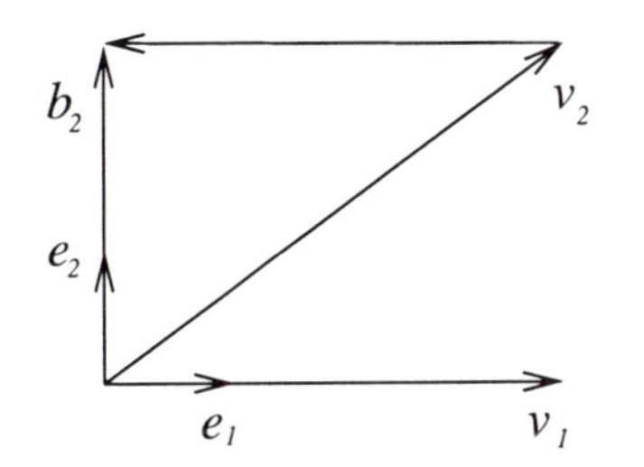

Fig. 3.1. Gram–Schmidt algorithm, at stage $k = 2$

Theorem 3.22

Let X be a k-dimensional inner product space and let $\{e_1, \ldots, e_k\}$ be an orthonormal basis for X. Then, for any numbers $\alpha_n \in \mathbb{F}$, $n = 1, \ldots, k$,

$$\|\sum_{n=1}^{k} \alpha_n e_n\|^2 = \sum_{n=1}^{k} |\alpha_n|^2.$$

Proof

By orthonormality and the algebraic properties of the inner product we have

$$\begin{aligned}\|\sum_{n=1}^{k} \alpha_n e_n\|^2 &= \Big(\sum_{m=1}^{k} \alpha_m e_m, \sum_{n=1}^{k} \alpha_n e_n \Big) \\ &= \sum_{m=1}^{k} \sum_{n=1}^{k} \alpha_m \overline{\alpha}_n (e_m, e_n) \\ &= \sum_{n=1}^{k} \alpha_n \overline{\alpha}_n = \sum_{n=1}^{k} |\alpha_n|^2.\end{aligned}$$

□

We saw, when discussing normed spaces, that completeness is an extremely important property and this is also true for inner product spaces. Complete normed spaces were called Banach spaces, and complete inner product spaces also have a special name.

Definition 3.23

An inner product space which is complete with respect to the metric associated with the norm induced by the inner product is called a *Hilbert space.*

From the preceding results we have the following examples of Hilbert spaces.

Example 3.24

(a) Every finite-dimensional inner product space is a Hilbert space.

(b) $L^2(X)$ with the standard inner product is a Hilbert space.

(c) ℓ^2 with the standard inner product is a Hilbert space.

Solution

Part (a) follows from Corollary 2.19, while parts (b) and (c) follow from Theorem 1.60. □

In general, infinite-dimensional inner product spaces need not be complete. We saw in Theorem 2.28 that a linear subspace of a Banach space is a Banach space if and only if it is closed. A similar result holds for Hilbert spaces.

Lemma 3.25

If $\mathcal{H}$ is a Hilbert space and $Y \subset \mathcal{H}$ is a linear subspace, then Y is a Hilbert space if and only if Y is closed in $\mathcal{H}$.

Proof

By the above definitions, Y is a Hilbert space if and only if it is complete. But a subset of a complete metric space is complete if and only if it is closed by Theorem 1.34. □

EXERCISES

3.8 Let X be an inner product space. Show that two vectors $x, y \in X$ are orthogonal if and only if $\|x + \alpha y\| = \|x - \alpha y\|$ for all $\alpha \in \mathbb{F}$. Draw a picture of this in $\mathbb{R}^2$.

3.9 If S is the linear span of $a = (1, 4, 1)$ and $b = (-1, 0, 1)$ in $\mathbb{R}^3$, use the Gram–Schmidt algorithm to find an orthonormal basis of $\mathbb{R}^3$ containing multiples of a and b.

3.10 Use the Gram–Schmidt algorithm to find an orthonormal basis for $\mathrm{Sp}\,\{1, x, x^2\}$ in $L^2[-1, 1]$.

3.11 Let $\mathcal{M}$ and $\mathcal{N}$ be Hilbert spaces with inner products $(\cdot\,,\cdot)_1$ and $(\cdot\,,\cdot)_2$ respectively, and let $\mathcal{H} = \mathcal{M} \times \mathcal{N}$ be the Cartesian product space, with the inner product defined in Example 3.10. Show that $\mathcal{H}$ is a Hilbert space.

3.3 Orthogonal Complements

We have already seen that the idea of orthogonality of two vectors is an important concept. We now extend this idea to consider the set of all vectors orthogonal to a given vector, or even the set of all vectors orthogonal to a given set of vectors.

Definition 3.26

Let X be an inner product space and let A be a subset of X. The *orthogonal complement of* A is the set

$$A^{\perp} = \{x \in X : (x, a) = 0 \text{ for all } a \in A\}.$$

Thus the set $A^{\perp}$ consists of those vectors in X which are orthogonal to every vector in A (if $A = \emptyset$ then $A^{\perp} = X$). Note that $A^{\perp}$ is *not* the set-theoretic complement of A.

The link between A and $A^{\perp}$ is given by the condition $(x, a) = 0$ for all $a \in A$, and this has to be used to obtain $A^{\perp}$, as shown in the following example.

Example 3.27

If $X = \mathbb{R}^3$ and $A = \{(a_1, a_2, 0) : a_1,\, a_2 \in \mathbb{R}\}$, then $A^{\perp} = \{(0, 0, x_3) : x_3 \in \mathbb{R}\}$.

Solution

By definition, a given vector $x = (x_1 x_2, x_3)$ belongs to $A^{\perp}$ if and only if, for any $a = (a_1, a_2, 0)$ with $a_1,\, a_2 \in \mathbb{R}$, we have

$$(x, a) = x_1 a_1 + x_2 a_2 = 0.$$

Putting $a_1 = x_1$, $a_2 = x_2$, we see that if $x \in A^{\perp}$ then we must have $x_1 = x_2 = 0$. Furthermore, if $x_1 = x_2 = 0$, then it follows that $x \in A^{\perp}$. □

Although the above example is quite simple it is a useful one to keep in mind when considering the concept of orthogonal complement. It can be generalized to the following example, whose solution we leave to Exercise 3.14.

Example 3.28

Suppose that X is a k-dimensional inner product space, and $\{e_1, \ldots, e_k\}$ is an orthonormal basis for X. If $A = \operatorname{Sp}\{e_1, \ldots, e_p\}$, for some $1 \le p < k$, then $A^\perp = \operatorname{Sp}\{e_{p+1}, \ldots, e_k\}$.

Using the above example, the computation of $A^\perp$ for a subset A of a finite-dimensional inner product space is relatively easy. We now turn to general inner product spaces and list some of the principal properties of the orthogonal complement.

Lemma 3.29

If X is an inner product space and $A \subset X$ then:

(a) $0 \in A^\perp$.

(b) If $0 \in A$ then $A \cap A^\perp = \{0\}$, otherwise $A \cap A^\perp = \emptyset$.

(c) $\{0\}^\perp = X$; $X^\perp = \{0\}$.

(d) If A contains an open ball $B_a(r)$, for some $a \in X$ and some positive $r > 0$, then $A^\perp = \{0\}$; in particular, if A is a non-empty open set then $A^\perp = \{0\}$.

(e) If $B \subset A$ then $A^\perp \subset B^\perp$.

(f) $A^\perp$ is a closed linear subspace of X.

(g) $A \subset (A^\perp)^\perp$.

Proof

(a) Since $(0, a) = 0$ for all $a \in A$, we have $0 \in A^\perp$.

(b) Suppose that $x \in A \cap A^\perp$. Then $(x, x) = 0$, and so $x = 0$ (by part (b) of the definition of an inner product).

(c) If $A = \{0\}$, then for any $x \in X$ we have trivially $(x, a) = (x, 0) = 0$ for all $a \in A$, so $x \in A^\perp$ and hence $A^\perp = X$. If $A = X$ and $x \in A^\perp$ then $(x, a) = 0$ for all $a \in X$. In particular, putting $a = x$ gives $(x, x) = 0$, which implies that $x = 0$. Hence, $A^\perp = \{0\}$.

(d) Suppose that $x \in A^\perp$. If $a \in A$ then by definition we have $(x, a) = 0$. But also, since $a + \frac{1}{2}rx \in A$, we must have

$$0 = (x, a + \tfrac{1}{2}rx) = (x, a) + \tfrac{1}{2}r(x, x),$$

which implies that $(x, x) = 0$ and hence $x = 0$.

(e) Let $x \in B^\perp$ and $a \in A$. Then $a \in B$ (since $A \subset B$), so $(x, a) = 0$. Since this holds for arbitrary $a \in A$, we have $x \in A^\perp$. Hence, $B^\perp \subset A^\perp$.

(f) Let $x, y \in A^\perp$, $\alpha, \beta \in \mathbb{F}$ and $a \in A$. Then

$$(\alpha x + \beta y, a) = \alpha(x, a) + \beta(y, a) = 0,$$

so $\alpha x + \beta y \in A^\perp$, and hence $A^\perp$ is a linear subspace of X. Next, let $\{x_n\}$ be a sequence in $A^\perp$ converging to $x \in X$. Then by Lemma 3.12(a) and Lemma 3.17, for any $a \in A$ we have

$$0 = \lim_{n\to\infty}(x - x_n, a) = (x, a) - \lim_{n\to\infty}(x_n, a) = (x, a)$$

(since $(x_n, a) = 0$ for all n). Thus $x \in A^\perp$ and so $A^\perp$ is closed, by Theorem 1.25.

(g) Let $a \in A$. Then for all $x \in A^\perp$, $(a, x) = \overline{(x, a)} = 0$, so $a \in (A^\perp)^\perp$. Thus, $A \subset (A^\perp)^\perp$. □

Part (e) of Lemma 3.29 shows that as the set A gets bigger the orthogonal complement of A gets smaller, and this is consistent with part (c) of the lemma.

There is an alternative characterization of the orthogonal complement for linear subspaces.

Lemma 3.30

Let Y be a linear subspace of an inner product space X. Then

$$x \in Y^\perp \iff \|x - y\| \geq \|x\|, \quad \forall y \in Y.$$

Proof

($\Rightarrow$) From part (c) of Lemma 3.12, the following identity

$$\|x - \alpha y\|^2 = (x - \alpha y, x - \alpha y) = \|x\|^2 - \overline{\alpha}(x, y) - \alpha(y, x) + |\alpha|^2\|y\|^2 \tag{3.3}$$

holds for all $x \in X$, $y \in Y$ and $\alpha \in \mathbb{F}$. Now suppose that $x \in Y^\perp$ and $y \in Y$. Then $(x, y) = (y, x) = 0$, so by putting $\alpha = 1$ in (3.3) we obtain

$$\|x - y\|^2 = \|x\|^2 + \|y\|^2 \geq \|x\|^2.$$

($\Leftarrow$) Now suppose that $\|x - y\|^2 \geq \|x\|^2$ for all $y \in Y$. Then since Y is a linear subspace, $\alpha y \in Y$ for all $y \in Y$ and $\alpha \in \mathbb{F}$, so by (3.3),

$$0 \leq -\overline{\alpha(y, x)} - \alpha(y, x) + |\alpha|^2\|y\|^2.$$

Now let

$$\beta = \begin{cases} \dfrac{|(x,y)|}{(y,x)}, & \text{if } (y,x) \neq 0, \\ 1, & \text{if } (y,x) = 0, \end{cases}$$

so that $\beta(y,x) = |(x,y)|$, and let $\alpha = t\beta$, where $t \in \mathbb{R}$ and $t > 0$. Then, $-t|(x,y)| - t|(x,y)| + t^2\|y\|^2 \geq 0$, so $|(x,y)| \leq \frac{1}{2}t\|y\|^2$ for all $t > 0$. Therefore,

$$|(x,y)| \leq \lim_{t \to 0+} \tfrac{1}{2}t\|y\|^2 = 0,$$

so $|(x,y)| = 0$ and hence $x \in Y^{\perp}$. □

The above result, together with the following theorem on convex sets, gives more information about A and $A^{\perp}$ in Hilbert spaces. We first recall the definition of a convex set.

Definition 3.31

A subset A of a vector space X is *convex* if, for all x, $y \in A$ and all $\lambda \in [0,1]$, $\lambda x + (1-\lambda)y \in A$.

In other words, A is convex if, for any two points x, y in A, the line segment joining x and y also lies in A, see Fig. 3.2. In particular, every linear subspace is a convex set.

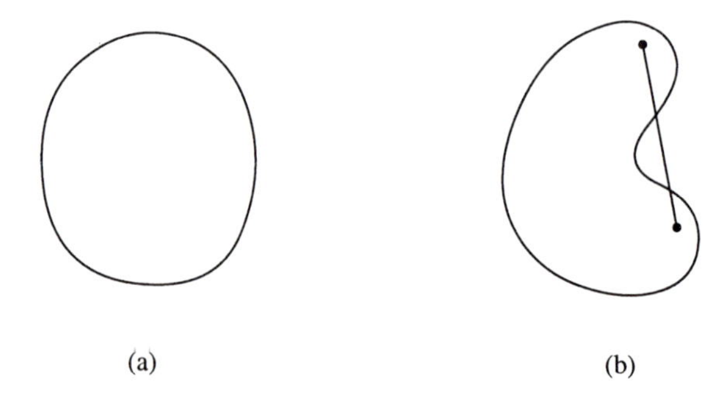

(a) (b)

Fig. 3.2. Convex and non-convex planar regions: (a) convex; (b) non-convex

Theorem 3.32

Let A be a non-empty, closed, convex subset of a Hilbert space $\mathcal{H}$ and let $p \in \mathcal{H}$. Then there exists a unique $q \in A$ such that

$$\|p - q\| = \inf\{\|p - a\| : a \in A\}.$$

Proof

Let $\gamma = \inf\{\|p-a\| : a \in A\}$ (note that the set in this definition is non-empty and bounded below, so γ is well-defined). We first prove the existence of q. By the definition of γ, for each $n \in \mathbb{N}$ there exists $q_n \in A$ such that

$$\gamma^2 \le \|p-q_n\|^2 < \gamma^2 + n^{-1}. \tag{3.4}$$

We will show that the sequence $\{q_n\}$ is a Cauchy sequence. Applying the parallelogram rule to $p-q_n$ and $p-q_m$ we obtain

$$\|(p-q_n)+(p-q_m)\|^2 + \|(p-q_n)-(p-q_m)\|^2 = 2\|p-q_n\|^2 + 2\|p-q_m\|^2,$$

and so

$$\|2p-(q_n+q_m)\|^2 + \|q_n-q_m\|^2 < 4\gamma^2 + 2(n^{-1}+m^{-1}).$$

Since $q_m, q_n \in A$ and A is convex, $\frac{1}{2}(q_m+q_n) \in A$, so

$$\|2p-(q_n+q_m)\|^2 = 4\|p-\tfrac{1}{2}(q_m+q_n)\|^2 \ge 4\gamma^2$$

(by the definition of γ), and hence

$$\|q_n-q_m\|^2 < 4\gamma^2 + 2(n^{-1}+m^{-1}) - 4\gamma^2 = 2(n^{-1}+m^{-1}).$$

Therefore the sequence $\{q_n\}$ is Cauchy and hence must converge to some point $q \in \mathcal{H}$ since $\mathcal{H}$ is a complete metric space. Since A is closed, $q \in A$. Also, by (3.4),

$$\gamma^2 \le \lim_{n\to\infty} \|p-q_n\|^2 = \|p-q\|^2 \le \lim_{n\to\infty}(\gamma^2+n^{-1}) = \gamma^2,$$

and so $\|p-q\| = \gamma$. Thus the required q exists.

Next we prove the uniqueness of q. Suppose that $w \in A$ and $\|p-w\| = \gamma$. Then $\frac{1}{2}(q+w) \in A$ since A is convex, and so $\|p-\frac{1}{2}(q+w)\| \ge \gamma$. Applying the parallelogram rule to $p-w$ and $p-q$ we obtain

$$\|(p-w)+(p-q)\|^2 + \|(p-w)-(p-q)\|^2 = 2\|p-w\|^2 + 2\|p-q\|^2,$$

and so

$$\begin{aligned}\|q-w\|^2 &= 2\gamma^2 + 2\gamma^2 - 4\|p-\tfrac{1}{2}(q+w)\|^2 \\ &\le 4\gamma^2 - 4\gamma^2 = 0.\end{aligned}$$

Thus $w = q$, which proves uniqueness. □

Remark 3.33

Theorem 3.32 shows that if A is a non-empty, closed, convex subset of a Hilbert space $\mathcal{H}$ and p is a point in $\mathcal{H}$, then there is a unique point q in A which is the closest point in A to p. In finite dimensions, even if the set A is not convex the existence of the point q can be proved in a similar manner (using the compactness of closed bounded sets to give the necessary convergent sequence). However, in this case the point q need not be unique (for example, let A be a circle in the plane and p be its centre, then q can be any point on A). In infinite dimensions, closed bounded sets need not be compact (see Theorem 2.26) so the existence question is more difficult and q may not exist if A is not convex.

Theorem 3.34

Let Y be a closed linear subspace of a Hilbert space $\mathcal{H}$. For any $x \in \mathcal{H}$, there exists a unique $y \in Y$ and $z \in Y^\perp$ such that $x = y+z$. Also, $\|x\|^2 = \|y\|^2+\|z\|^2$.

Proof

Since Y is a non-empty, closed, convex set it follows from Theorem 3.32 that there exists $y \in Y$ such that for all $u \in Y$, $\|x-y\| \leq \|x-u\|$. Let $z = x-y$ (so $x = y+z$). Then, for all $u \in Y$,

$$\|z-u\| = \|x-(y+u)\| \geq \|x-y\| = \|z\|$$

(since $y+u \in Y$). Thus by Lemma 3.30, $z \in Y^\perp$. This shows that the desired y, z exist. To prove uniqueness, suppose that $x = y_1+z_1 = y_2+z_2$, where $y_1, y_2 \in Y$, $z_1, z_2 \in Y^\perp$. Then $y_1-y_2 = z_2-z_1$. But $y_1-y_2 \in Y$ and $z_2-z_1 \in Y^\perp$ (since both Y and $Y^\perp$ are linear subspaces), so $y_1-y_2 \in Y \cap Y^\perp = \{0\}$ (by part (f) of Lemma 3.29). Hence $y_1 = y_2$ and $z_1 = z_2$.

Finally,

$$\begin{aligned}\|x\|^2 = \|y+z\|^2 = (y+z, y+z) &= \|y\|^2 + (y,z) + (z,y) + \|z\|^2 \\ &= \|y\|^2 + \|z\|^2.\end{aligned}$$

□

The simplest example of the above result is $\mathcal{H} = \mathbb{R}^2$, $Y = \{(x,0) : x \in \mathbb{R}\}$, $Y^\perp = \{(0,y) : y \in \mathbb{R}\}$. In this case the result is Pythagoras' theorem in the plane. Hence Theorem 3.34 can be regarded as another generalization of Pythagoras' theorem. The result of the theorem will be so useful that we give it a name.

Notation

Suppose that Y is a closed linear subspace of a Hilbert space $\mathcal{H}$ and $x \in \mathcal{H}$. The decomposition $x = y+z$, with $y \in Y$ and $z \in Y^\perp$, will be called the *orthogonal decomposition* of x with respect to Y.

Now that we have studied the orthogonal complement $Y^\perp$ of a subspace Y, a natural question is: what is the orthogonal complement of $Y^\perp$? We will write this as $Y^{\perp\perp} = (Y^\perp)^\perp$. It follows from Examples 3.27 and 3.28 that if $Y = \{u = (u_1, u_2, 0) : u_1, u_2 \in \mathbb{R}\}$, then $Y^\perp = \{x = (0, 0, x_3) : x_3 \in \mathbb{R}\}$ and $Y^{\perp\perp} = Y$. A similar result holds more generally.

Corollary 3.35

If Y is a closed linear subspace of a Hilbert space $\mathcal{H}$ then $Y^{\perp\perp} = Y$.

Proof

By part (g) of Lemma 3.29 we have $Y \subset Y^{\perp\perp}$. Now suppose that $x \in Y^{\perp\perp}$. Then by Theorem 3.34, $x = y + z$, where $y \in Y$ and $z \in Y^\perp$. Since $y \in Y$ and $x \in Y^{\perp\perp}$, $(x, z) = 0 = (y, z)$. Thus

$$0 = (x, z) = (y + z, z) = (y, z) + (z, z) = \|z\|^2,$$

so $z = 0$ and $x = y \in Y$. Therefore $Y^{\perp\perp} \subset Y$, which completes the proof. □

Note that since $Y^{\perp\perp}$ is always closed (by part (f) of Lemma 3.29) the above result cannot hold for subspaces which are not closed. However, we do have the following result.

Corollary 3.36

If Y is any linear subspace of a Hilbert space $\mathcal{H}$ then $Y^{\perp\perp} = \overline{Y}$.

Proof

Since $Y \subset \overline{Y}$, it follows from part (e) of Lemma 3.29 that $\overline{Y}^\perp \subset Y^\perp$ and hence $Y^{\perp\perp} \subset \overline{Y}^{\perp\perp}$. But $\overline{Y}$ is closed so by Corollary 3.35, $\overline{Y}^{\perp\perp} = \overline{Y}$, and hence $Y^{\perp\perp} \subset \overline{Y}$. Next, by part (g) of Lemma 3.29, $Y \subset Y^{\perp\perp}$, but $Y^{\perp\perp}$ is closed so $\overline{Y} \subset Y^{\perp\perp}$. Therefore $Y^{\perp\perp} = \overline{Y}$. □

EXERCISES

3.12 If S is as in Exercise 3.9, use the result of that exercise to find $S^\perp$.

3.13 If $X = \mathbb{R}^k$ and $A = \{a\}$, for some non-zero vector $a = (a_1, \ldots, a_k)$, show that

$$A^\perp = \{(x_1, \ldots, x_k) \in \mathbb{R}^k : \sum_{j=1}^k a_j x_j = 0\}.$$

3.14 Give the solution of Example 3.28.

3.15 If $A = \{\{x_n\} \in \ell^2 : x_{2n} = 0 \text{ for all } n \in \mathbb{N}\}$, find $A^\perp$.

3.16 Let X be an inner product space and let $A \subset X$. Show that $A^\perp = \overline{A}^\perp$.

3.17 Let X and Y be linear subspaces of a Hilbert space $\mathcal{H}$. Recall that $X+Y = \{x+y : x \in X,\ y \in Y\}$. Prove that $(X+Y)^\perp = X^\perp \cap Y^\perp$.

3.18 Let $\mathcal{H}$ be a Hilbert space, let $y \in \mathcal{H} \setminus \{0\}$ and let $S = \operatorname{Sp}\{y\}$. Show that $\{x \in \mathcal{H} : (x, y) = 0\}^\perp = S$.

3.19 Let Y be a closed linear subspace of a Hilbert space $\mathcal{H}$. Show that if $Y \neq \mathcal{H}$ then $Y^\perp \neq \{0\}$. Is this always true if Y is not closed? [Hint: consider dense, non-closed linear subspaces. Show that the subspace in Exercise 2.10 is such a subspace.]

3.20 Let X be an inner product space and let $A \subset X$ be non-empty. Show that:

(a) $A^{\perp\perp} = \overline{\operatorname{Sp}}\, A$;

(b) $A^{\perp\perp\perp} = A^\perp$ (where $A^{\perp\perp\perp} = ((A^\perp)^\perp)^\perp$).

3.4 Orthonormal Bases in Infinite Dimensions

We now wish to extend the idea of an orthonormal basis, which we briefly discussed in Section 3.2 for finite-dimensional spaces, to infinite-dimensional spaces.

Definition 3.37

Let X be an inner product space. A sequence $\{e_n\} \subset X$ is said to be an *orthonormal sequence* if $\|e_n\| = 1$ for all $n \in \mathbb{N}$, and $(e_m, e_n) = 0$ for all $m, n \in \mathbb{N}$ with $m \neq n$.

The following example is easy to check.

Example 3.38

The sequence $\{\widetilde{e}_n\}$ (see Definition 1.59) is an orthonormal sequence in ℓ^2. Note that each of the elements of this sequence (in ℓ^2) is itself a sequence (in $\mathbb{F}$). This can be a source of confusion, so it is important to keep track of what space a sequence lies in.

A rather more complicated, but extremely important, example is the following.

Example 3.39

The set of functions $\{e_n\}$, where $e_n(x) = (2\pi)^{-1/2}e^{inx}$ for $n \in \mathbb{Z}$, is an orthonormal sequence in the space $L^2_{\mathbb{C}}[-\pi, \pi]$.

Solution

This follows from

$$(e_m, e_n) = \frac{1}{2\pi}\int_{-\pi}^{\pi} e^{i(m-n)x}\,dx = \begin{cases} 1, & \text{if } m = n, \\ 0, & \text{if } m \neq n. \end{cases} \qquad \square$$

The orthonormal sequence in Example 3.39, and related sequences of trigonometric functions, will be considered in more detail in Section 3.5 on Fourier series. We also note that, strictly speaking, to use the word "sequence" in Example 3.39 we should choose an ordering of the functions so that they are indexed by $n \in \mathbb{N}$, rather than $n \in \mathbb{Z}$, but this is a minor point.

It follows immediately from part (a) of Lemma 3.20 that an orthonormal sequence is linearly independent and if a space X contains an orthonormal sequence then it must be infinite-dimensional. A converse result also holds.

Theorem 3.40

Any infinite-dimensional inner product space X contains an orthonormal sequence.

Proof

Using the construction in the proof of Theorem 2.26 we obtain a linearly independent sequence of unit vectors $\{x_n\}$ in X. Now, by inductively applying the Gram–Schmidt algorithm (see the proof of part (b) of Lemma 3.20) to the sequence $\{x_n\}$ we can construct an orthonormal sequence $\{e_n\}$ in $\mathcal{H}$. $\square$

For a general orthonormal sequence $\{e_n\}$ in an infinite-dimensional inner product space X and an element $x \in X$, the obvious generalization of the expansion (3.2) is the formula

$$x = \sum_{n=1}^{\infty} (x, e_n) e_n \tag{3.5}$$

(convergence of infinite series was defined in Definition 2.29). However, in the infinite-dimensional setting there are two major questions associated with this formula.

(a) Does the series converge?

(b) Does it converge to x?

We will answer these questions in a series of lemmas.

Lemma 3.41 (Bessel's inequality)

Let X be an inner product space and let $\{e_n\}$ be an orthonormal sequence in X. For any $x \in X$ the (real) series $\sum_{n=1}^{\infty} |(x, e_n)|^2$ converges and

$$\sum_{n=1}^{\infty} |(x, e_n)|^2 \leq \|x\|^2.$$

Proof

For each $k \in \mathbb{N}$ let $y_k = \sum_{n=1}^{k} (x, e_n) e_n$. Then,

$$\begin{aligned}
\|x - y_k\|^2 &= (x - y_k, x - y_k) \\
&= \|x\|^2 - \sum_{n=1}^{k} \overline{(x, e_n)} (x, e_n) - \sum_{n=1}^{k} (x, e_n)(e_n, x) + \|y_k\|^2 \\
&= \|x\|^2 - \sum_{n=1}^{k} |(x, e_n)|^2
\end{aligned}$$

(applying Theorem 3.22 to $\|y_k\|^2$). Thus,

$$\sum_{n=1}^{k} |(x, e_n)|^2 = \|x\|^2 - \|x - y_k\|^2 \leq \|x\|^2,$$

and hence this sequence of partial sums is increasing and bounded above so the result follows. □

The next result gives conditions on the coefficients $\{\alpha_n\}$ which guarantee the convergence of a general series $\sum_{n=1}^{\infty} \alpha_n e_n$.

Theorem 3.42

Let $\mathcal{H}$ be a Hilbert space and let $\{e_n\}$ be an orthonormal sequence in $\mathcal{H}$. Let $\{\alpha_n\}$ be a sequence in $\mathbb{F}$. Then the series $\sum_{n=1}^{\infty} \alpha_n e_n$ converges if and only if $\sum_{n=1}^{\infty} |\alpha_n|^2 < \infty$. If this holds, then

$$\|\sum_{n=1}^{\infty} \alpha_n e_n\|^2 = \sum_{n=1}^{\infty} |\alpha_n|^2.$$

Proof

($\Rightarrow$) Suppose that $\sum_{n=1}^{\infty} \alpha_n e_n$ converges and let $x = \sum_{n=1}^{\infty} \alpha_n e_n$. Then for any $m \in \mathbb{N}$,

$$(x, e_m) = \lim_{k\to\infty} \Big(\sum_{n=1}^{k} \alpha_n e_n, e_m \Big) = \alpha_m$$

(since, ultimately, $k \geq m$). Therefore, by Bessel's inequality,

$$\sum_{n=1}^{\infty} |\alpha_n|^2 = \sum_{n=1}^{\infty} |(x, e_n)|^2 \leq \|x\|^2 < \infty.$$

($\Leftarrow$) Suppose that $\sum_{n=1}^{\infty} |\alpha_n|^2 < \infty$ and, for each $k \in \mathbb{N}$, let $x_k = \sum_{n=1}^{k} \alpha_n e_n$. By Theorem 3.22, for any j, $k \in \mathbb{N}$ with $k > j$,

$$\|x_k - x_j\|^2 = \|\sum_{n=j+1}^{k} \alpha_n e_n\|^2 = \sum_{n=j+1}^{k} |\alpha_n|^2.$$

Since $\sum_{n=1}^{\infty} |\alpha_n|^2 < \infty$, the partial sums of this series converge and so form a Cauchy sequence. Therefore the sequence $\{x_k\}$ is a Cauchy sequence in $\mathcal{H}$, and hence converges.

Finally,

$$\begin{aligned}
\|\sum_{n=1}^{\infty} \alpha_n e_n\|^2 &= \lim_{k\to\infty} \|\sum_{n=1}^{k} (x, e_n) e_n\|^2 && \text{(by Theorem 2.11)} \\
&= \lim_{k\to\infty} \sum_{n=1}^{k} |\alpha_n|^2 && \text{(by Theorem 3.22)} \\
&= \sum_{n=1}^{\infty} |\alpha_n|^2.
\end{aligned}$$

□

Remark 3.43

The result of Theorem 3.42 can be rephrased as: the series $\sum_{n=1}^{\infty} \alpha_n e_n$ converges if and only if the sequence $\{\alpha_n\} \in \ell^2$.

We can now answer question (a) above.

Corollary 3.44

Let $\mathcal{H}$ be a Hilbert space and let $\{e_n\}$ be an orthonormal sequence in $\mathcal{H}$. For any $x \in \mathcal{H}$ the series $\sum_{n=1}^{\infty}(x, e_n)e_n$ converges.

Proof

By Bessel's inequality, $\sum_{n=1}^{\infty} |(x, e_n)|^2 < \infty$, so by Theorem 3.42 the series $\sum_{n=1}^{\infty}(x, e_n)e_n$ converges. □

We next consider question (b), that is, when does the series in (3.5) converge to x? Without further conditions on the sequence there is certainly no reason why the series should converge.

Example 3.45

In $\mathbb{R}^3$, consider the orthonormal set $\{\widehat{e}_1,\ \widehat{e}_2\}$, and let $x = (3, 0, 4)$, say. Then

$$(x, \widehat{e}_1)\widehat{e}_1 + (x, \widehat{e}_2)\widehat{e}_2 \neq x.$$

It is clear that in this example the problem arises because the orthonormal set $\{\widehat{e}_1,\ \widehat{e}_2\}$ does not have enough vectors to span the space $\mathbb{R}^3$, that is, it is not a basis. We have not so far defined the idea of a basis in infinite-dimensional spaces, but a similar problem can occur, even with orthonormal sequences having infinitely many vectors.

Example 3.46

Let $\{e_n\}$ be an orthonormal sequence in a Hilbert space $\mathcal{H}$, and let S be the subsequence $S = \{e_{2n}\}_{n \in \mathbb{N}}$ (that is, S consists of just the even terms in the sequence $\{e_n\}$). Then S is an orthonormal sequence in $\mathcal{H}$ with infinitely many elements, but, for instance, $e_1 \neq \sum_{n=1}^{\infty} \alpha_{2n} e_{2n}$, for any numbers α_{2n}.

Solution

Since S is a subset of an orthonormal sequence it is also an orthonormal sequence. Suppose now that the vector e_1 can be expressed as $e_1 = \sum_{n=1}^{\infty} \alpha_{2n} e_{2n}$, for some numbers α_{2n}. Then, by Lemma 3.17,

$$0 = (e_1, e_{2m}) = \lim_{k\to\infty} \Big(\sum_{n=1}^{k} \alpha_{2n} e_{2n}, e_{2m} \Big) = \alpha_{2m},$$

for all $m \in \mathbb{N}$, and hence $e_1 = \sum_{n=1}^{\infty} \alpha_{2n} e_{2n} = 0$, which contradicts the orthonormality of the sequence $\{e_n\}$. □

We now give various conditions which ensure that (3.5) holds for all $x \in \mathcal{H}$.

Theorem 3.47

Let $\mathcal{H}$ be a Hilbert space and let $\{e_n\}$ be an orthonormal sequence in $\mathcal{H}$. The following conditions are equivalent:

(a) $\{e_n : n \in \mathbb{N}\}^\perp = \{0\}$;

(b) $\overline{\mathrm{Sp}}\,\{e_n : n \in \mathbb{N}\} = \mathcal{H}$;

(c) $\|x\|^2 = \sum_{n=1}^{\infty} |(x, e_n)|^2$ for all $x \in \mathcal{H}$;

(d) $x = \sum_{n=1}^{\infty} (x, e_n) e_n$ for all $x \in \mathcal{H}$.

Proof

(a) ⇒ (d) Let $x \in \mathcal{H}$ and let $y = x - \sum_{n=1}^{\infty}(x, e_n)e_n$. For each $m \in \mathbb{N}$,

$$\begin{aligned}(y, e_m) &= (x, e_m) - \lim_{k\to\infty} \Big(\sum_{n=1}^{k} (x, e_n) e_n, e_m \Big) \qquad \text{(by Lemma 3.13)}\\ &= (x, e_m) - \lim_{k\to\infty} \sum_{n=1}^{k} (x, e_n)(e_n, e_m)\\ &= (x, e_m) - (x, e_m) = 0.\end{aligned}$$

Therefore property (a) implies that $y = 0$, and so $x = \sum_{n=1}^{\infty}(x, e_n)e_n$ for any $x \in \mathcal{H}$. Hence, property (d) holds.

(d) ⇒ (b) For any $x \in \mathcal{H}$, $x = \lim_{k\to\infty} \sum_{n=1}^{k}(x, e_n)e_n$. But $\sum_{n=1}^{k}(x, e_n)e_n \in \mathrm{Sp}\,\{e_1, \ldots, e_k\}$, so $x \in \overline{\mathrm{Sp}}\,\{e_n : n \in \mathbb{N}\}$, which is property (b).

(d) ⇒ (c) This follows immediately from Theorem 3.22.

(b) ⇒ (a) Suppose that $y \in \{e_n : n \in \mathbb{N}\}^\perp$. Then $(y, e_n) = 0$ for all $n \in \mathbb{N}$, and so $e_n \in \{y\}^\perp$, for all $n \in \mathbb{N}$. But by part (f) of Lemma 3.29, $\{y\}^\perp$ is a closed linear subspace, so this shows that $\mathcal{H} = \overline{\mathrm{Sp}}\,\{e_n\} \subset \{y\}^\perp$. It follows that $y \in \{y\}^\perp$, hence $(y, y) = 0$, and so $y = 0$.

(c) $\Rightarrow$ (a) If $(x, e_n) = 0$ for all $n \in \mathbb{N}$, then by (c), $\|x\|^2 = \sum_{n=1}^{\infty} |(x, e_n)|^2 = 0$, so $x = 0$. □

Remark 3.48

The linear span ($\mathrm{Sp}\,\{e_n\}$) of the set $\{e_n\}$ consist of all possible *finite* linear combinations of the vectors in $\{e_n\}$, that is, all possible finite sums in the expansion (3.5). However, for (3.5) to hold for all $x \in \mathcal{H}$ it is necessary to also consider infinite sums in (3.5). This corresponds to considering the closed linear span ($\overline{\mathrm{Sp}}\,\{e_n\}$). In finite dimensions the linear span is necessarily closed, and so equals the closed linear span, so it is not necessary to distinguish between these concepts in finite-dimensional linear algebra.

Definition 3.49

Let $\mathcal{H}$ be a Hilbert space and let $\{e_n\}$ be an orthonormal sequence in $\mathcal{H}$. Then $\{e_n\}$ is called an *orthonormal basis for* $\mathcal{H}$ if any of the conditions in Theorem 3.47 hold.

Remark 3.50

Some books call an orthonormal basis $\{e_n\}$ a *complete orthonormal sequence* – the sequence is "complete" in the sense that there are enough vectors in it to span the space (as in Theorem 3.47). We prefer not to use the term "complete" in this sense to avoid confusion with the previous use of "complete" to describe spaces in which all Cauchy sequences converge.

The result in part (c) of Theorem 3.47 is sometimes called *Parseval's theorem*. Comparing Lemma 3.41 with Theorem 3.47 we see that Parseval's theorem corresponds to the case when equality holds in Bessel's inequality. Parseval's theorem holds for orthonormal bases while Bessel's inequality holds for general orthonormal sequences. In fact, if an orthonormal sequence $\{e_n\}$ is not an orthonormal basis then there must be vectors $x \in \mathcal{H}$ for which Bessel's inequality holds with strict inequality (see Exercise 3.21).

It is usually relatively easy to decide whether a given sequence is orthonormal, but in general it is usually rather difficult to decide whether a given orthonormal sequence is actually a basis. In Examples 3.38 and 3.39 we saw two orthonormal sequences in infinite-dimensional spaces. Both are in fact orthonormal bases. It is easy to check this for the first, see Exercise 3.51, but the second is rather more complicated and will be dealt with in Section 3.5, where we discuss Fourier series. However, we note that in infinite dimensions the components (x, e_n) in the expansion (3.5) are often called the *Fourier coefficients* of x with respect to the basis $\{e_n\}$ (by analogy with the theory of Fourier series).

Example 3.51

The orthonormal sequence $\{\widetilde{e}_n\}$ in ℓ^2 given in Example 3.38 is an orthonormal basis. This basis will be called the *standard orthonormal basis* in ℓ^2.

Solution

Let $x = \{x_n\} \in \ell^2$. Then by definition,

$$\|x\|^2 = \sum_{n=1}^{\infty} |x_n|^2 = \sum_{n=1}^{\infty} |(x, e_n)|^2.$$

Thus by part (c) of Theorem 3.47, $\{e_n\}$ is an orthonormal basis. □

We emphasize that although the sequence $\{\widetilde{e}_n\}$ is an orthonormal basis in the space ℓ^2, this says nothing about this sequence being a basis in the space ℓ^p when $p \neq 2$. In fact, we have given no definition of the idea of a basis in an infinite dimensional space other than an orthonormal basis in a Hilbert space.

Theorem 3.40 showed that any infinite-dimensional Hilbert space $\mathcal{H}$ contains an orthonormal sequence. However, there is no reason to suppose that this sequence is a basis. The question now arises – do all Hilbert spaces have orthonormal bases? The answer is no. There exist Hilbert spaces which are too "large" to be spanned by the countable collection of vectors in a sequence. In this section we will show that a Hilbert space has an orthonormal basis if and only if it is separable (that is, it contains a countable, dense subset). In a sense, the countability condition in the definition of separability ensures that such spaces are "small enough" to be spanned by a countable orthonormal basis.

We begin with some illustrations of the idea of separability in the context of vector spaces. As noted in Chapter 1, the space $\mathbb{R}$ is separable since the set of rational numbers is countable and dense in $\mathbb{R}$. Similarly, $\mathbb{C}$ is separable since the set of complex numbers of the form $p+iq$, with p and q rational, is countable and dense in $\mathbb{C}$ (for convenience, we will call numbers of this form *complex rationals*). A very common method of constructing countable dense sets is to take a general element of the space, expressed in terms of certain naturally occurring real or complex coefficients, and restrict these coefficients to be rationals or complex rationals. Also, in the vector space context, infinite sums are replaced by finite sums of arbitrarily large length. The proof of the following theorem illustrates these ideas. The theorem is an extremely important result on separable Hilbert spaces.

Theorem 3.52

(a) Finite dimensional normed vector spaces are separable.

(b) An infinite-dimensional Hilbert space $\mathcal{H}$ is separable if and only if it has an orthonormal basis.

Proof

(a) Let X be a finite-dimensional, real normed vector space, and let $\{e_1, \ldots, e_k\}$ be a basis for X. Then the set of vectors having the form $\sum_{n=1}^{k} \alpha_n e_n$, with α_n a rational is countable and dense (the proof of density is similar to the proof below in part (b)), so X is separable. In the complex case we use complex rational coefficients α_n.

(b) Suppose that $\mathcal{H}$ is infinite-dimensional and separable, and let $\{x_n\}$ be a countable, dense sequence in $\mathcal{H}$. We construct a new sequence $\{y_n\}$ by omitting every member of the sequence $\{x_n\}$ which is a linear combination of the preceding members of the sequence. By this construction the sequence $\{y_n\}$ is linearly independent. Now, by inductively applying the Gram–Schmidt algorithm (see the proof of part (b) of Lemma 3.20) to the sequence $\{y_n\}$ we can construct an orthonormal sequence $\{e_n\}$ in $\mathcal{H}$ with the property that for each $k \geq 1$, $\operatorname{Sp}\{e_1, \ldots, e_k\} = \operatorname{Sp}\{y_1, \ldots, y_k\}$. Thus,

$$\operatorname{Sp}\{e_n : n \in \mathbb{N}\} = \operatorname{Sp}\{y_n : n \in \mathbb{N}\} = \operatorname{Sp}\{x_n : n \in \mathbb{N}\}.$$

Since the sequence $\{x_n\}$ is dense in $\mathcal{H}$ it follows that $\overline{\operatorname{Sp}}\{e_n : n \in \mathbb{N}\} = \mathcal{H}$, and so, by Theorem 3.47, $\{e_n\}$ is an orthonormal basis for $\mathcal{H}$.

Now suppose that $\mathcal{H}$ has an orthonormal basis $\{e_n\}$. The set of elements $x \in \mathcal{H}$ expressible as a finite sum of the form $x = \sum_{n=1}^{k} \alpha_n e_n$, with $k \in \mathbb{N}$ and rational (or complex rational) coefficients α_n, is clearly countable, so to show that $\mathcal{H}$ is separable we must show that this set is dense. To do this, choose arbitrary $y \in \mathcal{H}$ and $\epsilon > 0$. Then y can be written in the form $y = \sum_{n=1}^{k} \beta_n e_n$, with $\sum_{n=1}^{\infty} |\beta_n|^2 < \infty$, so there exists an integer N such that $\sum_{n=N+1}^{\infty} |\beta_n|^2 < \epsilon^2/2$. Now, for each $n = 1, \ldots, N$, we choose rational coefficients α_n, such that $|\beta_n - \alpha_n|^2 < \epsilon^2/2N$, and let $x = \sum_{n=1}^{k} \alpha_n e_n$. Then

$$\|y - x\|^2 = \sum_{n=1}^{N} |\beta_n - \alpha_n|^2 + \sum_{n=N+1}^{\infty} |\beta_n|^2 < \epsilon^2,$$

which shows that the above set is dense, by part (f) of Theorem 1.25. □

Example 3.53

The Hilbert space ℓ^2 is separable.

It will be shown in Section 3.5 that the space $L^2[a,b]$, a, $b \in \mathbb{R}$, has an orthonormal basis, so is also separable. In addition, by an alternative argument it will be shown that $C[a,b]$ is separable. In fact, most spaces which arise in applications are separable.

EXERCISES

3.21 Use Example 3.46 to find an orthonormal sequence in a Hilbert space $\mathcal{H}$ and a vector $x \in \mathcal{H}$ for which Bessel's inequality holds with strict inequality.

3.22 Let $\mathcal{H}$ be a Hilbert space and let $\{e_n\}$ be an orthonormal sequence in $\mathcal{H}$. Determine whether the following series converge in $\mathcal{H}$:

$$\text{(a)} \sum_{n=1}^{\infty} n^{-1} e_n; \quad \text{(b)} \sum_{n=1}^{\infty} n^{-1/2} e_n.$$

3.23 Let $\mathcal{H}$ be a Hilbert space and let $\{e_n\}$ be an orthonormal basis in $\mathcal{H}$. Let $\rho : \mathbb{N} \to \mathbb{N}$ be a permutation of $\mathbb{N}$ (so that for all $x \in \mathcal{H}$, $\sum_{n=1}^{\infty} |(x, e_{\rho(n)})|^2 = \sum_{n=1}^{\infty} |(x, e_n)|^2$). Show that:

$$\text{(a)} \sum_{n=1}^{\infty} (x, e_{\rho(n)}) e_n \text{ converges for all } x \in \mathcal{H};$$

$$\text{(b)} \ \|\sum_{n=1}^{\infty} (x, e_{\rho(n)}) e_n\|^2 = \|x\|^2 \text{ for all } x \in \mathcal{H}.$$

3.24 Let $\mathcal{H}$ be a Hilbert space and let $\{e_n\}$ be an orthonormal basis in $\mathcal{H}$. Prove that the *Parseval relation*

$$(x, y) = \sum_{n=1}^{\infty} (x, e_n)(e_n, y)$$

holds for all x, $y \in \mathcal{H}$.

3.25 Show that a metric space M is separable if and only if M has a countable subset A with the property: for every integer $k \geq 1$ and every $x \in X$ there exists $a \in A$ such that $d(x, a) < 1/k$.

Show that any subset N of a separable metric space M is separable. [Note: separability of M ensures that there is a countable dense subset of M, but none of the elements of this set need belong to N. Thus it is necessary to construct a countable dense subset of N.]

3.26 Suppose that $\mathcal{H}$ is a separable Hilbert space and $Y \subset \mathcal{H}$ is a closed linear subspace. Show that there is an orthonormal basis for $\mathcal{H}$ consisting only of elements of Y and $Y^{\perp}$.

3.5 Fourier Series

In this section we will prove that the orthonormal sequence in Example 3.39 is a basis for $L^2_{\mathbb{C}}[-\pi, \pi]$, and we will also consider various related bases consisting of sets of trigonometric functions.

Theorem 3.54

The set of functions

$$C = \left\{ c_0(x) = (1/\pi)^{1/2},\ c_n(x) = (2/\pi)^{1/2} \cos nx : n \in \mathbb{N} \right\}$$

is an orthonormal basis in $L^2[0, \pi]$.

Proof

We first consider $L^2_{\mathbb{R}}[0, \pi]$. It is easy to check that C is orthonormal. Thus by Theorem 3.47 we must show that $\operatorname{Sp} C$ is dense in $L^2_{\mathbb{R}}[0, \pi]$. We will combine the approximation properties in Theorems 1.39 and 1.61 to do this.

Firstly, by Theorem 1.61 there is a function $g_1 \in C_{\mathbb{R}}[0, \pi]$ with $\|g_2 - f\| < \epsilon/2$ (here, $\|\cdot\|$ denotes the $L^2_{\mathbb{R}}[0, \pi]$ norm). Thus it is sufficient to show that for any function $g_1 \in C_{\mathbb{R}}[0, \pi]$ there is a function $g_2 \in \operatorname{Sp} C$ with $\|g_2 - g_1\| < \epsilon/2$ (it will then follow that there is a function $g_2 \in \operatorname{Sp} C$ such that $\|g_2 - f\| < \epsilon$).

Now suppose that $g_1 \in C_{\mathbb{R}}[0, \pi]$ is arbitrary. We recall that the function $\cos^{-1} : [-1, 1] \to [0, \pi]$ is a continuous bijection, so we may define a function $h \in C_{\mathbb{R}}[-1, 1]$ by $h(s) = g_1(\cos^{-1} s)$ for $s \in [-1, 1]$. It follows from Theorem 1.39 that there is a polynomial p such that $|h(s) - p(s)| < \epsilon/2\sqrt{\pi}$ for all $s \in [-1, 1]$, and hence, writing $g_2(x) = p(\cos x)$, we have $|g_2(x) - g_1(x)| < \epsilon/2\sqrt{\pi}$ for all $x \in [0, \pi]$, and so $\|g_2 - g_1\| < \epsilon/2$. But standard trigonometry now shows that any polynomial in $\cos x$ of the form $\sum_{n=0}^{m} \alpha_n (\cos x)^n$ can be rewritten in the form $\sum_{n=0}^{m} \beta_n \cos nx$, which shows that $g_2 \in \operatorname{Sp} C$, and so completes the proof in the real case.

In the complex case, for any $f \in L^2_{\mathbb{C}}[0, \pi]$ we let $f_{\mathbb{R}}, f_{\mathbb{C}} \in L^2_{\mathbb{R}}[0, \pi]$ denote the functions obtained by taking the real and imaginary parts of f, and we apply the result just proved to these functions to obtain

$$f = f_{\mathbb{R}} + i f_{\mathbb{C}} = \sum_{n=0}^{\infty} \alpha_n c_n + i \sum_{n=0}^{\infty} \beta_n c_n = \sum_{n=0}^{\infty} (\alpha_n + i\beta_n) c_n,$$

which proves the result in the complex case. □

From Theorems 3.52 and 3.54 we have the following result.

Corollary 3.55

The space $L^2[0, \pi]$ is separable.

Theorem 3.56

The set of functions

$$S = \left\{ s_n(x) = (2/\pi)^{1/2} \sin nx : n \in \mathbb{N} \right\}$$

is an orthonormal basis in $L^2[0, \pi]$.

Proof

The proof is similar to the proof of the previous theorem so we will merely sketch it. This time we first approximate f (in $L^2_{\mathbb{R}}[0, \pi]$) by a function f_δ, with $\delta > 0$, defined by

$$f_\delta(x) = \begin{cases} 0, & \text{if } x \in [0, \delta], \\ f(x), & \text{if } x \in (\delta, \pi] \end{cases}$$

(clearly, $\|f - f_\delta\|$ can be made arbitrarily small by choosing δ sufficiently small). Then the function $g_1(x)/\sin x$ belongs to $L^2_{\mathbb{R}}[0, \pi]$, so by the previous proof it can be approximated by functions of the form $\sum_{n=0}^{m} \alpha_n \cos nx$, and hence $g_1(x)$ can be approximated by functions of the form

$$\sum_{n=0}^{m} \alpha_n \cos nx \, \sin x = \frac{1}{2} \sum_{n=0}^{m} \alpha_n (\sin(n+1)x - \sin(n-1)x).$$

The latter function is an element of $\operatorname{Sp} S$, which completes the proof. □

It follows from Theorems 3.54, 3.56, and 3.47, that an arbitrary function $f \in L^2[0, \pi]$ can be represented in either of the forms

$$f = \sum_{n=0}^{\infty} (f, c_n) c_n, \qquad \sum_{n=1}^{\infty} (f, s_n) s_n, \tag{3.6}$$

where the convergence is in the $L^2[0,\pi]$ sense. These series are called, respectively, *Fourier cosine* and *sine series* expansions of f. Other forms of Fourier series expansions can be obtained from the following corollary.

Corollary 3.57

The sets of functions

$$\begin{aligned} E &= \{e_n(x) = (2\pi)^{-1/2}e^{inx} : n \in \mathbb{Z}\}, \\ F &= \{2^{-1/2}c_0,\, 2^{-1/2}c_n,\, 2^{-1/2}s_n : n \in \mathbb{N}\}, \end{aligned}$$

are orthonormal bases in the space $L^2_{\mathbb{C}}[-\pi,\pi]$. The set F is also an orthonormal basis in the space $L^2_{\mathbb{R}}[-\pi,\pi]$ (the set E is clearly not appropriate for the space $L^2_{\mathbb{R}}[-\pi,\pi]$ since the functions in E are complex).

Proof

Again it is easy to check that the set F is orthonormal in $L^2_{\mathbb{F}}[-\pi,\pi]$. Suppose that F is not a basis for $L^2_{\mathbb{F}}[-\pi,\pi]$. Then, by part (a) of Theorem 3.47, there exists a non-zero function $f \in L^2_{\mathbb{F}}[-\pi,\pi]$ such that $(f,c_0) = 0$, $(f,c_n) = 0$ and $(f,s_n) = 0$, for all $n \in \mathbb{N}$, which can be rewritten as,

$$\begin{aligned} 0 &= \int_{-\pi}^{\pi} f(x)\,dx = \int_0^{\pi} (f(x) + f(-x))\,dx, \\ 0 &= \int_{-\pi}^{\pi} f(x)\cos nx\,dx = \int_0^{\pi} (f(x) + f(-x))\cos nx\,dx, \\ 0 &= \int_{-\pi}^{\pi} f(x)\sin nx\,dx = \int_0^{\pi} (f(x) - f(-x))\sin nx\,dx. \end{aligned}$$

Thus, by part (a) of Theorem 3.47 and Theorems 3.54 and 3.56, it follows that for a.e. $x \in [0,\pi]$,

$$\begin{aligned} f(x) + f(-x) &= 0, \\ f(x) - f(-x) &= 0, \end{aligned}$$

and hence $f(x) = 0$ for a.e. $x \in [-\pi,\pi]$. But this contradicts the assumption that $f \neq 0$ in $L^2_{\mathbb{F}}[-\pi,\pi]$, so F must be a basis. Next, it was shown in Example 3.39 that the set E is orthonormal in $L^2_{\mathbb{C}}[-\pi,\pi]$, and it follows from the formula $e^{in\theta} = \cos n\theta + i\sin n\theta$ that $\operatorname{Sp} E$ is equal to $\operatorname{Sp} F$, so E is also an orthonormal basis. □

The above results give the basic theory of Fourier series in an L^2 setting. This theory is simple and elegant, but there is much more to the theory of Fourier series than this. For instance, one could consider the convergence of the various series in the pointwise sense (that is, for each x in the interval concerned), or uniformly, for all x in the interval. A result in this area will be obtained in Corollary 7.29, but we will not consider these topics further here.

Finally, we note that there is nothing special about the interval $[0, \pi]$ (and $[-\pi, \pi]$) used above. By the change of variables $x \to \tilde{x} = a + (b - a)x/\pi$ in the above proofs we see that they are valid for a general interval $[a, b]$.

EXERCISES

3.27 Show that for any $b > a$ the set of polynomials with rational (or complex rational) coefficients is dense in the spaces: (a) $C[a, b]$; (b) $L^2[a, b]$.
Deduce that the space $C[a, b]$ is separable.

3.28 (Legendre polynomials) For each integer $n \geq 0$, define the polynomials

$$u_n(x) = (x^2 - 1)^n, \quad P_n(x) = \frac{1}{2^n n!} \frac{d^n u_n}{dx^n}$$

(clearly, u_n has order $2n$, while P_n has order n). The polynomials P_n are called *Legendre polynomials.* We consider these polynomials on the interval $[-1, 1]$, and let $\mathcal{H} = L^2[-1, 1]$, with the standard inner product $(\cdot\,,\cdot)$. Prove the following results.

(a) $d^{2n}u_n/dx^{2n}(x) \equiv (2n)!$.

(b) $(P_m, P_n) = 0$, for m, $n \geq 0$, $m \neq n$.

(c) $\|P_n\|^2 = (2^n n!)^2 \frac{2}{2n+1}$, for $n \geq 0$.

(d) $\left\{ e_n = \sqrt{\frac{2n+1}{2}} P_n : n \geq 0 \right\}$ is an orthonormal basis for $\mathcal{H}$.

[Hint: use integration by parts, noting that u_n, and its derivatives to order $n - 1$, are zero at ± 1.]

4 Linear Operators

4.1 Continuous Linear Transformations

Now that we have studied some of the properties of normed spaces we turn to look at functions which map one normed space into another. The simplest maps between two vector spaces are the ones which respect the linear structure, that is, the linear transformations. We recall the convention introduced in Chapter 1 that if we have two vector spaces X and Y and a linear transformation from X to Y it is taken for granted that X and Y are vector spaces over the same scalar field. Since normed vector spaces have a metric associated with the norm, and continuous functions between metric spaces are in general more important than functions which are not continuous, the important maps between normed vector spaces will be the *continuous linear transformations.*

After giving examples of these, we fix two normed spaces X and Y and look at the space of all continuous linear transformations from X to Y. We show this space is also a normed vector space and then study in more detail the cases when $Y = \mathbb{F}$ and when $Y = X$. In the latter case we will see that it is possible to define the product of continuous linear transformations and therefore, for some continuous linear transformations, the inverse of a continuous linear transformation. The final section of this chapter is devoted to determining when a continuous linear transformation has an inverse.

We start by studying continuous linear transformations. Before we look at examples of continuous linear transformations, it is convenient to give alternative characterizations of continuity for linear transformations. A notational convention should be clarified here. If X and Y are normed spaces and $T : X \to Y$ is a linear transformation, the norm of an element of X and the norm of an element of Y will frequently occur in the same equation. We should therefore introduce notation which distinguishes between these norms. In practice we just use the symbol $\|\cdot\|$ for the norm on both spaces as it is usually easy to determine which space an element is in and therefore, implicitly, to which norm we are referring. We recall also that if we write down one of the spaces in Examples 2.2, 2.4, 2.5 and 2.6 without explicitly mentioning a norm, it is assumed that the norm on this space is the standard norm.

Lemma 4.1

Let X and Y be normed linear spaces and let $T : X \to Y$ be a linear transformation. The following are equivalent:

(a) T is uniformly continuous;

(b) T is continuous;

(c) T is continuous at 0;

(d) there exists a positive real number k such that $\|T(x)\| \leq k$ whenever $x \in X$ and $\|x\| \leq 1$;

(e) there exists a positive real number k such that $\|T(x)\| \leq k\|x\|$ for all $x \in X$.

Proof

The implications (a) $\Rightarrow$ (b) and (b) $\Rightarrow$ (c) hold in more generality so all that is required to be proved is (c) $\Rightarrow$ (d), (d) $\Rightarrow$ (e) and (e) $\Rightarrow$ (a).

(c) $\Rightarrow$ (d). As T is continuous at 0, taking $\epsilon = 1$ there exists a $\delta > 0$ such that $\|T(x)\| < 1$ when $x \in X$ and $\|x\| < \delta$. Let $w \in X$ with $\|w\| \leq 1$. As

$$\|\frac{\delta w}{2}\| = \frac{\delta}{2}\|w\| \leq \frac{\delta}{2} < \delta,$$

$\|T\left(\frac{\delta w}{2}\right)\| < 1$ and as T is a linear transformation $T\left(\frac{\delta w}{2}\right) = \frac{\delta}{2}T(w)$. Thus $\frac{\delta}{2}\|T(w)\| < 1$ and so $\|T(w)\| < \frac{2}{\delta}$. Therefore condition (d) holds with $k = \frac{2}{\delta}$.

(d) $\Rightarrow$ (e). Let k be such that $\|T(x)\| \leq k$ whenever $x \in X$ and $\|x\| \leq 1$. Since $T(0) = 0$ it is clear that $\|T(0)\| \leq k\|0\|$. Let $y \in X$ with $y \neq 0$. As $\left\|\frac{y}{\|y\|}\right\| = 1$ we have $\left\|T\left(\frac{y}{\|y\|}\right)\right\| \leq k$. Since T is a linear transformation

$$\frac{1}{\|y\|}\|T(y)\| = \left\|\left(\frac{1}{\|y\|}\right)T(y)\right\| = \left\|T\left(\frac{y}{\|y\|}\right)\right\| \leq k,$$

and so $\|T(y)\| \leq k\|y\|$. Hence $\|T(x)\| \leq k\|x\|$ for all $x \in X$.

(e) $\Rightarrow$ (a). Since T is a linear transformation,

$$\|T(x) - T(y)\| = \|T(x - y)\| \leq k\|x - y\|$$

for all $x, y \in X$. Let $\epsilon > 0$ and let $\delta = \frac{\epsilon}{k}$. Then when $x, y \in X$ and $\|x - y\| < \delta$

$$\|T(x) - T(y)\| \leq k\|x - y\| < k\left(\frac{\epsilon}{k}\right) < \epsilon.$$

Therefore T is uniformly continuous. □

Having obtained these alternative characterizations of continuity of linear transformations, we can now look at some examples. It will normally be clear that the maps we are considering are linear transformations so we shall just concentrate on showing that they are continuous. It is usual to check continuity of linear transformations using either of the equivalent conditions (d) or (e) of Lemma 4.1.

Example 4.2

The linear transformation $T : C_{\mathbb{F}}[0, 1] \to \mathbb{F}$ defined by

$$T(f) = f(0)$$

is continuous.

Solution

Let $f \in C_{\mathbb{C}}[0, 1]$. Then

$$|T(f)| = |f(0)| \leq \sup\{|f(x)| : x \in [0, 1]\} = \|f\|.$$

Hence T is continuous by condition (e) of Lemma 4.1 with $k = 1$. □

Before starting to check that a linear transformation T is continuous it is sometimes first necessary to check that T is well defined. Lemma 4.3 will be used to check that the following examples of linear transformations are well defined.

Lemma 4.3

If $\{c_n\} \in \ell^\infty$ and $\{x_n\} \in \ell^p$, where $1 \leq p < \infty$, then $\{c_n x_n\} \in \ell^p$ and

$$\sum_{n=1}^{\infty} |c_n x_n|^p \leq \|\{c_n\}\|_\infty^p \sum_{n=1}^{\infty} |x_n|^p. \tag{4.1}$$

Proof

Since $\{c_n\} \in \ell^\infty$ and $\{x_n\} \in \ell^p$ we have $\lambda = \|\{c_n\}\|_\infty = \sup\{|c_n| : n \in \mathbb{N}\} < \infty$ and $\sum_{n=1}^\infty |x_n|^p < \infty$. Since, for all $n \in \mathbb{N}$

$$|c_n x_n|^p \leq \lambda^p |x_n|^p,$$

$\sum_{n=1}^\infty |c_n x_n|^p$ converges by the comparison test. Thus $\{c_n x_n\} \in \ell^p$ and the above inequality also verifies (4.1). □

Example 4.4

If $\{c_n\} \in \ell^\infty$ then the linear transformation $T : \ell^1 \to \mathbb{F}$ defined by

$$T(\{x_n\}) = \sum_{n=1}^\infty c_n x_n$$

is continuous.

Solution

Since $\{c_n x_n\} \in \ell^1$ by Lemma 4.3, it follows that T is well defined. Moreover,

$$|T(\{x_n\})| = |\sum_{n=1}^\infty c_n x_n| \leq \sum_{n=1}^\infty |c_n x_n| \leq \|\{c_n\}\|_\infty \sum_{n=1}^\infty |x_n| = \|\{c_n\}\|_\infty \|\{x_n\}\|_1.$$

Hence T is continuous by condition (e) of Lemma 4.1 with $k = \|\{c_n\}\|_\infty$. □

Example 4.5

If $\{c_n\} \in \ell^\infty$ then the linear transformation $T : \ell^2 \to \ell^2$ defined by

$$T(\{x_n\}) = \{c_n x_n\}$$

is continuous.

Solution

Let $\lambda = \|\{c_n\}\|_\infty$. Since $\{c_n x_n\} \in \ell^2$ by Lemma 4.3, it follows that T is well defined. Moreover

$$\|T(\{x_n\})\|_2^2 = \sum_{n=1}^\infty |c_n x_n|^2 \leq \lambda^2 \sum_{n=1}^\infty |x_n|^2 = \lambda^2 \|\{x_n\}\|_2^2.$$

Hence T is continuous by condition (e) of Lemma 4.1 with $k = \|\{c_n\}\|_\infty$. □

There is another notation for maps which satisfy condition (e) of Lemma 4.1 for some k.

Definition 4.6

Let X and Y be normed linear spaces and let $T : X \to Y$ be a linear transformation. T is said to be *bounded* if there exists a positive real number k such that $\|T(x)\| \leq k\|x\|$ for all $x \in X$.

By Lemma 4.1 we can use the words *continuous* and *bounded* interchangeably for linear transformations. Note, however, that this is a different use of the word bounded from that used for functions from $\mathbb{R}$ to $\mathbb{R}$. For example if $T : \mathbb{R} \to \mathbb{R}$ is the linear transformation defined by $T(x) = x$ then T is bounded in the sense given in Definition 4.6 but, of course, is not bounded in the usual sense of a bounded function. Despite this apparent conflict of usage there is not a serious problem since apart from the zero linear transformation, linear transformations are never bounded in the usual sense of bounded functions so the word may be used in an alternative way. Since the term "bounded" gives a good indication of what has to be shown this compensates for the disadvantage of potential ambiguity. The use of this term also explains the abbreviation used in the following notation for the set of all continuous linear transformations between two normed spaces.

Notation

Let X and Y be normed linear spaces. The set of all continuous linear transformations from X to Y is denoted by $B(X, Y)$. Elements of $B(X, Y)$ are also called *bounded linear operators* or *linear operators* or sometimes just *operators*.

If X and Y are normed linear spaces then $B(X, Y) \subseteq L(X, Y)$.

Example 4.7

Let $a, b \in \mathbb{R}$, let $k : [a, b] \times [a, b] \to \mathbb{C}$ be continuous and let

$$M = \sup\{|k(s, t)| : (s, t) \in [a, b] \times [a, b]\}.$$

(a) If $g \in C[a, b]$, then $f : [a, b] \to \mathbb{C}$ defined by

$$f(s) = \int_a^b k(s, t)\, g(t)\, dt$$

is in $C[a, b]$.

(b) If the linear transformation $K : C[a,b] \to C[a,b]$ is defined by

$$(K(g))(s) = \int_a^b k(s,t)\, g(t)\, dt$$

then $K \in B(C[a,b], C[a,b])$ and

$$\|K(g)\| \leq M(b-a)\|g\|.$$

Solution

(a) Suppose that $\epsilon > 0$ and $s \in [a,b]$. We let $k_s \in C[a,b]$ be the function $k_s(t) = k(s,t)$, $t \in [a,b]$. Since the square $[a,b] \times [a,b]$ is a compact subset of $\mathbb{R}^2$, the function k is uniformly continuous and so there exists $\delta > 0$ such that if $|s - s'| < \delta$ then $|k_s(t) - k_{s'}(t)| < \epsilon$ for all $t \in [a,b]$. Hence

$$|f(s) - f(s')| \leq \int_a^b |k(s,t) - k(s',t)||g(t)|\, dt \leq \epsilon(a-b)\|g\|.$$

Therefore f is continuous.

(b) For all $s \in [a,b]$,

$$|(K(g))(s)| \leq \int_a^b |k(s,t)\, g(t)|\, dt \leq \int_a^b M\|g\|\, dt = M(b-a)\|g\|.$$

Hence $\|K(g)\| \leq M(b-a)\|g\|$ and so $K \in B(C[a,b], C[a,b])$. □

In Example 4.7 there are lots of brackets. To avoid being overwhelmed by these, if $T \in B(X,Y)$ and $x \in X$ it is usual to write Tx rather than $T(x)$.

The examples presented so far may give the impression that all linear transformations are continuous. Unfortunately, this is not the case as the following example shows.

Example 4.8

Let $\mathcal{P}$ be the linear subspace of $C_{\mathbb{C}}[0,1]$ consisting of all polynomial functions. If $T : \mathcal{P} \to \mathcal{P}$ is the linear transformation defined by

$$T(p) = p',$$

where p' is the derivative of p, then T is not continuous.

Solution

Let $p_n \in \mathcal{P}$ be defined by $p_n(t) = t^n$. Then

$$\|p_n\| = \sup\{|p_n(t)| : t \in [0,1]\} = 1,$$

while

$$\|T(p_n)\| = \|p_n'\| = \sup\{|p_n'(t)| : t \in [0,1]\} = \sup\{|nt^{n-1}| : t \in [0,1]\} = n.$$

Therefore there does not exist $k \geq 0$ such that $\|T(p)\| \leq k\|p\|$ for all $p \in \mathcal{P}$, and so T is not continuous. □

The space $\mathcal{P}$ in Example 4.8 was not finite-dimensional, so it is natural to ask whether all linear transformations between finite-dimensional normed spaces are continuous. The answer is given in Theorem 4.9.

Theorem 4.9

Let X be a finite-dimensional normed space, let Y be any normed linear space and let $T : X \to Y$ be a linear transformation. Then T is continuous.

Proof

To show this we first define a new norm on X. Since this will be different from the original norm, in this case we have to use notation which will distinguish between the two norms. Let $\|\cdot\|_1 : X \to \mathbb{R}$ be defined by $\|x\|_1 = \|x\| + \|T(x)\|$. We will show that $\|\cdot\|_1$ is a norm for X. Let $x, y \in X$ and let $\lambda \in \mathbb{F}$.

(i) $\|x\|_1 = \|x\| + \|T(x)\| \geq 0$.

(ii) If $\|x\|_1 = 0$ then $\|x\| = \|T(x)\| = 0$ and so $x = 0$ while if $x = 0$ then $\|x\| = \|T(x)\| = 0$ and so $\|x\|_1 = 0$.

(iii) $\|\lambda x\|_1 = \|\lambda x\| + \|T(\lambda x)\| = |\lambda|\|x\| + |\lambda|\|T(x)\| = |\lambda|(\|x\| + \|T(x)\|)$.
$= |\lambda|\|x\|_1$

(iv)
$$\begin{aligned}\|x+y\|_1 &= \|x+y\| + \|T(x+y)\| \\ &= \|x+y\| + \|T(x) + T(y)\| \\ &\leq \|x\| + \|y\| + \|T(x)\| + \|T(y)\| \\ &= \|x\|_1 + \|y\|_1.\end{aligned}$$

Hence $\|\cdot\|_1$ is a norm on X. Now, as X is finite-dimensional, $\|\cdot\|$ and $\|\cdot\|_1$ are equivalent and so there exists $K > 0$ such that $\|x\|_1 \leq K\|x\|$ for all $x \in X$ by Corollary 2.17. Hence $\|T(x)\| \leq \|x\|_1 \leq K\|x\|$ for all $x \in X$ and so T is bounded. □

If the domain of a linear transformation is finite-dimensional then the linear transformation is continuous by Theorem 4.9. Unfortunately, if the range is finite-dimensional instead, then the linear transformation need not be continuous as we see in Example 4.10, whose solution is left as an exercise.

Example 4.10

Let $\mathcal{P}$ be the linear subspace of $C_{\mathbb{C}}[0,1]$ consisting of all polynomial functions. If $T : \mathcal{P} \to \mathbb{C}$ is the linear transformation defined by

$$T(p) = p'(1),$$

where p' is the derivative of p, then T is not continuous.

Now that we have seen how to determine whether a given linear transformation is continuous, we give some elementary properties of continuous linear transformations. We should remark here that although the link between matrices and linear transformations between finite-dimensional vector spaces given in Theorem 1.15 can be very useful, any possible extension to linear transformations between infinite-dimensional spaces is not so straightforward since both bases in infinite-dimensional spaces and infinite-sized matrices are much harder to manipulate. We will therefore only use the matrix representation of a linear transformation between finite-dimensional vector spaces.

Lemma 4.11

If X and Y are normed linear spaces and $T : X \to Y$ is a continuous linear transformation then $\mathrm{Ker}\,(T)$ is closed.

Proof

Since T is continuous, $\mathrm{Ker}\,(T) = \{x \in X : T(x) = 0\}$ and $\{0\}$ is closed in Y it follows that $\mathrm{Ker}\,(T)$ is closed, by Theorem 1.28. □

Before our next definition we recall that if X and Y are normed spaces then the Cartesian product $X \times Y$ is a normed space by Example 2.8.

Definition 4.12

If X and Y are normed spaces and $T : X \to Y$ is a linear transformation, the *graph* of T is the linear subspace $\mathcal{G}(T)$ of $X \times Y$ defined by

$$\mathcal{G}(T) = \{(x, Tx) : x \in X\}.$$

Lemma 4.13

If X and Y are normed spaces and $T : X \to Y$ is a continuous linear transformation then $\mathcal{G}(T)$ is closed.

Proof

Let $\{(x_n, y_n)\}$ be a sequence in $\mathcal{G}(T)$ which converges to (x, y) in $X \times Y$. Then $\{x_n\}$ converges to x in X and $\{y_n\}$ converges in y in Y by Exercise 2.5. However, $y_n = T(x_n)$ for all $n \in \mathbb{N}$ since $(x_n, y_n) \in \mathcal{G}(T)$. Hence, as T is continuous,

$$y = \lim_{n\to\infty} y_n = \lim_{n\to\infty} T(x_n) = T(x).$$

Therefore $(x, y) = (x, T(x)) \in \mathcal{G}(T)$ and so $\mathcal{G}(T)$ is closed. □

We conclude this section by showing that if X and Y are fixed normed spaces the set $B(X, Y)$ is a vector space. This will be done by showing that $B(X, Y)$ is a linear subspace of $L(X, Y)$, which is a vector space under the algebraic operations given in Definition 1.7.

Lemma 4.14

Let X and Y be normed linear spaces and let $S, T \in B(X, Y)$ with $\|S(x)\| \leq k_1\|x\|$ and $\|T(x)\| \leq k_2\|x\|$ for all $x \in X$. Let $\lambda \in \mathbb{F}$. Then

(a) $\|(S + T)(x)\| \leq (k_1 + k_2)\|x\|$ for all $x \in X$;

(b) $\|(\lambda S)(x)\| \leq |\lambda|k_1\|x\|$ for all $x \in X$;

(c) $B(X, Y)$ is a linear subspace of $L(X, Y)$ and so $B(X, Y)$ is a vector space.

Proof

(a) If $x \in X$ then

$$\|(S + T)(x)\| \leq \|S(x)\| + \|T(x)\| \leq k_1\|x\| + k_2\|x\| = (k_1 + k_2)\|x\|.$$

(b) If $x \in X$ then

$$\|(\lambda S)(x)\| = |\lambda|\|S(x)\| \leq |\lambda|k_1\|x\|.$$

(c) By parts (a) and (b), $S + T$ and λS are in $B(X, Y)$ so $B(X, Y)$ is a linear subspace of $L(X, Y)$. Hence $B(X, Y)$ is a vector space. □

EXERCISES

4.1 If $T : C_{\mathbb{R}}[0,1] \to \mathbb{R}$ is the linear transformation defined by

$$T(f) = \int_0^1 f(x)dx$$

show that T is continuous.

4.2 Let $h \in L^\infty[0,1]$.

(a) If f is in $L^2[0,1]$, show that $fh \in L^2[0,1]$.

(b) Let $T : L^2[0,1] \to L^2[0,1]$ be the linear transformation defined by $T(f) = hf$. Show that T is continuous.

4.3 Let $\mathcal{H}$ be a complex Hilbert space and let $y \in \mathcal{H}$. Show that the linear transformation $f : \mathcal{H} \to \mathbb{C}$ defined by

$$f(x) = (x, y)$$

is continuous.

4.4 (a) If $(x_1, x_2, x_3, x_4, \ldots) \in \ell^2$, show that

$$(0, 4x_1, x_2, 4x_3, x_4, \ldots) \in \ell^2.$$

(b) Let $T : \ell^2 \to \ell^2$ be the linear transformation defined by

$$T(x_1, x_2, x_3, x_4, \ldots) = (0, 4x_1, x_2, 4x_3, x_4, \ldots).$$

Show that T is continuous.

4.5 Give the solution to Example 4.10.

4.2 The Norm of a Bounded Linear Operator

If X and Y are normed linear spaces we showed in Lemma 4.14 that $B(X, Y)$ is a vector space. We now show that $B(X, Y)$ is also a normed space. While doing this we often have as many as three different norms, from three different spaces, in the same equation, and so we should in principle distinguish between these norms. In practice we simply use the symbol $\|\cdot\|$ for the norm on all three spaces as it is still usually easy to determine which space an element is in and therefore, implicitly, to which norm we are referring. To check the axioms for the norm on $B(X,Y)$ that we will define in Lemma 4.15 we need the following consequence of Lemma 4.1

$$\sup\{\|T(x)\| : \|x\| \leq 1\} = \inf\{k : \|T(x)\| \leq k\|x\| \text{ for all } x \in X\}$$

and so in particular $\|T(y)\| \leq \sup\{\|T(x)\| : \|x\| \leq 1\}\|y\|$ for all $y \in X$.

Lemma 4.15

Let X and Y be normed spaces. If $\|\cdot\| : B(X,Y) \to \mathbb{R}$ is defined by

$$\|T\| = \sup\{\|T(x)\| : \|x\| \leq 1\}$$

then $\|\cdot\|$ is a norm on $B(X,Y)$.

Proof

Let $S, T \in B(X,Y)$ and let $\lambda \in \mathbb{F}$.

(i) Clearly $\|T\| \geq 0$ for all $T \in B(X,Y)$.

(ii) Recall that the zero linear transformation R satisfies $R(x) = 0$ for all $x \in X$. Hence,

$$\begin{aligned} \|T\| = 0 &\iff \|Tx\| = 0 \text{ for all } \ x \in X \\ &\iff Tx = 0 \text{ for all } \ x \in X \\ &\iff T \text{ is the zero linear transformation.} \end{aligned}$$

(iii) As $\|T(x)\| \leq \|T\|\|x\|$ we have $\|(\lambda T)(x)\| \leq |\lambda|\|T\|\|x\|$ for all $x \in X$ by Lemma 4.14(b). Hence

$$\|\lambda T\| \leq |\lambda|\|T\|.$$

If $\lambda = 0$ then $\|\lambda T\| = |\lambda|\,\|T\|$ while if $\lambda \neq 0$ then

$$\|T\| = \|\lambda^{-1}\lambda T\| \leq \left|\lambda^{-1}\right| \|\lambda T\| \leq |\lambda|^{-1}\,|\lambda|\,\|T\| = \|T\|.$$

Hence $\|T\| = |\lambda|^{-1}\,\|\lambda T\|$ and so

$$\|\lambda T\| = |\lambda|\,\|T\|.$$

(iv) The final property to check is the triangle inequality.

$$\begin{aligned} \|(S+T)(x)\| &\leq \|S(x)\| + \|T(x)\| \\ &\leq \|S\|\|x\| + \|T\|\|x\| \\ &= (\|S\| + \|T\|)\|x\|. \end{aligned}$$

Therefore $\|S+T\| \leq \|S\| + \|T\|$.

Hence $B(X, Y)$ is a normed vector space. □

Definition 4.16

Let X and Y be normed linear spaces and let $T \in B(X, Y)$. The norm of T is defined by $\|T\| = \sup\{\|T(x)\| : \|x\| \leq 1\}$.

Using the link between matrices and linear transformations between finite-dimensional spaces we can use the definition of the norm of a bounded linear transformation to give a norm on the vector space of $m \times n$ matrices.

Definition 4.17

Let $\mathbb{F}^p$ have the standard norm and let A be a $m \times n$ matrix with entries in $\mathbb{F}$. If $T : \mathbb{F}^n \to \mathbb{F}^m$ is the bounded linear transformation defined by $T(x) = Ax$ then the norm of the matrix A is defined by $\|A\| = \|T\|$.

Let us now see how to compute the norm of a bounded linear transformation. Since the norm of an operator is the supremum of a set, the norm can sometimes be hard to find. Even if X is a finite-dimensional normed linear space and there is an element y with $\|y\| = 1$ in X such that $\|T\| = \sup\{\|T(x)\| : \|x\| \leq 1\} = \|T(y)\|$ it might not be easy to find this element y. In the infinite-dimensional case there is also the possibility that the supremum may not be attained. Hence there is no general procedure for finding the norm of a bounded linear transformation. Nevertheless there are some cases when the norm can easily be found. As a first example consider the norm of the linear transformation given in Example 4.2.

Example 4.18

If $T : C_{\mathbb{F}}[0, 1] \to \mathbb{F}$ is the bounded linear operator defined by

$$T(f) = f(0)$$

then $\|T\| = 1$.

Solution

It was shown in Example 4.2 that $|T(f)| \leq \|f\|$ for all $f \in C_{\mathbb{F}}[0, 1]$. Hence

$$\|T\| = \inf\{k : \|T(x)\| \leq k\|x\| \text{ for all } x \in X\} \leq 1.$$

On the other hand, if $g : [0, 1] \to \mathbb{C}$ is defined by $g(x) = 1$ for all $x \in X$ then $g \in C_{\mathbb{C}}[0, 1]$ with $\|g\| = \sup\{|g(x)| : x \in [0, 1]\} = 1$ and $|T(g)| = |g(0)| = 1$.

Hence

$$1 = |T(g)| \leq \|T\|\|g\| = \|T\|.$$

Therefore $\|T\| = 1$. □

Sometimes it is possible to use the norm of one operator to find the norm of another. We illustrate this in Theorem 4.19.

Theorem 4.19

Let X be a normed linear space and let W be a dense subspace of X. Let Y be a Banach space and let $S \in B(W, Y)$.

(a) If $x \in X$ and $\{x_n\}$ and $\{y_n\}$ are sequences in W such that $\lim_{n\to\infty} x_n = \lim_{n\to\infty} y_n = x$ then $\{S(x_n)\}$ and $\{S(y_n)\}$ both converge and $\lim_{n\to\infty} S(x_n) = \lim_{n\to\infty} S(y_n)$.

(b) There exists $T \in B(X, Y)$ such that $\|T\| = \|S\|$ and $Tx = Sx$ for all $x \in W$.

Proof

(a) Since $\{x_n\}$ converges it is a Cauchy sequence. Therefore, as

$$\|S(x_n) - S(x_m)\| = \|S(x_n - x_m)\| \leq \|S\|\|x_n - x_m\|,$$

$\{S(x_n)\}$ is also a Cauchy sequence and hence, since Y is a Banach space, $\{S(x_n)\}$ converges.

As $\lim_{n\to\infty} x_n = \lim_{n\to\infty} y_n = x$ we have $\lim_{n\to\infty} (x_n - y_n) = 0$. Since

$$\|S(x_n) - S(y_n)\| = \|S(x_n - y_n)\| \leq \|S\|\|x_n - y_n\|,$$

$\lim_{n\to\infty} S(x_n) - S(y_n) = 0$ and so $\lim_{n\to\infty} S(x_n) = \lim_{n\to\infty} S(y_n)$.

(b) We now define $T : X \to Y$ as follows: for any $x \in X$ there exists a sequence $\{x_n\}$ in W such that $\lim_{n\to\infty} x_n = x$ (since W is dense in X) and we define $T : X \to Y$ by

$$T(x) = \lim_{n\to\infty} S(x_n)$$

(T is well defined since the value of the limit is independent of the choice of sequence $\{x_n\}$ converging to x by part (a)). In this case it is perhaps not clear that T is a linear transformation so the first step in this part is to show that T is linear.

Let $x, y \in X$ and let $\lambda \in \mathbb{F}$. Let $\{x_n\}$ and $\{y_n\}$ be sequences in W such that $\lim_{n\to\infty} x_n = x$ and $\lim_{n\to\infty} y_n = y$. Then $\{x_n\}$ and $\{y_n\}$ are sequences in W such that $\lim_{n\to\infty} (x_n + y_n) = x + y$ and $\lim_{n\to\infty} \lambda x_n = \lambda x$. Hence

$$\begin{aligned} T(x+y) &= \lim_{n\to\infty} S(x_n + y_n) \\ &= \lim_{n\to\infty} (S(x_n) + S(y_n)) \\ &= \lim_{n\to\infty} S(x_n) + \lim_{n\to\infty} S(y_n) \\ &= T(x) + T(y) \end{aligned}$$

and

$$T(\lambda x) = \lim_{n\to\infty} S(\lambda x_n) = \lim_{n\to\infty} \lambda S(x_n) = \lambda \lim_{n\to\infty} S(x_n) = \lambda T(x).$$

Hence T is a linear transformation.

Now suppose that $x \in X$ with $\|x\| = 1$ and let $\{x_n\}$ be a sequence in W such that $\lim_{n\to\infty} x_n = x$. Since $\lim_{n\to\infty} \|x_n\| = \|x\| = 1$, if we let $w_n = \dfrac{x_n}{\|x_n\|}$ then $\{w_n\}$ is a sequence in W such that $\lim_{n\to\infty} w_n = \lim_{n\to\infty} \dfrac{x_n}{\|x_n\|} = x$ and $\|w_n\| = \dfrac{\|x_n\|}{\|x_n\|} = 1$ for all $n \in \mathbb{N}$. As

$$\begin{aligned} \|Tx\| &= \lim_{n\to\infty} \|Sw_n\| \\ &\leq \sup\{\|Sw_n\| : n \in \mathbb{N}\} \\ &\leq \sup\{\|S\|\|w_n\| : n \in \mathbb{N}\} \\ &= \|S\|, \end{aligned}$$

T is bounded and $\|T\| \leq \|S\|$. Moreover if $w \in W$ then the constant sequence $\{w\}$ is a sequence in W converging to w and so

$$Tw = \lim_{n\to\infty} Sw = Sw.$$

Thus $\|Sw\| = \|Tw\| \leq \|T\|\|w\|$ so $\|S\| \leq \|T\|$. Hence $\|S\| = \|T\|$ and we have already shown that if $x \in W$ then $Tx = Sx$. □

The operator T in Theorem 4.19 can be thought of as an extension of the operator S to the larger space X.

We now consider a type of operator whose norm is easy to find.

Definition 4.20

Let X and Y be normed linear spaces and let $T \in L(X, Y)$. If $\|T(x)\| = \|x\|$ for all $x \in X$ then T is called an *isometry*.

On every normed space there is at least one isometry.

Example 4.21

If X is a normed space and I is the identity linear transformation on X then I is an isometry.

Solution

If $x \in X$ then $I(x) = x$ and so $\|I(x)\| = \|x\|$. Hence I is an isometry. □

As another example of an isometry consider the following linear transformation.

Example 4.22

(a) If $x = (x_1, x_2, x_3, \ldots) \in \ell^2$ then $y = (0, x_1, x_2, x_3, \ldots) \in \ell^2$.

(b) The linear transformation $S : \ell^2 \to \ell^2$ defined by

$$S(x_1, x_2, x_3, \ldots) = (0, x_1, x_2, x_3, \ldots) \tag{4.2}$$

is an isometry.

Solution

(a) Since $x \in \ell^2$,

$$|0|^2 + |x_1|^2 + |x_2|^2 + |x_3|^2 + \ldots = |x_1|^2 + |x_2|^2 + |x_3|^2 + \ldots < \infty,$$

and so $y \in \ell^2$.

(b)
$$\begin{aligned} \|S(x)\|^2 &= |0|^2 + |x_1|^2 + |x_2|^2 + |x_3|^2 + \ldots \\ &= |x_1|^2 + |x_2|^2 + |x_3|^2 + \ldots \\ &= \|x\|^2, \end{aligned}$$

and hence S is an isometry. □

The operator defined in Example 4.22 will be referred to frequently in the next chapter so it will be useful to have a name for it.

Notation

The isometry $S : \ell^2 \to \ell^2$ defined by

$$S(x_1, x_2, x_3, \ldots) = (0, x_1, x_2, x_3, \ldots)$$

is called the *unilateral shift*.

It is easy to see that the unilateral shift does not map ℓ^2 onto ℓ^2. This contrasts with the finite-dimensional situation where if X is a normed linear space and T is an isometry of X into X then T maps X onto X, by Lemma 1.12.

We leave as an exercise the proof of Lemma 4.23 which shows that the norm of an isometry is 1.

Lemma 4.23

Let X and Y be normed linear spaces and let $T \in L(X, Y)$. If T is an isometry then T is bounded and $\|T\| = 1$.

The converse of Lemma 4.23 is not true. In Example 4.18, although $\|T\| = 1$ it is not true that $|T(h)| = \|h\|$ for all $h \in C_{\mathbb{F}}[0,1]$. For example, if $h : [0,1] \to \mathbb{F}$ is defined by $h(x) = x$ for all $x \in [0,1]$ then $\|h\| = 1$ while $|T(h)| = 0$. Therefore, saying that a linear transformation is an isometry asserts more than that it has norm 1.

Definition 4.24

If X and Y are normed linear spaces and T is an isometry from X onto Y then T is called an *isometric isomorphism* and X and Y are called *isometrically isomorphic.*

If two spaces are isometrically isomorphic it means that as far as the vector space and norm properties are concerned they have essentially the same structure. However, it can happen that one way of looking at a space gives more insight into the space. For instance, $\ell^2_{\mathbb{F}}$ is a simple example of a Hilbert space and we will show in Corollary 4.26 that any infinite-dimensional, separable Hilbert spaces over $\mathbb{F}$ is isomorphic to $\ell^2_{\mathbb{F}}$. We recall that $\{\widetilde{e}_n\}$ is the standard orthonormal basis in $\ell^2_{\mathbb{F}}$.

Theorem 4.25

Let $\mathcal{H}$ be an infinite-dimensional Hilbert space over $\mathbb{F}$ with an orthonormal basis $\{e_n\}$. Then there is an isometry T of $\mathcal{H}$ onto $\ell^2_{\mathbb{F}}$ such that $T(e_n) = \widetilde{e}_n$ for all $n \in \mathbb{N}$.

Proof

Let $x \in \mathcal{H}$. Then $x = \sum_{n=1}^{\infty}(x, e_n)e_n$ by Theorem 3.47 as $\{e_n\}$ is an orthonormal basis for $\mathcal{H}$. Moreover, if $\alpha_n = (x, e_n)$ then $\{\alpha_n\} \in \ell^2_{\mathbb{F}}$ by Lemma 3.41 (Bessel's inequality) so we can define a linear transformation $T : \mathcal{H} \to \ell^2_{\mathbb{F}}$ by $T(x) = \{\alpha_n\}$. Now

$$\|T(x)\|^2 = \sum_{n=1}^{\infty} |\alpha_n|^2 = \sum_{n=1}^{\infty} |(x, e_n)|^2 = \|x\|^2$$

for all $x \in \mathcal{H}$ by Theorem 3.47, so T is an isometry and by definition $T(e_n) = \widetilde{e}_n$ for all $n \in \mathbb{N}$.

Finally, if $\{\beta_n\}$ is in $\ell^2_{\mathbb{F}}$ then by Theorem 3.42 the series $\sum_{n=1}^{\infty} \beta_n e_n$ converges to a point $y \in \mathcal{H}$. Since $(y, e_n) = \beta_n$ we have $T(y) = \{\beta_n\}$. Hence T is an isometry of $\mathcal{H}$ onto $\ell^2_{\mathbb{F}}$. □

Corollary 4.26

Any infinite-dimensional, separable Hilbert space $\mathcal{H}$ over $\mathbb{F}$ is isometrically isomorphic to $\ell^2_{\mathbb{F}}$.

Proof

$\mathcal{H}$ has an orthonormal basis $\{e_n\}$ by Theorem 3.52, so $\mathcal{H}$ is isometrically isomorphic to $\ell^2_{\mathbb{F}}$ by Theorem 4.25. □

EXERCISES

4.6 Let $T : C_{\mathbb{R}}[0, 1] \to \mathbb{R}$ be the bounded linear transformation defined by

$$T(f) = \int_0^1 f(x)dx.$$

(a) Show that $\|T\| \leq 1$.

(b) If $g \in C_{\mathbb{R}}[0, 1]$ is defined by $g(x) = 1$ for all $x \in [0, 1]$, find $|T(g)|$ and hence find $\|T\|$.

4.7 Let $h \in L^\infty[0, 1]$ and let $T : L^2[0, 1] \to L^2[0, 1]$ be the bounded linear transformation defined by $T(f) = hf$. Show that

$$\|T\| \leq \|h\|_\infty.$$

4.8 Let $T : \ell^2 \to \ell^2$ be the bounded linear transformation defined by

$$T(x_1, x_2, x_3, x_4, \ldots) = (0, 4x_1, x_2, 4x_3, x_4, \ldots).$$

Find the norm of T.

4.9 Prove Lemma 4.23.

4.10 Let $\mathcal{H}$ be a complex Hilbert space and let $y \in \mathcal{H}$. Find the norm of the bounded linear transformation $f : \mathcal{H} \to \mathbb{C}$ defined by

$$f(x) = (x, y).$$

4.11 Let $\mathcal{H}$ be a Hilbert space and let $y, z \in \mathcal{H}$. If T is the linear transformation defined by $T(x) = (x, y)z$, show that T is bounded and that $\|T\| \leq \|y\|\|z\|$.

4.3 The Space $B(X, Y)$ and Dual Spaces

Now that we have seen some examples of norms of individual operators let us look in more detail at the space $B(X, Y)$ where X and Y are normed linear spaces. Since many of the deeper properties of normed linear spaces are obtained only for Banach spaces it is natural to ask when $B(X, Y)$ is a Banach space. An initial guess might suggest that it would be related to completeness of X and Y. This is only half correct. In fact, it is only the completeness of Y which matters.

Theorem 4.27

If X is a normed linear spaces and Y is a Banach space then the normed space $B(X, Y)$ is a Banach space.

Proof

We have to show that $B(X, Y)$ is a complete metric space. Let $\{T_n\}$ be a Cauchy sequence in $B(X, Y)$. In any metric space a Cauchy sequence is bounded so there exists $M > 0$ such that $\|T_n\| \leq M$ for all $n \in \mathbb{N}$. Let $x \in X$. As

$$\|T_n(x) - T_m(x)\| = \|(T_n - T_m)(x)\| \leq \|T_n - T_m\|\|x\|$$

and $\{T_n\}$ is Cauchy, it follows that $\{T_n(x)\}$ is a Cauchy sequence in Y. Since Y is complete $\{T_n(x)\}$ converges, so we may define $T : X \to Y$ by

$$T(x) = \lim_{n \to \infty} T_n(x).$$

It is perhaps not clear in this case that T is a linear transformation so the first step to show that T is the required limit is to show that it is linear. As

$$T(x+y) = \lim_{n\to\infty} T_n(x+y) = \lim_{n\to\infty} (T_nx+T_ny) = \lim_{n\to\infty} T_nx + \lim_{n\to\infty} T_ny = Tx+Ty$$

and

$$T(\alpha x) = \lim_{n\to\infty} T_n(\alpha x) = \lim_{n\to\infty} \alpha T_nx = \alpha \lim_{n\to\infty} T_nx = \alpha T(x),$$

T is a linear transformation.

Next we show that T is bounded. As $\|T(x)\| = \lim_{n\to\infty} \|T_n(x)\|$,

$$\|T(x)\| \leq \sup\{\|T_n(x)\| : n \in \mathbb{N}\} \leq \sup\{\|T_n\|\|x\| : n \in \mathbb{N}\} \leq M\|x\|.$$

Therefore T is bounded and so $T \in B(X, Y)$.

Finally, we have to show that $\lim_{n\to\infty} T_n = T$. Let $\epsilon > 0$. There exists $N \in \mathbb{N}$ such that when $m, n \geq N$

$$\|T_n - T_m\| < \frac{\epsilon}{2}.$$

Then for any x with $\|x\| \leq 1$

$$\|T_n(x) - T_m(x)\| \leq \|T_n - T_m\|\|x\| < \frac{\epsilon}{2},$$

for $m, n \geq N$. As $T(x) = \lim_{n\to\infty} T_n(x)$, there exists $N_1 \in \mathbb{N}$ such that when $m \geq N_1$,

$$\|T(x) - T_m(x)\| < \frac{\epsilon}{2}.$$

Then when $n \geq N$ and $m \geq N_1$

$$\|T(x) - T_n(x)\| \leq \|T(x) - T_m(x)\| + \|T_n(x) - T_m(x)\| < \frac{\epsilon}{2} + \frac{\epsilon}{2}\|x\| \leq \epsilon.$$

Thus $\|T - T_n\| \leq \epsilon$ when $n \geq N$ and so $\lim_{n\to\infty} T_n = T$. Hence $B(X, Y)$ is a Banach space. □

One case of the above which occurs sufficiently often to warrant separate notation is when $Y = \mathbb{F}$.

Definition 4.28

Let X be a normed space over $\mathbb{F}$. The space $B(X, \mathbb{F})$ is called the *dual space* of X and is denoted by X'.

The dual space is sometimes also denoted by X^*, but to avoid confusion with notation we will use in Chapter 5, we will not use X^* for the dual space.

Corollary 4.29

If X is a normed vector space then X' is a Banach space.

Proof

The space $\mathbb{F}$ is complete so X' is a Banach space by Theorem 4.27. □

In general it is relatively easy to produce some elements of a dual space but less easy to identify all of them. As an example of what is involved we find all the elements of the dual space of a Hilbert space. We have already identified some of the elements of this dual space in Exercise 4.3.

Example 4.30

Let $\mathcal{H}$ be a Hilbert space over $\mathbb{F}$ and let $y \in \mathcal{H}$. If $f : \mathcal{H} \to \mathbb{F}$ is defined by $f(x) = (x, y)$ then f is an element of $\mathcal{H}'$.

The harder part of the identification of the dual space of a Hilbert space is to show that all elements of the dual space are of the above form.

Theorem 4.31 (Riesz–Fréchet theorem)

If $\mathcal{H}$ is a Hilbert space and $f \in \mathcal{H}'$ then there is a unique $y \in \mathcal{H}$ such that $f(x) = (x, y)$ for all $x \in \mathcal{H}$. Moreover $\|f\| = \|y\|$.

Proof

(a) (Existence). If $f(x) = 0$ for all $x \in \mathcal{H}$ then $y = 0$ will be a suitable choice. Otherwise, $\operatorname{Ker} f = \{x \in \mathcal{H} : f(x) = 0\}$ is a proper closed subspace of $\mathcal{H}$ so that $(\operatorname{Ker} f)^\perp \neq 0$ by Theorem 3.34. Therefore there exists $z \in (\operatorname{Ker} f)^\perp$ such that $f(z) = 1$. In particular, $z \neq 0$ so we may define $y = \dfrac{z}{\|z\|^2}$. Since f is a linear transformation

$$f(x - f(x)z) = f(x) - f(x)f(z) = 0,$$

and hence $x - f(x)z \in \operatorname{Ker} f$. However, $z \in (\operatorname{Ker} f)^\perp$ so

$$(x - f(x)z, z) = 0.$$

Therefore, $(x, z) - f(x)(z, z) = 0$ and so $(x, z) = f(x)\|z\|^2$. Hence

$$f(x) = (x, \frac{z}{\|z\|^2}) = (x, y).$$

Moreover, if $\|x\| \leq 1$ then by the Cauchy–Schwarz inequality

$$|f(x)| = |(x, y)| \leq \|x\|\|y\| \leq \|y\|.$$

Hence $\|f\| \leq \|y\|$.

On the other hand if $x = \frac{y}{\|y\|}$ then $\|x\| = \left\|\frac{y}{\|y\|}\right\| = 1$ and so

$$\|f\| \geq |f(x)| = \frac{|f(y)|}{\|y\|} = \frac{(y, y)}{\|y\|} = \|y\|.$$

Therefore $\|f\| \geq \|y\|$.

(b) (Uniqueness). If y and w are such that

$$f(x) = (x, y) = (x, w)$$

for all $x \in \mathcal{H}$, then $(x, y - w) = 0$ for all $x \in \mathcal{H}$. Hence by Exercise 3.1 we have $y - w = 0$ and so $y = w$ as required. □

Another Banach space whose dual is relatively easy to identify is ℓ^1.

Theorem 4.32

(a) If $c = \{c_n\} \in \ell^\infty$ and $\{x_n\} \in \ell^1$ then $\{c_n x_n\} \in \ell^1$ and if the linear transformation $f_c : \ell^1 \to \mathbb{F}$ is defined by $f_c(\{x_n\}) = \sum_{n=1}^\infty c_n x_n$ then $f_c \in (\ell^1)'$ with

$$\|f_c\| \leq \|c\|_\infty.$$

(b) If $f \in (\ell^1)'$ then there exists $c \in \ell^\infty$ such that $f = f_c$ and $\|c\|_\infty \leq \|f\|$.

(c) The space $(\ell^1)'$ is isometrically isomorphic to ℓ^∞.

Proof

(a) This follows from Example 4.4.

(b) Let $\{\widetilde{e}_n\}$ be the sequence in ℓ^1 given in Definition 1.59. Let $c_n = f(\widetilde{e}_n)$ for all $n \in \mathbb{N}$. Then

$$|c_n| = |f(\widetilde{e}_n)| \leq \|f\|\|\widetilde{e}_n\| = \|f\|,$$

for all $n \in \mathbb{N}$, and hence $c = \{c_n\} \in \ell^\infty$ and $\|c\|_\infty \leq \|f\|$.

If $\mathcal{S}$ is the linear subspace of ℓ^1 consisting of sequences with only finitely many non-zero terms then $\mathcal{S}$ is dense in ℓ^1. Let x be an element of $\mathcal{S}$ where $x = (x_1, x_2, x_3, \ldots, x_n, 0, 0, \ldots)$. Then

$$f(x) = f(\sum_{j=1}^{n} x_j \widetilde{e}_j) = \sum_{j=1}^{n} x_j f(\widetilde{e}_j) = \sum_{j=1}^{n} x_j c_j = f_c(x).$$

Since the continuous functions f and f_c agree on the dense subset $\mathcal{S}$ it follows that $f = f_c$, by Corollary 1.29.

(c) The map $T : \ell^\infty \to (\ell^1)'$ defined by $T(c) = f_c$ is a linear transformation which maps ℓ^∞ onto $(\ell^1)'$ by part (a). From the inequalities $\|f_c\| \le \|c\|_\infty$ and $\|c\|_\infty \le \|f\| = \|f_c\|$,

$$\|T(c)\| = \|f_c\| = \|c\|_\infty.$$

Hence T is an isometry. □

It is possible to identify many of the dual spaces of other standard normed spaces introduced in Chapter 2 in a similar manner, but the proofs of these identifications in many cases require more knowledge of measure theory than we are assuming as prerequisite, so we shall not go into more details. Instead we shall return to the investigation of the algebraic structure of $B(X, Y)$.

Lemma 4.33

If X, Y and Z are normed linear spaces and $T \in B(X, Y)$ and $S \in B(Y, Z)$ then $S \circ T \in B(X, Z)$ and

$$\|S \circ T\| \le \|S\|\|T\|.$$

Proof

As $S \circ T \in L(X, Z)$ by Lemma 1.8 and

$$\|(S \circ T)(x)\| = \|S(T(x))\| \le \|S\|\|T(x)\| \le \|S\|\|T\|\|x\|,$$

$S \circ T \in B(X, Z)$ and $\|S \circ T\| \le \|S\|\|T\|$. □

If X, Y and Z are three finite-dimensional spaces with bases $\mathbf{x}$, $\mathbf{y}$ and $\mathbf{z}$ respectively and $T \in L(X, Y)$ and $S \in L(Y, Z)$ then $M_{\mathbf{z}}^{\mathbf{x}}(S \circ T) = M_{\mathbf{z}}^{\mathbf{y}}(S) M_{\mathbf{y}}^{\mathbf{x}}(T)$ by Theorem 1.15. A natural candidate for the product of bounded linear operators is therefore function composition.

Definition 4.34

Let X, Y and Z be normed linear spaces and $T \in B(X, Y)$ and $S \in B(Y, Z)$. The composition $S \circ T$ of S and T will be denoted by ST and called the *product* of S and T.

Let X, Y and Z be normed linear spaces and $T \in B(X, Y)$ and $S \in B(Y, Z)$. If the spaces X, Y and Z are not all the same the fact that we can define the product ST does not mean we can define the product TS. However, if $X = Y = Z$ then TS and ST will both be defined. We note, however, that even if all the spaces are finite-dimensional and $X = Y = Z$, in general $TS \neq ST$.

As the situation when $X = Y = Z$ arises frequently it is convenient to have the following notation.

Notation

If X is a normed linear space the set $B(X, X)$ of all bounded linear operators from X to X will be denoted by $B(X)$.

If X is a normed space then we have seen that $B(X)$ has a lot of algebraic structure. This is summarized in part (a) of Lemma 4.35.

Lemma 4.35

Let X be a normed linear space.

(a) $B(X)$ is an algebra with identity and hence a ring with identity.

(b) If $\{T_n\}$ and $\{S_n\}$ are sequences in $B(X)$ such that $\lim_{n\to\infty} T_n = T$ and $\lim_{n\to\infty} S_n = S$ then $\lim_{n\to\infty} S_n T_n = ST$.

Proof

(a) The identities given in Lemma 1.9 hold for all linear transformations and hence in particular hold for those in $B(X)$. Therefore $B(X)$ is an algebra with identity (and hence a ring with identity) under the operations of addition and scalar multiplication given in Definition 1.7 and multiplication given in Definition 4.34.

(b) As $\{T_n\}$ is convergent it is bounded so there exists $K > 0$ such that $\|T_n\| \leq K$ for all $n \in \mathbb{N}$. Let $\epsilon > 0$. There exists $N_1 \in \mathbb{N}$ such that when $n \geq N_1$

$$\|S_n - S\| < \frac{\epsilon}{2K},$$

and $N_2 \in \mathbb{N}$ such that when $n \geq N_2$

$$\|T_n - T\| < \frac{\epsilon}{2(\|S\| + 1)}.$$

As

$$\|S_nT_n - ST\| \leq \|S_nT_n - ST_n\| + \|ST_n - ST\| \leq K\|S_n - S\| + \|S\|\|T_n - T\|,$$

when $n \geq \max(N_1, N_2)$ we have

$$\|S_nT_n - ST\| \leq K\|S_n - S\| + \|S\|\|T_n - T\| < \epsilon.$$

Thus $\lim_{n\to\infty} S_nT_n = ST$. □

Notation

Let X be a normed space and let $T \in B(X)$.

(a) TT will be denoted by T^2, TTT will be denoted by T^3, and more generally the product of T with itself n times will be denoted by T^n.

(b) If $a_0, a_1, \ldots, a_n \in \mathbb{F}$ and $p : \mathbb{F} \to \mathbb{F}$ is the polynomial defined by $p(x) = a_0 + a_1x + \ldots + a_nx^n$, then we define $p(T)$ by $p(T) = a_0I + a_1T + \ldots + a_nT^n$.

Since $B(X)$ is a ring the above notation is consistent with using the power notation for products of an element with itself in any ring. The following is such an easy consequence of the definitions that we omit the proof.

Lemma 4.36

Let X be a normed linear space and let $T \in B(X)$. If p and q are polynomials and $\lambda, \mu \in \mathbb{C}$ then

(a) $(\lambda p + \mu q)(T) = \lambda p(T) + \mu q(T)$;

(b) $(pq)(T) - p(T)q(T)$.

EXERCISES

4.12 Let $\mathcal{H}$ be a complex Hilbert space and let $\mathcal{M}$ be a closed linear subspace of $\mathcal{H}$. If $f \in \mathcal{M}'$, show that there is $g \in \mathcal{H}'$ such that $g(x) = f(x)$ for all $x \in \mathcal{M}$ and $\|f\| = \|g\|$.

4.13 Let c_0 be the linear subspace of ℓ^∞ consisting of all sequences which converge to 0.

(a) If $a = \{a_n\} \in \ell^1$ and $\{x_n\} \in c_0$, show that $\{a_n x_n\} \in \ell^1$, and that the linear transformation $f_a : c_0 \to \mathbb{C}$ defined by

$$f_a(\{x_n\}) = \sum_{n=1}^{\infty} a_n x_n$$

is continuous with $\|f_a\| \leq \|\{a_n\}\|_1$.

(b) Show that for any $f \in (c_0)'$ there exists a unique $a = \{a_n\} \in \ell^1$ such that $f = f_a$ and show also that $\|\{a_n\}\|_1 \leq \|f\|$.

(c) Show that the linear transformation $T : \ell^1 \to (c_0)'$ defined by $T(a) = f_a$ is an isometry.

4.14 Let X be a normed linear space and let $P, Q \in B(X)$. Show that the linear transformation $T : B(X) \to B(X)$ defined by $T(R) = PRQ$ is bounded.

4.15 Let $T : \ell^2 \to \ell^2$ be the bounded linear operator defined by

$$T(x_1, x_2, x_3, x_4, \ldots) = (0, 4x_1, x_2, 4x_3, x_4, \ldots).$$

(a) Find T^2.

(b) Hence find $\|T^2\|$ and compare this with $\|T\|^2$.

4.16 Let X, Y and Z be a normed linear spaces and let $T : X \to Y$ and $S : Y \to Z$ be isometries. Show that $S \circ T$ is an isometry.

4.4 Inverses of Operators

One way to solve a matrix equation

$$Ax = y$$

is to find the inverse matrix A^{-1} (if it exists) and then the solution is $x = A^{-1}y$. In this section we will see to what extent this can be generalized to the infinite-dimensional case. For some operators it will be possible to define negative powers, namely the operators T for which we can make sense of the inverse T^{-1} of T. If X is a normed linear space, we can use the standard ring-theoretic definition of invertibility for our definition of invertibility of operators, as we have seen that $B(X)$ is a ring.

Definition 4.37

Let X be a normed linear space. An operator $T \in B(X)$ is said to be *invertible* if there exists $S \in B(X)$ such that $ST = I = TS$.

From elementary ring theory we know that if $T \in B(X)$ is invertible then there is at most one element $S \in B(X)$ such that $ST = I = TS$. Not surprisingly therefore, we use the following notation.

Notation

Let X be a normed linear space and let $T \in B(X)$ be invertible. The element $S \in B(X)$ such that $ST = I = TS$ is called the *inverse* of T and is denoted by T^{-1}.

The rest of this section is devoted to studying properties of inverses, characterizing invertible operators and giving examples of these operators. The following lemma on invertibility is true in any ring so we leave it as an exercise.

Lemma 4.38

If X is a normed linear space and T_1, T_2 are invertible elements of $B(X)$ then

(a) T_1^{-1} is invertible with inverse T_1;

(b) $T_1 T_2$ is invertible with inverse $T_2^{-1} T_1^{-1}$.

In Example 4.39 we check directly from the definition that an operator is invertible.

Example 4.39

For any $h \in C[0,1]$ let $T_h \in B(L^2[0,1])$ be defined by

$$(T_h g)(t) = h(t)g(t). \tag{4.3}$$

If $f \in C[0,1]$ is defined by $f(t) = 1 + t$ then T_f is invertible.

Solution

Note that we showed in Exercise 4.2 that T_h is bounded for any $h \in C[0,1]$. Let $k(t) = \dfrac{1}{1+t}$. Then $k \in C[0,1]$ and

$$(T_k T_f g)(t) = (T_k f g)(t) = k(t)f(t)g(t) = g(t),$$

for all $t \in [0,1]$. Thus $(T_k T_f)g = g$ for all $g \in L^2[0,1]$ so

$$T_k T_f = I.$$

Similarly $T_f T_k = I$. Hence T_f is invertible with $T_f^{-1} = T_k$. □

Once we had a suitable candidate for the inverse it was easy to determine that T_f was invertible in Example 4.39. Usually it is not so easy to determine whether an inverse exists. Even if we consider the same space and a very similar function $g \in C[0,1]$ defined by $g(t) = t$, the same solution as in Example 4.39 would not work as $\dfrac{1}{g(t)}$ is not in $C[0,1]$. Unfortunately, we cannot immediately conclude from this that T_g is not invertible as it could have an inverse not of the form T_h for any $h \in C[0,1]$ and we do not have the techniques available yet to check this. Therefore we have to look for other ways of determining whether an operator is invertible. The determinant of a matrix can be used to determine if the matrix is invertible, but there is no obvious generalization of this to the infinite-dimensional case.

We therefore have to approach the problem in a different way to try to find other methods to determine whether an operator is invertible. The first is based on knowledge of the norm of an operator.

Theorem 4.40

Let X be a Banach space. If $T \in B(X)$ is an operator with $\|T\| < 1$ then $I - T$ is invertible and the inverse is given by

$$(I-T)^{-1} = \sum_{n=0}^{\infty} T^n.$$

Proof

Because X is a Banach space, so is $B(X)$ by Theorem 4.27. Since $\|T\| < 1$ the series $\sum_{n=0}^{\infty} \|T\|^n$ converges and hence, as $\|T^n\| \le \|T\|^n$ for all $n \in \mathbb{N}$, the series $\sum_{n=0}^{\infty} \|T^n\|$ also converges. Therefore $\sum_{n=0}^{\infty} T^n$ converges by Theorem 2.30. Let $S = \sum_{n=0}^{\infty} T^n$ and let $S_k = \sum_{n=0}^{k} T^n$. Then the sequence $\{S_k\}$ converges to S in $B(X)$. Now

$$\|(I-T)S_k - I\| = \|I - T^{k+1} - I\| = \|-T^{k+1}\| \le \|T\|^{k+1}.$$

Since $\|T\| < 1$ we deduce that $\lim\limits_{k\to\infty} (I-T)S_k = I$. Therefore,

$$(I-T)S = (I-T)\lim_{k\to\infty} S_k = \lim_{k\to\infty} (I-T)S_k = I,$$

by Lemma 4.35. Similarly, $S(I-T) = I$ so $I-T$ is invertible and $(I-T)^{-1} = S$. □

Notation

The series in Theorem 4.40 is sometimes called the *Neumann series.*

As an application of the above method let us show how to obtain a solution to an integral equation.

Example 4.41

Let $A \in \mathbb{C}$ and let $k : [a,b] \times [a,b] \to \mathbb{R}$, a, $b \in \mathbb{R}$, be defined by

$$k(x,y) = A\sin(x-y).$$

Show that if $|A| < 1$ then for any $f \in C[a,b]$ there exists $g \in C[a,b]$ such that

$$g(x) = f(x) + \int_0^1 k(x,y)g(y)\,dy.$$

Solution

In Example 4.7(b) we showed that the linear transformation $K : C[a,b] \to C[a,b]$ defined by

$$(K(g))(s) = \int_a^b k(s,t)\,g(t)\,dt$$

is bounded and $\|K(g)\| \leq |A|\|g\|$. Hence $\|K\| \leq A$. Since the integral equation can be written as

$$(I-K)g = f$$

and $I-K$ is invertible by Theorem 4.40, the integral equation has the (unique) solution

$$g = (I-K)^{-1}f.$$

□

Theorem 4.40 can also be used to derive a metric space property of the invertible operators.

Corollary 4.42

Let X be a Banach space. The set $\mathcal{A}$ of invertible elements of $B(X)$ is open.

Proof

Let $T \in \mathcal{A}$ and let $\eta = \|T^{-1}\|^{-1}$. To show that $\mathcal{A}$ is open it suffices to show that if $\|T - S\| < \eta$ then $S \in \mathcal{A}$. Accordingly, let $\|T - S\| < \eta$. Then

$$\|(T - S)T^{-1}\| \leq \|T - S\|\|T^{-1}\| < \|T^{-1}\|^{-1}\|T^{-1}\| = 1.$$

Hence $I - (T - S)T^{-1}$ is invertible by Theorem 4.40. However,

$$I - (T - S)T^{-1} = I - (I - ST^{-1}) = ST^{-1}.$$

Therefore ST^{-1} is invertible and so $S = ST^{-1}T$ is invertible by Lemma 4.38. Hence $S \in \mathcal{A}$ as required. □

Let us now turn to a second method of determining invertibility of an operator. If X and Y are normed spaces and $T \in B(X, Y)$ is an invertible operator then T is bijective and so $\mathrm{Ker}\,(T) = \{0\}$ and $\mathrm{Im}(T) = Y$ by Lemma 1.12. Conversely if T is a one-to-one bounded linear transformation of X onto Y then there is a unique linear transformation $S : Y \to X$ such that $S \circ T = I_X$ and $T \circ S = I_Y$, again by Lemma 1.12. However, we do not know that S is bounded. The last part of the following result is therefore rather surprising.

Theorem 4.43 (Open mapping theorem)

Let X and Y be Banach spaces and $T \in B(X, Y)$ map X onto Y. Let

$$L = \{T(x) : x \in X \text{ and } \|x\| \leq 1\},$$

with closure $\overline{L}$. Then:

(a) there exists $r > 0$ such that $\{y \in Y : \|y\| \leq r\} \subseteq \overline{L}$;

(b) $\{y \in Y : \|y\| \leq \frac{r}{2}\} \subseteq L$;

(c) if, in addition, T is one-to-one then there exists $S \in B(X, Y)$ such that $S \circ T = I_X$ and $T \circ S = I_Y$.

Proof

To clarify which spaces various sets are in we will use the following notation in this proof. Let

$$B_{0,X}(r) = \{x \in X : \|x\| < r\}$$

be the open ball with centre 0 and radius r in the space X, with closure

$$\overline{B_{0,X}(r)} = \{x \in X : \|x\| \leq r\},$$

and let $B_{0,Y}(r)$ and $\overline{B_{0,Y}(r)}$ be the corresponding sets in the space Y.

(a) Since T maps X onto Y, for any $y \in Y$ there exists $x \in X$ such that $T(x) = y$. Thus $y \in \|x\|L$ and so

$$X = \bigcup_{n=1}^{\infty} n\overline{L}.$$

Therefore, by Theorem 1.32 (the Baire category theorem), there is $N \in \mathbb{N}$ such that $N\overline{L}$ contains an open ball. Hence $\overline{L}$ also contains an open ball, as the operation of multiplying by a non-zero scalar is a continuous function with a continuous inverse. Therefore, there exists $p \in \overline{L}$ and $t > 0$ such that

$$p + B_{0,Y}(t) \subseteq \overline{L}.$$

Let $y \in B_{0,Y}(t)$. Then $p + y$ and $y - p$ are both in $\overline{L}$ and so

$$y = \frac{1}{2}((p+y) + (y-p)) \in \overline{L}.$$

Thus $\overline{L}$ contains $B_{0,Y}(t)$ and so the result follows by taking $r = t/2$.

(b) Let $y \in \overline{B_{0,Y}(r/2)}$. Since $\|2y\| \leq r$ and $\overline{B_{0,Y}(r)} \subseteq \overline{L}$, there exists $w_1 \in L$ such that

$$\|2y - w_1\| < r/2.$$

Since $2^2y - 2w_1 \in \overline{B_{0,Y}(r)}$ and $\overline{B_{0,Y}(r)} \subseteq \overline{L}$ there exists $w_2 \in L$ such that

$$\|2^2y - 2w_1 - w_2\| < r/2.$$

Continuing in this way we obtain a sequence $\{w_n\}$ in L such that for all $n \in \mathbb{N}$

$$\|2^n y - 2^{n-1}w_1 - 2^{n-2}w_2 - \cdots - w_n\| < r/2.$$

Therefore, for all $n \in \mathbb{N}$,

$$\|y - \sum_{j=1}^{n} 2^{-j}w_j\| < 2^{-n-1}r,$$

and hence $y = \sum_{j=1}^{\infty} 2^{-j}w_j$. Since $w_n \in L$ for all $n \in \mathbb{N}$, there exists $x_n \in \overline{B_{0,X}(1)}$ such that $w_n = T(x_n)$. Now as

$$\|\sum_{j=1}^{\infty} 2^{-j}x_j\| \leq \sum_{j=1}^{\infty} 2^{-j} = 1,$$

the sequence of partial sums of $\sum_{j=1}^{\infty} 2^{-j}x_j$ is a Cauchy sequence and so converges to $x \in \overline{B_{0,X}(1)}$. Also

$$T(x) = T(\sum_{j=1}^{\infty} 2^{-j}x_j) = \sum_{j=1}^{\infty} 2^{-j}T(x_j) = \sum_{j=1}^{\infty} 2^{-j}w_j = y.$$

Therefore $y \in L$ so $\overline{B_{0,Y}(r/2)} \subseteq L$.

(c) There is a unique linear transformation $S : Y \to X$ such that $S \circ T = I_X$ and $T \circ S = I_Y$ by Lemma 1.12. Let $y \in Y$ with $\|y\| \leq 1$ and let $w = \frac{ry}{2}$. As $\|w\| \leq \frac{r}{2}$, we have $w = T(x)$ for some $x \in X$ with $\|x\| \leq 1$ by part (b). Therefore $y = T\left(\frac{2x}{r}\right)$ so

$$\|S(y)\| = \left\|\frac{2x}{r}\right\| \leq \frac{2}{r}.$$

Thus S is bounded. □

As an application of Theorem 4.43 we give a converse to Lemma 4.13 for operators between Banach spaces.

Corollary 4.44 (Closed graph theorem)

If X and Y are Banach space and T is a linear transformation from X into Y such that $\mathcal{G}(T)$, the graph of T, is closed, then T is continuous.

Proof

As $X \times Y$ is a Banach space by Exercise 2.13, $\mathcal{G}(T)$ is also a Banach space since it is a closed linear subspace of $X \times Y$. Let $R : \mathcal{G}(T) \to X$ be defined by $R(x, Tx) = x$. Then R is a linear transformation from $\mathcal{G}(T)$ to X which is one-to-one and onto. Since

$$\|R(x, Tx)\| = \|x\| \leq \|x\| + \|Tx\| = \|(x, Tx)\|,$$

R is bounded and $\|R\| \leq 1$. Hence there is a bounded linear operator $S : X \to \mathcal{G}(T)$ such that $S \circ R = I_{\mathcal{G}(T)}$ and $R \circ S = I_X$ by Theorem 4.43. In particular, $Sx = (x, Tx)$ for all $x \in X$. As

$$\|Tx\| \leq \|x\| + \|Tx\| = \|(x, Tx)\| = \|Sx\| \leq \|S\|\|x\|$$

for all $x \in X$, it follows that T is continuous. □

If we take $X = Y$ in Theorem 4.43 we obtain the following characterization of invertible operators on a Banach space.

Corollary 4.45 (Banach's isomorphism theorem)

If X is a Banach space and $T \in B(X)$ is one-to-one and maps X onto X then T is invertible.

Despite this neat characterization of invertibility of an operator T, if we try to use it to solve the equation $T(x) = y$ we observe that if we can check from the definition that T maps X onto X then we could already have solved the equation! To avoid this cyclic argument we have to try to find another way of characterizing invertibility.

Lemma 4.46

If X is a normed linear space and $T \in B(X)$ is invertible then, for all $x \in X$,

$$\|T(x)\| \geq \|T^{-1}\|^{-1}\|x\|.$$

Proof

For all $x \in X$ we have $\|x\| = \|T^{-1}(T(x))\| \leq \|T^{-1}\|\|T(x)\|$ and so

$$\|T(x)\| \geq \|T^{-1}\|^{-1}\|x\|. \qquad \square$$

If T is an invertible operator, then T has the property that there exists $\alpha > 0$ such that $\|T(x)\| \geq \alpha\|x\|$ for all $x \in X$ by Lemma 4.46. This property will be part of our next invertibility criteria.

Lemma 4.47

If X is a Banach space and $T \in B(X)$ has the property that there exists $\alpha > 0$ such that $\|T(x)\| \geq \alpha\|x\|$ for all $x \in X$, then $\mathrm{Im}(T)$ is a closed set.

Proof

We use the sequential characterization of closed sets, so let $\{y_n\}$ be a sequence in $\mathrm{Im}(T)$ which converges to $y \in B(X)$. As $y_n \in \mathrm{Im}(T)$ there exists $x_n \subset X$ such that $T(x_n) = y_n$. As $\{y_n\}$ converges, it is a Cauchy sequence so since

$$\|y_n - y_m\| = \|T(x_n) - T(x_m)\| = \|T(x_n - x_m)\| \geq \alpha\|x_n - x_m\|,$$

for all m and $n \in \mathbb{N}$, we deduce that the sequence $\{x_n\}$ is a Cauchy sequence. As X is complete there exists $x \in X$ such that $\{x_n\}$ converges to x. Therefore, by the continuity of T,

$$T(x) = \lim_{n\to\infty} T(x_n) = \lim_{n\to\infty} y_n = y.$$

Hence $y = T(x) \in \mathrm{Im}(T)$ and so $\mathrm{Im}(T)$ is closed. $\square$

From Lemmas 4.46 and 4.47 we can now give another characterization of invertible operators.

Theorem 4.48

Let X be a Banach space and let $T \in B(X)$. The following are equivalent:

(a) T is invertible;

(b) $\mathrm{Im}(T)$ is dense in X and there exists $\alpha > 0$ such that $\|T(x)\| \geq \alpha\|x\|$ for all $x \in X$.

Proof

(a) $\Rightarrow$ (b). This is a consequence of Lemmas 1.12 and 4.46.
(b) $\Rightarrow$ (a). By hypothesis $\mathrm{Im}(T)$ is dense in X and so, since $\mathrm{Im}(T)$ is also closed by Lemma 4.47, we have $\mathrm{Im}(T) = X$. If $x \in \mathrm{Ker}\,(T)$ then $T(x) = 0$ so

$$0 = \|T(x)\| \geq \alpha\|x\|,$$

and thus $x = 0$. Hence $\mathrm{Ker}\,(T) = \{0\}$ and so T is invertible by Corollary 4.45. □

It is worthwhile stating explicitly how this result can be used to show that an operator is not invertible. The following corollary is therefore just a reformulation of Theorem 4.48.

Corollary 4.49

Let X be a Banach space and let $T \in B(X)$. The operator T is not invertible if and only if $\mathrm{Im}(T)$ is not dense in X or there exists a sequence $\{x_n\} \in X$ with $\|x_n\| = 1$ for all $n \in \mathbb{N}$ but $\lim_{n\to\infty} T(x_n) = 0$.

Not surprisingly Corollary 4.49 is a useful way of showing non-invertibility of an operator. Example 4.50 resolves a question left open after Example 4.39.

Example 4.50

For any $h \in C[0,1]$, let $T_h \in B(L^2[0,1])$ be defined by (4.3). If $f \in C[0,1]$ is defined by $f(t) = t$, then T_f is not invertible.

Solution

For each $n \in \mathbb{N}$ let $g_n = \sqrt{n}\chi_{[0,1/n]}$. Then $g_n \in L^2[0,1]$ and

$$\|g_n\|^2 = \int_0^1 \sqrt{n}\chi_{[0,1/n]}(t)\,\sqrt{n}\chi_{[0,1/n]}(t)\,dt = \frac{n}{n} = 1$$

for all $n \in \mathbb{N}$. However

$$\|T_f(g_n)\|^2 = \int_0^1 t\,\sqrt{n}\chi_{[0,1/n]}(t)\,t\,\sqrt{n}\chi_{[0,1/n]}(t)\,dt = n\int_0^{1/n} t^2\,dt = \frac{n}{3n^3}.$$

Therefore, $\lim_{n\to\infty} T_f(g_n) = 0$ and so T_f is not invertible by Corollary 4.49. □

It is also possible to use Theorem 4.48 to show that an operator is invertible. We use it to give an alternative solution to Example 4.39.

Example 4.51

For any $h \in C[0,1]$, let $T_h \in B(L^2[0,1])$ be defined by (4.3). If $f \in C[0,1]$ is defined by $f(t) = 1 + t$, then T_f is invertible.

Solution

We verify that the two conditions of Theorem 4.48 are satisfied for this operator. Let $k \in C[0,1]$ and let $g = \dfrac{k}{f}$. Then $g \in C[0,1]$ and

$$(T_f(g))(t) = f(t)g(t) = k(t)$$

for all $t \in [0,1]$. Hence $T_f(g) = k$ and so $\operatorname{Im}(T_f) = C[0,1]$. Moreover,

$$\|T_f(g)\|^2 = \int_0^1 f(t)g(t)\overline{f}(t)\overline{g}(t)dt = \int_0^1 |f(t)|^2|g(t)|^2 dt \geq \int_0^1 |g(t)|^2 dt = \|g\|^2$$

for all $g \in C[0,1]$. Therefore T_f is invertible by Theorem 4.48. □

It could be argued that to show that $\operatorname{Im}(T_f)$ was dense in the solution to Example 4.51 was still just about as hard as finding the inverse formula explicitly. However, in the case of operators in $B(\mathcal{H})$ where $\mathcal{H}$ is a Hilbert space, there is some additional structure in $B(\mathcal{H})$, which we shall study in the next chapter, which will enable us to obtain an alternative characterization of this density condition which is much easier to study.

EXERCISES

4.17 Prove Lemma 4.38.

4.18 Let X be a vector space on which there are two norms $\|\cdot\|_1$ and $\|\cdot\|_2$. Suppose that under both norms X is a Banach space and that there exists $k > 0$ such that

$$\|x\|_1 \leq k\|x\|_2,$$

for all $x \in X$. Show that $\|\cdot\|_1$ and $\|\cdot\|_2$ are equivalent norms.

4.19 Let $c = \{c_n\} \in \ell^\infty$ and let $T_c \in B(\ell^2)$ be defined by $T_c(\{x_n\}) = \{c_n x_n\}$.

(a) If $\inf\{|c_n| : n \in \mathbb{N}\} > 0$ and $d_n = \dfrac{1}{c_n}$ show that $d = \{d_n\} \in \ell^\infty$ and that $T_c T_d = T_d T_c = I$.

(b) If $\lambda \notin \{c_n : n \in \mathbb{N}\}^-$ show that $T_c - \lambda I$ is invertible.

4.20 Let $c = \{c_n\} \in \ell^\infty$ and let $T_c \in B(\ell^2)$ be defined by $T_c(\{x_n\}) = \{c_n x_n\}$. If $c_n = \dfrac{1}{n}$ show that T_c is not invertible.

4.21 Let X be a Banach space and suppose that $\{T_n\}$ is a sequence of invertible operators in $B(X)$ which converges to $T \in B(X)$. Suppose also that $\|T_n^{-1}\| < 1$ for all $n \in \mathbb{N}$. Show that T is invertible.

4.22 (The uniform boundedness principle) Let X and Y be Banach spaces over $\mathbb{F}$, let S be a non-empty set and let $F_b(S, X)$ be the Banach space defined in Exercises 2.2 and 2.14. Suppose that for each $s \in S$ we have $T_s \in B(Y, X)$ and suppose that for each $y \in Y$ the set $\{\|T_s(y)\| : s \in S\}$ is bounded.

(a) For each $y \in Y$, define $f^y : S \to X$ by $f^y(s) = T_s(y)$. Show that $f^y \in F_b(S, X)$.

(b) If the linear transformation $\phi : Y \to F_b(S, X)$ is defined by $\phi(y) = f^y$ show that $\mathcal{G}(\phi)$, the graph of ϕ, is closed.

(c) Show that $\{\|T_s\| : s \in S\}$ is bounded.

5

Linear Operators on Hilbert Spaces

5.1 The Adjoint of an Operator

At the end of the previous chapter we stated that there is an additional structure on the space of all operators on a Hilbert space which enables us to obtain a simpler characterization of invertibility. This is the "adjoint" of an operator and we start this chapter by showing what this is and giving some examples to show how easy it is to find adjoints. We describe some of the the properties of adjoints and show how they are used to give the desired simpler characterization of invertibility. We then use the adjoint to define three important types of operators (normal, self-adjoint and unitary operators) and give properties of these. The set of eigenvalues of a matrix have so many uses in finite-dimensional applications that it is not surprising that its analogue for operators on infinite-dimensional spaces is also important. So, for any operator T, we try to find as much as possible about the set $\{\lambda \in \mathbb{C} : T - \lambda I \text{ is not invertible}\}$ which is called the spectrum of T and which generalizes the set of eigenvalues of a matrix. We conclude the chapter by investigating the properties of those self-adjoint operators whose spectrum lies in $[0, \infty)$. Although some of the earlier results in this chapter have analogues for real spaces, when we deal with the spectrum it is necessary to use complex spaces, so for simplicity we shall only consider complex spaces throughout this chapter.

We start by defining the adjoint and showing its existence and uniqueness. As in Chapter 4 with normed spaces, if we have two or more inner product spaces we should, in principle, use notation which distinguishes the inner products. However, in practice we just use the symbol $(\cdot\,,\cdot)$ for the inner product on all the spaces as it is usually easy to determine which space elements are in and therefore, implicitly, to which inner product we are referring.

Theorem 5.1

Let $\mathcal{H}$ and $\mathcal{K}$ be complex Hilbert spaces and let $T \in B(\mathcal{H}, \mathcal{K})$. There exists a unique operator $T^* \in B(\mathcal{K}, \mathcal{H})$ such that

$$(Tx, y) = (x, T^*y)$$

for all $x \in \mathcal{H}$ and all $y \in \mathcal{K}$.

Proof

Let $y \in \mathcal{K}$ and let $f : \mathcal{H} \to \mathbb{C}$ be defined by $f(x) = (Tx, y)$. Then f is a linear transformation and by the Cauchy–Schwarz inequality (Lemma 3.13(a)) and the boundedness of T

$$|f(x)| = |(Tx, y)| \le \|Tx\|\|y\| \le \|T\|\|x\|\|y\|.$$

Hence f is bounded and so by the Riesz–Fréchet theorem (Theorem 4.31) there exists a unique $z \in \mathcal{H}$ such that $f(x) = (x, z)$ for all $x \in \mathcal{H}$. We define $T^*(y) = z$, so that T^* is a function from $\mathcal{K}$ to $\mathcal{H}$ such that

$$(T(x), y) = (x, T^*(y)) \tag{5.1}$$

for all $x \in \mathcal{H}$ and all $y \in \mathcal{K}$. Thus T^* is a function which satisfies the equation in the statement of the theorem but we have yet to show that it is in $B(\mathcal{K}, \mathcal{H})$. It is perhaps not even clear yet that T^* is a linear transformation so the first step is to show this.

Let $y_1, y_2 \in \mathcal{K}$, let $\lambda, \mu \in \mathbb{C}$ and let $x \in \mathcal{H}$. Then by (5.1),

$$\begin{aligned}(x, T^*(\lambda y_1 + \mu y_2)) &= (T(x), \lambda y_1 + \mu y_2)\\ &= \overline{\lambda}(T(x), y_1) + \overline{\mu}(T(x), y_2)\\ &= \overline{\lambda}(x, T^*(y_1)) + \overline{\mu}(x, T^*(y_2))\\ &= (x, \lambda T^*(y_1) + \mu T^*(y_2)).\end{aligned}$$

Hence $T^*(\lambda y_1 + \mu y_2) = \lambda T^*(y_1) + \mu T^*(y_2)$, by Exercise 3.1, and so T^* is a linear transformation.

Next we show that T^* is bounded. Using the Cauchy–Schwarz inequality

$$\|T^*(y)\|^2 = (T^*(y), T^*(y)) = (TT^*(y), y) \le \|TT^*(y)\|\|y\| \le \|T\|\|T^*(y)\|\|y\|.$$

If $\|T^*(y)\| > 0$ then we can divide through the above inequality by $\|T^*(y)\|$ to get $\|T^*(y)\| \leq \|T\|\|y\|$, while if $\|T^*(y)\| = 0$ then trivially $\|T^*(y)\| \leq \|T\|\|y\|$. Hence for all $y \in \mathcal{K}$,

$$\|T^*(y)\| \leq \|T\|\|y\|$$

and so T^* is bounded and $\|T^*\| \leq \|T\|$.

Finally, we have to show that T^* is unique. Suppose that B_1 and B_2 are in $B(\mathcal{K}, \mathcal{H})$ and that for all $x \in \mathcal{H}$ and all $y \in \mathcal{K}$,

$$(Tx, y) = (x, B_1 y) = (x, B_2 y).$$

Therefore $B_1 y = B_2 y$ for all $y \in \mathcal{K}$, by Exercise 3.1, so $B_1 = B_2$ and hence T^* is unique. □

Definition 5.2

If $\mathcal{H}$ and $\mathcal{K}$ are complex Hilbert spaces and $T \in B(\mathcal{H}, \mathcal{K})$ the operator T^* constructed in Theorem 5.1 is called the *adjoint* of T.

The uniqueness part of Theorem 5.1 is very useful when finding the adjoint of an operator. In the notation of Theorem 5.1, if we find an operator S which satisfies the equation $(Tx, y) = (x, Sy)$ for all x and y then $S = T^*$. In practice, finding an adjoint often boils down to solving an equation.

The first example of an adjoint of an operator we will find is that of an operator between finite-dimensional spaces. Specifically, we will find the matrix representation of the adjoint of an operator in terms of the matrix representation of the operator itself (recall that matrix representations of operators were discussed at the end of Section 1.1). In the solution, given a matrix $A = [\,a_{i,j}\,] \in M_{mn}(\mathbb{F})$ the notation $[\,\overline{a_{j,i}}\,]$ denotes the matrix obtained by taking the complex conjugates of the entries of the transpose A^T of A.

Example 5.3

Let $\mathbf{u} = \{\widehat{e}_1, \widehat{e}_2\}$ be the standard basis of $\mathbb{C}^2$ and let $T \in B(\mathbb{C}^2)$. If $M_{\mathbf{u}}^{\mathbf{u}}(T) = [\,a_{i,j}\,]$ then $M_{\mathbf{u}}^{\mathbf{u}}(T^*) = [\,\overline{a_{j,i}}\,]$.

Solution

Let $M_{\mathbf{u}}^{\mathbf{u}}(T^*) = [\,b_{i,j}\,]$. Then using the equation which defines the adjoint, for all $\begin{bmatrix} x_1 \\ x_2 \end{bmatrix}$ and $\begin{bmatrix} y_1 \\ y_2 \end{bmatrix}$ in $\mathbb{C}^2$ we have

$$\left(\begin{bmatrix} a_{1,1} & a_{1,2} \\ a_{2,1} & a_{2,2} \end{bmatrix} \begin{bmatrix} x_1 \\ x_2 \end{bmatrix}, \begin{bmatrix} y_1 \\ y_2 \end{bmatrix} \right) = \left(\begin{bmatrix} x_1 \\ x_2 \end{bmatrix}, \begin{bmatrix} b_{1,1} & b_{1,2} \\ b_{2,1} & b_{2,2} \end{bmatrix} \begin{bmatrix} y_1 \\ y_2 \end{bmatrix} \right).$$

Therefore,

$$\left(\begin{bmatrix} a_{1,1}x_1 + a_{1,2}x_2 \\ a_{2,1}x_1 + a_{2,2}x_2 \end{bmatrix}, \begin{bmatrix} y_1 \\ y_2 \end{bmatrix}\right) = \left(\begin{bmatrix} x_1 \\ x_2 \end{bmatrix}, \begin{bmatrix} b_{1,1}y_1 + b_{1,2}y_2 \\ b_{2,1}y_1 + b_{2,2}y_2 \end{bmatrix}\right)$$

and so

$$a_{1,1}x_1\overline{y_1} + a_{1,2}x_2\overline{y_1} + a_{2,1}x_1\overline{y_2} + a_{2,2}x_2\overline{y_2}$$

$$= x_1\overline{b_{1,1}}\,\overline{y_1} + x_1\overline{b_{1,2}}\,\overline{y_2} + x_2\overline{b_{2,1}}\,\overline{y_1} + x_2\overline{b_{2,2}}\,\overline{y_2}.$$

Since this is true for all $\begin{bmatrix} x_1 \\ x_2 \end{bmatrix}$ and $\begin{bmatrix} y_1 \\ y_2 \end{bmatrix}$ in $\mathbb{C}^2$ we deduce that

$$a_{1,1} = \overline{b_{1,1}}\ ,\ a_{1,2} = \overline{b_{2,1}}\ ,\ a_{2,1} = \overline{b_{1,2}} \text{ and } a_{2,2} = \overline{b_{2,2}}.$$

Hence $[\,b_{i,j}\,] = [\,\overline{a_{j,i}}\,]$. □

More generally, if $\mathbf{u}$ is the standard basis for $\mathbb{C}^n$ and $\mathbf{v}$ is the standard basis for $\mathbb{C}^m$ and $T \in B(\mathbb{C}^n, \mathbb{C}^m)$ with $M_{\mathbf{v}}^{\mathbf{u}}(T) = [\,a_{i,j}\,]$ then it can be shown similarly that $M_{\mathbf{u}}^{\mathbf{v}}(T^*) = [\,\overline{a_{j,i}}\,]$. Because of this we shall use the following notation.

Definition 5.4

If $A = [\,a_{i,j}\,] \in M_{mn}(\mathbb{F})$ then the matrix $[\,\overline{a_{j,i}}\,]$ is called the *adjoint* of A and is denoted by A^*.

The next two examples illustrate ways of finding adjoints of operators between infinite-dimensional spaces.

Example 5.5

For any $k \in C[0,1]$ let $T_k \in B(L^2[0,1])$ be defined by

$$(T_k g)(t) = k(t)g(t). \tag{5.2}$$

If $f \in C[0,1]$, then $(T_f)^* = T_{\overline{f}}$.

Solution

Let $g, h \in L^2[0,1]$ and let $k = (T_f)^*h$. Then $(T_f g, h) = (g, k)$ by definition of the adjoint and so

$$\int_0^1 f(t)g(t)\overline{h(t)}\,dt = \int_0^1 g(t)\overline{k(t)}\,dt.$$

Now this equation is true if $\overline{k(t)} = \overline{h(t)}f(t)$, that is, if $k(t) = \overline{f(t)}h(t)$. Hence by the uniqueness of the adjoint, we deduce that $(T_f)^*h = k = \overline{f}h$ and so $(T_f)^* = T_{\overline{f}}$. □

Example 5.6

The adjoint of the unilateral shift $S \in B(\ell^2)$ (see (4.2)) is S^* where

$$S^*(y_1, y_2, y_3, \ldots) = (y_2, y_3, y_4, \ldots).$$

Solution

Let $x = \{x_n\}$ and $y = \{y_n\} \in \ell^2$ and let $z = \{z_n\} = S^*(y)$. Then $(Sx, y) = (x, S^*y)$ by the definition of the adjoint and so

$$((0, x_1, x_2, x_3, \ldots), (y_1, y_2, y_3, \ldots)) = ((x_1, x_2, x_3, \ldots), (z_1, z_2, z_3, \ldots)).$$

Therefore

$$x_1\overline{y_2} + x_2\overline{y_3} + x_3\overline{y_4} + \ldots = x_1\overline{z_1} + x_2\overline{z_2} + x_3\overline{z_3} + \ldots.$$

Now if $z_1 = y_2$, $z_2 = y_3$, $\ldots$, then this equation is true for all $x_1, x_2, x_3, \ldots$ and hence by the uniqueness of the adjoint

$$S^*(y_1, y_2, y_3, \ldots) = (z_1, z_2, z_3, \ldots) = (y_2, y_3, y_4, \ldots).$$ □

The adjoint of the unilateral shift found in Example 5.6 is another type of operator which shifts the entries of each sequence while maintaining their original order. If S is called a forward shift as the entries in the sequence "move" forward then S^* could be called a backward shift as the entries in the sequence "move" backwards.

It is possible that the adjoint of an operator could be the operator itself.

Example 5.7

Let $\mathcal{H}$ be a complex Hilbert space. If I is the identity operator on $\mathcal{H}$ then $I^* = I$.

Solution

If $x, y \in \mathcal{H}$ then

$$(Ix, y) = (x, y) = (x, Iy).$$

Therefore by the uniqueness of adjoint, $I^* = I$. □

Having seen how easy it is to compute adjoints of operators in the above examples we now consider the adjoints of linear combinations and products of operators. Starting with the matrix case, if A and B are matrices and λ, $\mu \in \mathbb{C}$ then it follows easily from the standard rule for matrix transpositions, $(AB)^T =$

$B^T A^T$, that $(AB)^* = B^* A^*$ and $(\lambda A + \mu B)^* = \overline{\lambda} A^* + \overline{\mu} B^*$. This matrix result has the following analogue for operators. The proof is left as an exercise.

Lemma 5.8

Let $\mathcal{H}$, $\mathcal{K}$ and $\mathcal{L}$ be complex Hilbert spaces, let R, $S \in B(\mathcal{H}, \mathcal{K})$ and let $T \in B(\mathcal{K}, \mathcal{L})$. Let $\lambda, \mu \in \mathbb{C}$. Then:

(a) $(\mu R + \lambda S)^* = \overline{\mu} R^* + \overline{\lambda} S^*$;

(b) $(TR)^* = R^* T^*$.

Remark 5.9

Many of the results which follow hold for both linear operators and for matrices (as in Lemma 5.8), with only minor differences between the two cases. Thus, from now on we will normally only write out each result for the linear operator case, the corresponding result for matrices being an obvious modification of that for operators.

Further properties of adjoints are given in Theorem 5.10.

Theorem 5.10

Let $\mathcal{H}$ and $\mathcal{K}$ be complex Hilbert spaces and let $T \in B(\mathcal{H}, \mathcal{K})$.

(a) $(T^*)^* = T$.

(b) $\|T^*\| = \|T\|$.

(c) The function $f : B(\mathcal{H}, \mathcal{K}) \to B(\mathcal{K}, \mathcal{H})$ defined by $f(R) = R^*$ is continuous.

(d) $\|T^* T\| = \|T\|^2$.

Proof

(a) We have to show that the adjoint of the adjoint is the original operator.

$$\begin{aligned}
(y, (T^*)^* x) &= (T^* y, x) && \text{by definition of } (T^*)^* \\
&= \overline{(x, T^* y)} && \text{by definition of an inner product} \\
&= \overline{(Tx, y)} && \text{by definition of } T^* \\
&= (y, Tx) && \text{by definition of an inner product.}
\end{aligned}$$

Hence $(T^*)^* x = Tx$ for all $x \in \mathcal{H}$ so $(T^*)^* = T$.

(b) By Theorem 5.1 we have $\|T^*\| \leq \|T\|$. Applying this result to $(T^*)^*$ and using part (a) we have

$$\|T\| = \|(T^*)^*\| \leq \|T^*\| \leq \|T\|,$$

and so $\|T^*\| = \|T\|$.

(c) We cannot apply Lemma 4.1 here as f is not a linear transformation, by Lemma 5.8. Let $\epsilon > 0$ and let $\delta = \epsilon$. Then, when $\|R - S\| < \delta$,

$$\|f(R) - f(S)\| = \|R^* - S^*\| = \|(R - S)^*\| = \|(R - S)\| < \epsilon$$

by part (b). Hence f is continuous.

(d) Since $\|T\| = \|T^*\|$, we have

$$\|T^*T\| \leq \|T^*\|\|T\| = \|T\|^2.$$

On the other hand

$$\begin{aligned} \|Tx\|^2 &= (Tx, Tx) \\ &= (T^*Tx, x) && \text{by the definition of } T^* \\ &\leq \|T^*Tx\|\|x\| && \text{by the Cauchy–Schwarz inequality} \\ &\leq \|T^*T\|\|x\|^2. \end{aligned}$$

Therefore $\|T\|^2 \leq \|T^*T\|$, which proves the result. □

Part (d) of the Theorem 5.10 will later be used to find norms of operators. However, the following lemma is the important result as far as getting a characterization of invertibility is concerned.

Lemma 5.11

Let $\mathcal{H}$ and $\mathcal{K}$ be complex Hilbert spaces and let $T \in B(\mathcal{H}, \mathcal{K})$.

(a) $\operatorname{Ker} T = (\operatorname{Im} T^*)^\perp$;

(b) $\operatorname{Ker} T^* = (\operatorname{Im} T)^\perp$;

(c) $\operatorname{Ker} T^* = \{0\}$ if and only if $\operatorname{Im} T$ is dense in $\mathcal{K}$.

Proof

(a) First we show that $\operatorname{Ker} T \subseteq (\operatorname{Im} T^*)^\perp$. To this end, let $x \in \operatorname{Ker} T$ and let $z \in \operatorname{Im} T^*$. As $z \in \operatorname{Im} T^*$ there exists $y \in \mathcal{K}$ such that $T^*y = z$. Thus

$$(x, z) = (x, T^*y) = (Tx, y) = 0.$$

Hence $x \in (\text{Im } T^*)^\perp$ so Ker $T \subseteq (\text{Im } T^*)^\perp$.

Next we show that $(\text{Im } T^*)^\perp \subseteq$ Ker T. In this case let $v \in (\text{Im } T^*)^\perp$. As $T^*Tv \in \text{Im } T^*$ we have

$$(Tv, Tv) = (v, T^*Tv) = 0.$$

Thus $Tv = 0$ and so $v \in$ Ker T.

Therefore Ker $T = (\text{Im } T^*)^\perp$.

(b) By part (a) and Theorem 5.10 we have

$$\text{Ker } T^* = (\text{Im } ((T^*)^*))^\perp = (\text{Im } T)^\perp.$$

(c) If Ker $T^* = \{0\}$ then

$$((\text{Im } T)^\perp)^\perp = (\text{Ker } T^*)^\perp = \{0\}^\perp = \mathcal{K}.$$

Hence Im T is dense in $\mathcal{K}$ by Corollary 3.36.

Conversely, if Im T is dense in $\mathcal{K}$ then $(\text{Im } T^\perp)^\perp = \mathcal{K}$ by Corollary 3.36. Therefore, by Corollary 3.35,

$$\text{Ker } T^* = \text{Im } T^\perp = ((\text{Im } T^\perp)^\perp)^\perp = \mathcal{K}^\perp = \{0\}. \qquad \square$$

We now have an immediate consequence of Theorem 4.48 and Lemma 5.11.

Corollary 5.12

Let $\mathcal{H}$ be a complex Hilbert space and let $T \in B(\mathcal{H})$. The following are equivalent:

(a) T is invertible;

(b) Ker $T^* = \{0\}$ and there exists $\alpha > 0$ such that $\|T(x)\| \geq \alpha\|x\|$ for all $x \in \mathcal{H}$.

Despite having to do one more step it is usually easier to find the adjoint of an operator T and then Ker T^* than to show that Im T is dense. Corollary 5.12 can of course also be used to show that an operator is not invertible as Example 5.13 illustrates.

Example 5.13

The unilateral shift $S \in B(\ell^2)$ defined in (4.2) is not invertible.

Solution

We showed that $S^*(y_1, y_2, y_3, \ldots) = (y_2, y_3, y_4, \ldots)$ in Example 5.6. Thus $(1, 0, 0, 0, \ldots) \in \text{Ker } S^*$ and hence S is not invertible by Corollary 5.12. □

There is also a link between the invertibility of an operator and the invertibility of its adjoint.

Lemma 5.14

If $\mathcal{H}$ is a complex Hilbert space and $T \in B(\mathcal{H})$ is invertible then T^* is invertible and $(T^*)^{-1} = (T^{-1})^*$.

Proof

As $TT^{-1} = T^{-1}T = I$, if we take the adjoint of this equation we obtain $(TT^{-1})^* = (T^{-1}T)^* = I^*$ and so $(T^{-1})^*T^* = T^*(T^{-1})^* = I$ by Lemma 5.8 and Example 5.7. Hence T^* is invertible and $(T^*)^{-1} = (T^{-1})^*$. □

EXERCISES

5.1 Let $c = \{c_n\} \in \ell^\infty$. Find the adjoint of the linear operator $T_c : \ell^2 \to \ell^2$ defined by

$$T_c(\{x_n\}) = \{c_n x_n\}.$$

5.2 Find the adjoint of the linear operator $T : \ell^2 \to \ell^2$ defined by

$$T(x_1, x_2, x_3, x_4, \ldots) = (0, 4x_1, x_2, 4x_3, x_4, \ldots).$$

5.3 Let $\mathcal{H}$ be a Hilbert space and let $y, z \in \mathcal{H}$. If T is the bounded linear operator defined by $T(x) = (x, y)z$, show that $T^*(w) = (w, z)y$.

5.4 Prove Lemma 5.8.

5.5 Let $\mathcal{H}$ be a complex Hilbert space and let $T \in B(\mathcal{H})$.

(a) Show that $\text{Ker } T = \text{Ker } (T^*T)$.

(b) Deduce that the closure of $\text{Im } T^*$ is equal to the closure of $\text{Im } (T^*T)$.

5.2 Normal, Self-adjoint and Unitary Operators

Although the adjoint enables us to obtain further information about all operators on a Hilbert space, it can also be used to define particular classes of operators which frequently arise in applications and for which much more is known. In this section we define three classes of operators which will occur in many later examples and investigate some of their properties.

The first such class of operators which we study is the class of normal operators.

Definition 5.15

(a) If $\mathcal{H}$ is a complex Hilbert space and $T \in B(\mathcal{H})$ then T is *normal* if

$$TT^* = T^*T.$$

(b) If A is a square matrix then A is *normal* if

$$AA^* = A^*A.$$

It is quite easy to check whether an operator is normal or not as the two steps required, namely finding the adjoint and working out the products, are themselves easy. As matrix adjoints and matrix multiplication are even easier to find and do than their operator analogues, we shall not give any specific examples to determine whether matrices are normal as this would not illustrate any new points.

Example 5.16

For any $k \in C[0,1]$, let $T_k \in B(L^2[0,1])$ be defined as in (5.2). If $f \in C[0,1]$ then T_f is normal.

Solution

From Example 5.5 we know that $(T_f)^* = T_{\overline{f}}$. Hence, for all $g \in L^2[0,1]$,

$$(T_f(T_f)^*)(g) = T_f(T_{\overline{f}}(g)) = T_f(\overline{f}g) = f\overline{f}g$$

and

$$((T_f)^*T_f)(g) = T_{\overline{f}}(T_f(g)) = T_{\overline{f}}(fg) = \overline{f}fg = f\overline{f}g.$$

Therefore $T_f(T_f)^* = (T_f)^*T_f$ and so T_f is normal. □

Example 5.17

The unilateral shift $S \in B(\ell^2)$ defined in (4.2) is not normal.

Solution

We know from Example 5.6 that, for all $\{y_n\} \in \ell^2$,

$$S^*(y_1, y_2, y_3, \ldots) = (y_2, y_3, y_4, \ldots).$$

For $\{x_n\} \in \ell^2$,

$$S^*(S(x_1, x_2, x_3, \ldots)) = S^*(0, x_1, x_2, x_3, \ldots) = (x_1, x_2, x_3, \ldots)$$

while

$$S(S^*(x_1, x_2, x_3, \ldots)) = S(x_2, x_3, \ldots) = (0, x_2, x_3, \ldots).$$

Thus $S^*(S(x_1, x_2, x_3, \ldots)) \neq S(S^*(x_1, x_2, x_3, \ldots))$ for all $\{x_n\} \in \ell^2$ and so S is not normal. □

Even in Example 5.18, which is slightly more abstract, it is easy to determine whether or not an operator is normal.

Example 5.18

If $\mathcal{H}$ is a complex Hilbert space, I is the identity operator on $\mathcal{H}$, $\lambda \in \mathbb{C}$ and $T \in B(\mathcal{H})$ is normal then $T - \lambda I$ is normal.

Solution

$(T - \lambda I)^* = T^* - \overline{\lambda} I$ by Lemma 5.8. Hence, using $T^*T = TT^*$,

$$\begin{aligned}
(T - \lambda I)(T - \lambda I)^* &= (T - \lambda I)(T^* - \overline{\lambda} I) \\
&= TT^* - \lambda T^* - \overline{\lambda} T - \lambda\overline{\lambda} I \\
&= T^*T - \lambda T^* - \overline{\lambda} T - \lambda\overline{\lambda} I \\
&= (T^* - \overline{\lambda} I)(T - \lambda I) \\
&= (T - \lambda I)^*(T - \lambda I).
\end{aligned}$$

Therefore $T - \lambda I$ is normal. □

We now give some properties of normal operators.

Lemma 5.19

Let $\mathcal{H}$ be a complex Hilbert space, let $T \in B(\mathcal{H})$ be normal and let $\alpha > 0$.

(a) $\|T(x)\| = \|T^*(x)\|$ for all $x \in \mathcal{H}$.

(b) If $\|T(x)\| \geq \alpha\|x\|$ for all $x \in \mathcal{H}$, then Ker $T^* = \{0\}$.

Proof

(a) Let $x \in \mathcal{H}$. As $T^*T = TT^*$,

$$\begin{aligned}\|T(x)\|^2 - \|T^*(x)\|^2 &= (Tx, Tx) - (T^*x, T^*x)\\ &= (T^*Tx, x) - (TT^*x, x)\\ &= (T^*Tx - TT^*x, x)\\ &= 0.\end{aligned}$$

Therefore $\|T(x)\| = \|T^*(x)\|$.

(b) Let $y \in$ Ker T^*. Then $T^*y = 0$, so by part (a)

$$0 = \|T^*y\| = \|Ty\| \geq \alpha\|y\|.$$

Therefore $\|y\| = 0$ and so $y = 0$. Hence Ker $T^* = \{0\}$. □

The following characterization of invertibility of normal operators is an immediate consequence of Corollary 5.12 and Lemma 5.19.

Corollary 5.20

Let $\mathcal{H}$ be a complex Hilbert space and let $T \in B(\mathcal{H})$ be normal. The following are equivalent:

(a) T is invertible;

(b) there exists $\alpha > 0$ such that $\|T(x)\| \geq \alpha\|x\|$ for all $x \in \mathcal{H}$.

It may not have been apparent why normal operators were important but Corollary 5.20 shows that it is easier to determine whether normal operators are invertible, compared with determining whether arbitrary operators are invertible. However, in addition, there are some subsets of the set of normal operators which seem natural objects of study if we consider them as generalizations of important sets of complex numbers. The set of 1×1 complex matrices is just the set of complex numbers and in this case the adjoint of a complex number z is $z^* = \overline{z}$. Two important subsets of $\mathbb{C}$ are the real numbers

$$\mathbb{R} = \{z \in \mathbb{C} : z = \overline{z}\},$$

and the circle centre 0 radius 1, which is the set

$$\{z \in \mathbb{C} : z\overline{z} = \overline{z}z = 1\}.$$

We look at the generalization of both these sets of numbers to sets of operators.

Definition 5.21

(a) If $\mathcal{H}$ is a complex Hilbert space and $T \in B(\mathcal{H})$ then T is *self-adjoint* if $T = T^*$.

(b) If A is a square matrix then A is *self-adjoint* if $A = A^*$.

Although they amount to the same thing, there are several ways of trying to show that an operator is self-adjoint. The first is to find the adjoint and show it is equal to the original operator. This approach is illustrated in Example 5.22.

Example 5.22

The matrix $A = \begin{bmatrix} 2 & i \\ -i & 3 \end{bmatrix}$ is self-adjoint.

Solution

As $A^* = \begin{bmatrix} 2 & i \\ -i & 3 \end{bmatrix} = A$ we conclude that A is self-adjoint. □

The second approach to show that an operator T is self-adjoint is to show directly that $(Tx, y) = (x, Ty)$ for all vectors x and y. The uniqueness of the adjoint ensures that if this equation is satisfied then T is self-adjoint. This was previously illustrated in Example 5.7 which, using the notation we have just introduced, can be rephrased as follows.

Example 5.23

Let $\mathcal{H}$ be a complex Hilbert space. The identity operator $I \in B(\mathcal{H})$ is self-adjoint.

If we already know the adjoint of an operator it is even easier to check whether it is self-adjoint as Example 5.24 shows.

Example 5.24

For any $k \in C[0,1]$, let $T_k \in B(L^2[0,1])$ be defined as in (5.2). If $f \in C[0,1]$ is real-valued then T_f is self-adjoint.

Solution

From Example 5.5 we know that $(T_f)^* = T_{\overline{f}}$. As f is real-valued, $\overline{f} = f$. Hence

$$(T_f)^* = T_{\overline{f}} = T_f$$

and so T_f is self-adjoint. □

More general algebraic properties of self-adjoint operators are given in the following results. The analogy between self-adjoint operators and real numbers is quite well illustrated here.

Lemma 5.25

Let $\mathcal{H}$ be a complex Hilbert space and let $\mathcal{S}$ be the set of self-adjoint operators in $B(\mathcal{H})$.

(a) If α and β are real numbers and T_1 and $T_2 \in \mathcal{S}$ then $\alpha T_1 + \beta T_2 \in \mathcal{S}$.

(b) $\mathcal{S}$ is a closed subset of $B(\mathcal{H})$.

Proof

(a) As T_1 and T_2 are self-adjoint, $(\alpha T_1 + \beta T_2)^* = \alpha T_1^* + \beta T_2^* = \alpha T_1 + \beta T_2$, by Lemma 5.8. Hence $\alpha T_1 + \beta T_2 \in \mathcal{S}$.

(b) Let $\{T_n\}$ be a sequence in $\mathcal{S}$ which converges to $T \in B(\mathcal{H})$. Then $\{T_n^*\}$ converges to T^* by Theorem 5.10. Therefore, $\{T_n\}$ converges to T^* as $T_n^* = T_n$ for all $n \in \mathbb{N}$. Hence $T = T^*$ and so $T \in \mathcal{S}$. Hence $\mathcal{S}$ is closed. □

An alternative way of stating Lemma 5.25 is that the set of self-adjoint operators in $B(\mathcal{H})$ is a real Banach space.

Lemma 5.26

Let $\mathcal{H}$ be a complex Hilbert space and let $T \in B(\mathcal{H})$.

(a) T^*T and TT^* are self-adjoint.

(b) $T = R + iS$ where R and S are self-adjoint.

Proof

(a) $(T^*T)^* = T^*T^{**} = T^*T$ by Lemma 5.8. Hence T^*T is self-adjoint. Similarly TT^* is self-adjoint.

(b) Let $R = \frac{1}{2}(T + T^*)$ and $S = \frac{1}{2i}(T - T^*)$. Then $T = R + iS$. Also

$$R^* = \frac{1}{2}(T + T^*)^* = \frac{1}{2}(T^* + T) = R,$$

so R is self-adjoint and

$$S^* = \frac{-1}{2i}(T - T^*)^* = -\frac{1}{2i}(T^* - T^{**}) = \frac{1}{2i}(T - T^*) = S,$$

so S is also self-adjoint. □

By analogy with complex numbers, the operators R and S defined in Lemma 5.26 are sometimes called the *real* and *imaginary* parts of the operator T. We now look at the generalization of the set $\{z \in \mathbb{C} : z\overline{z} = \overline{z}z = 1\}$.

Definition 5.27

(a) If $\mathcal{H}$ is a complex Hilbert space and $T \in B(\mathcal{H})$ then T is *unitary* if $TT^* = T^*T = I$.

(b) If A is a square matrix then A is *unitary* if $AA^* = A^*A = I$.

Hence, for a unitary operator or matrix the adjoint is the inverse. As it is again quite easy to check whether a matrix is unitary, we shall only give an example of an operator which is unitary.

Example 5.28

For any $k \in C[0,1]$, let $T_k \in B(L^2[0,1])$ be defined as in (5.2). If $f \in C[0,1]$ is such that $|f(t)| = 1$ for all $t \in [0,1]$ then T_f is unitary.

Solution

We know from Example 5.5 that $(T_f)^* = T_{\overline{f}}$. Thus

$$((T_f)^*T_f)g(t) = \overline{f(t)}f(t)g(t) = |f(t)|^2 g(t),$$

so $T_f^*T_f(g) = g$ and hence $T_f^*T_f = I$. Similarly, $T_fT_f^* = I$. Therefore T_f is unitary. □

There is an alternative characterization of unitary operators which is more geometric. We require the following lemma, whose proof is left as an exercise.

Lemma 5.29

If X is a complex inner product space and $S, T \in B(X)$ are such that $(Sz, z) = (Tz, z)$ for all $z \in X$, then $S = T$.

Theorem 5.30

Let $\mathcal{H}$ be a complex Hilbert spaces and let T, $U \in B(\mathcal{H})$.

(a) $T^*T = I$ if and only if T is an isometry.

(b) U is unitary if and only if U is an isometry of $\mathcal{H}$ onto $\mathcal{H}$.

Proof

(a) Suppose first that $T^*T = I$. Then

$$\|Tx\|^2 = (Tx, Tx) = (T^*Tx, x) = (Ix, x) = \|x\|^2.$$

Hence T is an isometry.
Conversely, suppose that T is an isometry. Then

$$(T^*Tx, x) = (Tx, Tx) = \|Tx\|^2 = \|x\|^2 = (Ix, x).$$

Hence $T^*T = I$ by Lemma 5.29.

(b) Suppose first that U is unitary. Then U is an isometry by part (a). Moreover if $y \in \mathcal{H}$ then $y = U(U^*y)$ so $y \in \operatorname{Im} U$. Hence U maps $\mathcal{H}$ onto $\mathcal{H}$.

Conversely, suppose that U is an isometry of $\mathcal{H}$ onto $\mathcal{H}$. Then $U^*U = I$ by part (a). If $y \in \mathcal{H}$, then since U maps $\mathcal{H}$ onto $\mathcal{H}$, there exists x in $\mathcal{H}$ such that $Ux = y$. Hence

$$UU^*y = UU^*(Ux) = U(U^*U)x = UIx = Ux = y.$$

Thus $UU^* = I$ and so U is unitary. □

We leave the proof of the following properties of unitary operators as an exercise.

Lemma 5.31

Let $\mathcal{H}$ be a complex Hilbert space and let $\mathcal{U}$ be the set of unitary operators in $B(\mathcal{H})$.

(a) If $U \in \mathcal{U}$ then $U^* \in \mathcal{U}$ and $\|U\| = \|U^*\| = 1$.

(b) If U_1 and $U_2 \in \mathcal{U}$ then U_1U_2 and U_1^{-1} are in $\mathcal{U}$.

(c) $\mathcal{U}$ is a closed subset of $B(\mathcal{H})$.

EXERCISES

5.6 Is the matrix $A = \begin{bmatrix} 1 & 1 \\ 0 & 1 \end{bmatrix}$ normal?

5.7 Are the operators defined in Exercises 5.1 and 5.2 normal?

5.8 Prove Lemma 5.29.
[Hint: use Lemma 3.14.]

5.9 If $\mathcal{H}$ is a complex Hilbert space and $T \in B(\mathcal{H})$ is such that $\|Tx\| = \|T^*x\|$ for all $x \in \mathcal{H}$, show that T is normal.

5.10 Let T_c be the operator defined in Exercise 5.1.

(a) If $c_n \in \mathbb{R}$ for all $n \in \mathbb{N}$ show that T_c is self-adjoint.

(b) If $|c_n| = 1$ for all $n \in \mathbb{N}$ show that T_c is unitary.

5.11 If $\mathcal{H}$ is a complex Hilbert space and $S, T \in B(\mathcal{H})$ with S self-adjoint, show that T^*ST is self-adjoint.

5.12 If $\mathcal{H}$ is a complex Hilbert space and $A \in B(\mathcal{H})$ is invertible and self-adjoint, show that A^{-1} is self-adjoint.

5.13 If $\mathcal{H}$ is a complex Hilbert space and $S, T \in B(\mathcal{H})$ are self-adjoint, show that ST is self-adjoint if and only if $ST = TS$.

5.14 Prove Lemma 5.31.

5.15 Let $\mathcal{H}$ be a complex Hilbert space and let $U \in B(\mathcal{H})$ be unitary. Show that the linear transformation $f : B(\mathcal{H}) \to B(\mathcal{H})$ defined by $f(T) = U^*TU$ is an isometry.

5.3 The Spectrum of an Operator

Given a square matrix A, an important set of complex numbers is the set $\mathcal{A} = \{\lambda \in \mathbb{C} : A - \lambda I \text{ is not invertible }\}$. In fact, $\mathcal{A}$ consists of the set of eigenvalues of A by Lemma 1.12. Since eigenvalues occur in so many applications of finite-dimensional linear algebra, it is natural to try to extend these notions to infinite-dimensional spaces. This is what we aim to do in this section. Since the adjoint is available to help with determining when an operator is invertible in Hilbert spaces, we will restrict consideration to Hilbert spaces, although the definitions we give can easily be extended to Banach spaces.

Definition 5.32

(a) Let $\mathcal{H}$ be a complex Hilbert space, let $I \in B(\mathcal{H})$ be the identity operator and let $T \in B(\mathcal{H})$. The *spectrum* of T, denoted by $\sigma(T)$, is defined to be

$$\sigma(T) = \{\lambda \in \mathbb{C} : T - \lambda I \text{ is not invertible }\}.$$

(b) If A is a square matrix then the *spectrum* of A, denoted by $\sigma(A)$, is defined to be

$$\sigma(A) = \{\lambda \in \mathbb{C} : A - \lambda I \text{ is not invertible }\}.$$

Our first example of the spectrum of an operator is perhaps the simplest possible example.

Example 5.33

Let $\mathcal{H}$ be a complex Hilbert space and let I be the identity operator on $\mathcal{H}$. If μ is any complex number then $\sigma(\mu I) = \{\mu\}$.

Solution

If $\tau \in \mathbb{C}$ then τI is invertible unless $\tau = 0$. Hence

$$\begin{aligned}\sigma(\mu I) &= \{\lambda \in \mathbb{C} : \mu I - \lambda I \text{ is not invertible }\}\\ &= \{\lambda \in \mathbb{C} : (\mu - \lambda) I \text{ is not invertible }\}\\ &= \{\mu\}.\end{aligned}$$

□

Finding the spectrum of other operators can be less straightforward. However, if an operator has any eigenvalues then these are in the spectrum by Lemma 5.34.

Lemma 5.34

Let $\mathcal{H}$ be a complex Hilbert space and let $T \in B(\mathcal{H})$. If λ is an eigenvalue of T then λ is in $\sigma(T)$.

Proof

As there is a non-zero vector $x \in \mathcal{H}$ such that $Tx = \lambda x$ it follows that $x \in \operatorname{Ker}(T - \lambda I)$ and so $T - \lambda I$ is not invertible by Lemma 1.12. □

If $\mathcal{H}$ is a finite-dimensional complex Hilbert space and $T \in B(\mathcal{H})$ then the spectrum of T consists solely of eigenvalues of T by Lemma 1.12. It might be hoped that the same would hold in the infinite-dimensional case. However, there are operators on infinite-dimensional spaces which have no eigenvalues at all.

Example 5.35

The unilateral shift $S \in B(\ell^2)$ has no eigenvalues.

Solution

Suppose that λ is an eigenvalue of S with corresponding non-zero eigenvector $x = \{x_n\}$. Then

$$(0, x_1, x_2, x_3, \ldots) = (\lambda x_1, \lambda x_2, \lambda x_3, \ldots).$$

If $\lambda = 0$, then the right-hand side of this equation is the zero vector so $0 = x_1 = x_2 = x_3 = \ldots = 0$, which is a contradiction as $x \neq 0$. If $\lambda \neq 0$, then since $\lambda x_1 = 0$, we have $x_1 = 0$. Hence, from $\lambda x_2 = x_1 = 0$, we have $x_2 = 0$. Continuing in this way, we again have $x_1 = x_2 = x_3 = \ldots = 0$, which is a contradiction. Hence S has no eigenvalues. □

Since the unilateral shift has no eigenvalues, how do we find its spectrum? The following two results can sometimes help.

Theorem 5.36

Let $\mathcal{H}$ be a complex Hilbert space and let $T \in B(\mathcal{H})$.

(a) If $|\lambda| > \|T\|$ then $\lambda \notin \sigma(T)$.

(b) $\sigma(T)$ is a closed set.

Proof

(a) If $|\lambda| > \|T\|$ then $\|\lambda^{-1}T\| < 1$ and so $I - \lambda^{-1}T$ is invertible by Theorem 4.40. Hence $\lambda I - T$ is invertible and therefore $\lambda \notin \sigma(T)$.

(b) Define $F : \mathbb{C} \to B(\mathcal{H})$ by $F(\lambda) = \lambda I - T$. Then as

$$\|F(\mu) - F(\lambda)\| = \|\mu I - T - (\lambda I - T)\| = |\mu - \lambda|,$$

F is continuous. Hence, $\sigma(T)$ is closed, as the set $\mathcal{C}$ of non-invertible elements is closed by Corollary 4.42 and

$$\sigma(T) = \{\lambda \in \mathbb{C} : F(\lambda) \in \mathcal{C}\}.$$

□

Theorem 5.36 states that the spectrum of an operator T is a closed bounded (and hence compact) subset of $\mathbb{C}$ which is contained in a circle centre the origin and radius $\|T\|$.

Lemma 5.37

If $\mathcal{H}$ is a complex Hilbert space and $T \in B(\mathcal{H})$ then $\sigma(T^*) = \{\overline{\lambda} : \lambda \in \sigma(T)\}$.

Proof

If $\lambda \notin \sigma(T)$ then $T - \lambda I$ is invertible and so $(T - \lambda I)^* = T^* - \overline{\lambda} I$ is invertible by Lemma 5.14. Hence $\overline{\lambda} \notin \sigma(T^*)$. A similar argument with T^* in place of T shows that if $\overline{\lambda} \notin \sigma(T^*)$ then $\lambda \notin \sigma(T)$. Therefore $\sigma(T^*) = \{\overline{\lambda} : \lambda \in \sigma(T)\}$. □

With these results we can now find the spectrum of the unilateral shift.

Example 5.38

If $S : \ell^2 \to \ell^2$ is the unilateral shift then:

(a) if $\lambda \in \mathbb{C}$ with $|\lambda| < 1$ then λ is an eigenvalue of S^*;

(b) $\sigma(S) = \{\lambda \in \mathbb{C} : |\lambda| \leq 1\}$.

Solution

(a) Let $\lambda \in \mathbb{C}$ with $|\lambda| < 1$. We have to find a non-zero vector $\{x_n\} \in \ell^2$ such that $S^*(\{x_n\}) = \lambda\{x_n\}$. As $S^*(x_1, x_2, x_3, \ldots) = (x_2, x_3, x_4, \ldots)$ by Example 5.6, this means that we need to find a non-zero $\{x_n\} \in \ell^2$ such that

$$(x_2, x_3, x_4, \ldots) = (\lambda x_1, \lambda x_2, \lambda x_3, \ldots),$$

that is, $x_{n+1} = \lambda x_n$ for all $n \in \mathbb{N}$. One solution to this set of equations is $\{x_n\} = \{\lambda^{n-1}\}$ which is non-zero. Moreover, as $|\lambda| < 1$,

$$\sum_{n=1}^{\infty} |x_n|^2 = \sum_{n=0}^{\infty} |\lambda^n|^2 = \sum_{n=0}^{\infty} |\lambda|^{2n} < \infty,$$

and so $\{x_n\} \in \ell^2$. Thus $S^*(\{x_n\}) = \lambda x$ and so λ is an eigenvalue of S^* with eigenvector $\{x_n\}$.

(b) We have $\{\lambda \in \mathbb{C} : |\lambda| < 1\} \subseteq \sigma(S^*)$ by part (a) and Lemma 5.34. Thus $\{\overline{\lambda} \in \mathbb{C} : |\lambda| < 1\} \subseteq \sigma(S)$, by Lemma 5.37. However, from elementary geometry

$$\{\overline{\lambda} \in \mathbb{C} : |\lambda| < 1\} = \{\lambda \in \mathbb{C} : |\lambda| < 1\}$$

and so $\{\lambda \in \mathbb{C} : |\lambda| < 1\} \subseteq \sigma(S)$. As $\sigma(S)$ is closed, by Theorem 5.36,

$$\{\lambda \in \mathbb{C} : |\lambda| \leq 1\} \subseteq \sigma(S).$$

On the other hand, if $|\lambda| > 1$ then $\lambda \notin \sigma(S)$ by Theorem 5.36, since $\|S\| = 1$. Hence $\sigma(S) = \{\lambda \in \mathbb{C} : |\lambda| \leq 1\}$. □

If we know the spectrum of an operator T it is possible to find the spectrum of powers of T and (if T is invertible) the inverse of T.

Theorem 5.39

Let $\mathcal{H}$ be a complex Hilbert space and let $T \in B(\mathcal{H})$.

(a) If p is a polynomial then $\sigma(p(T)) = \{p(\mu) : \mu \in \sigma(T)\}$.

(b) If T is invertible then $\sigma(T^{-1}) = \{\mu^{-1} : \mu \in \sigma(T)\}$.

Proof

(a) Let $\lambda \in \mathbb{C}$ and let $q(z) = \lambda - p(z)$. Then q is also a polynomial, so it has a factorization of the form $q(z) = c(z - \mu_1)(z - \mu_2)\ldots(z - \mu_n)$, where $c, \mu_1, \mu_2, \ldots, \mu_n \in \mathbb{C}$ with $c \neq 0$. Hence,

$$\begin{aligned}
\lambda \notin \sigma(p(T)) &\iff \lambda I - p(T) \text{ is invertible} \\
&\iff q(T) \text{ is invertible} \\
&\iff c(T - \mu_1 I)(T - \mu_2 I)\ldots(T - \mu_n I) \text{ is invertible} \\
&\iff (T - \mu_j I) \text{ is invertible for } 1 \leq j \leq n \text{ by Lemma 4.38} \\
&\iff \text{no zero of } q \text{ is in } \sigma(T) \\
&\iff q(\mu) \neq 0 \text{ for all } \mu \in \sigma(T) \\
&\iff \lambda \neq p(\mu) \text{ for all } \mu \in \sigma(T).
\end{aligned}$$

Hence $\sigma(p(T)) = \{p(\mu) : \mu \in \sigma(T)\}$.

(b) As T is invertible, $0 \notin \sigma(T)$. Hence any element of $\sigma(T^{-1})$ can be written as μ^{-1} for some $\mu \in \mathbb{C}$. Since

$$\mu^{-1} I - T^{-1} = -T^{-1}\mu^{-1}(\mu I - T)$$

and $-T^{-1}\mu^{-1}$ is invertible,

$$\begin{aligned}
\mu^{-1} \in \sigma(T^{-1}) &\iff \mu^{-1} I - T^{-1} \text{ is not invertible} \\
&\iff -T^{-1}\mu^{-1}(\mu I - T) \text{ is not invertible} \\
&\iff \mu I - T \text{ is not invertible} \\
&\iff \mu \in \sigma(T).
\end{aligned}$$

Thus $\sigma(T^{-1}) = \{\mu^{-1} : \mu \in \sigma(T)\}$. □

Notation

Let $\mathcal{H}$ be a complex Hilbert space and let $T \in B(\mathcal{H})$. If p is a polynomial, we denote the set $\{p(\mu) : \mu \in \sigma(T)\}$ by $p(\sigma(T))$.

As an application of Theorem 5.39 we can obtain information about the spectrum of unitary operators.

Lemma 5.40

If $\mathcal{H}$ is a complex Hilbert space and $U \in B(\mathcal{H})$ is unitary then

$$\sigma(U) \subseteq \{\lambda \in \mathbb{C} : |\lambda| = 1\}.$$

Proof

As U is unitary, $\|U\| = 1$ so $\sigma(U) \subseteq \{\lambda \in \mathbb{C} : |\lambda| \leq 1\}$, by Theorem 5.36. Also, $\sigma(U^*) \subseteq \{\lambda \in \mathbb{C} : |\lambda| \leq 1\}$, since U^* is also unitary. However, $U^* = U^{-1}$ so

$$\sigma(U) = \{\lambda^{-1} : \lambda \in \sigma(U^*)\} \subseteq \{\lambda \in \mathbb{C} : |\lambda| \geq 1\},$$

by Theorem 5.39, which proves the result. □

An obvious question now is whether anything can be said about the spectrum of self-adjoint operators. We first introduce some notation.

Definition 5.41

(a) Let $\mathcal{H}$ be a complex Hilbert space and let $T \in B(\mathcal{H})$.

(i) The *spectral radius* of T, denoted by $r_\sigma(T)$, is defined to be

$$r_\sigma(T) = \sup\{|\lambda| : \lambda \in \sigma(T)\}.$$

(ii) The *numerical range* of T, denoted by $V(T)$, is defined to be

$$V(T) = \{(Tx, x) : \|x\| = 1\}.$$

(b) Let A be a $n \times n$ matrix.

(i) The *spectral radius* of A, denoted by $r_\sigma(A)$, is defined to be

$$r_\sigma(A) = \sup\{|\lambda| : \lambda \in \sigma(A)\}.$$

(ii) The *numerical range* of A, denoted by $V(A)$, is defined to be

$$V(A) = \{(Ax, x) : x \in \mathbb{C}^n \text{ and } \|x\| = 1\}.$$

A link between the numerical range and the spectrum for normal operators is given in Lemma 5.42.

Lemma 5.42

If $\mathcal{H}$ is a complex Hilbert space and $T \in B(\mathcal{H})$ is normal, then $\sigma(T)$ is a subset of the closure of $V(T)$.

Proof

Let $\lambda \in \sigma(T)$. As $T - \lambda I$ is normal by Example 5.18, there exists a sequence $\{x_n\}$ in $\mathcal{H}$ with $\|x_n\| = 1$ for all $n \in \mathbb{N}$ and $\lim\limits_{n\to\infty} \|(T - \lambda I)x_n\| = 0$ by Corollary 5.20. Hence

$$\lim_{n\to\infty} ((T - \lambda I)x_n, x_n) = 0$$

by the Cauchy–Schwarz inequality. Thus

$$\lim_{n\to\infty} (Tx_n, x_n) - \lambda(x_n, x_n) = 0.$$

However, $(x_n, x_n) = 1$ for all $n \in \mathbb{N}$ and so

$$\lim_{n\to\infty} (Tx_n, x_n) = \lambda.$$

Therefore λ is in the closure of $V(T)$. □

Since it is relatively easy to find the numerical range for self-adjoint operators, we can use Lemma 5.42 to get information about the spectrum of self-adjoint operators.

Theorem 5.43

Let $\mathcal{H}$ be a complex Hilbert space and let $S \in B(\mathcal{H})$ be self-adjoint.

(a) $V(S) \subseteq \mathbb{R}$.

(b) $\sigma(S) \subseteq \mathbb{R}$.

(c) At least one of $\|S\|$ or $-\|S\|$ is in $\sigma(S)$.

(d) $r_\sigma(S) = \sup\{|\tau| : \tau \in V(S)\} = \|S\|$.

(e) $\inf\{\lambda : \lambda \in \sigma(S)\} \leq \mu \leq \sup\{\lambda : \lambda \in \sigma(S)\}$ for all $\mu \in V(S)$.

Proof

(a) As S is self-adjoint,

$$(Sx, x) = (x, Sx) = \overline{(Sx, x)},$$

for all $x \in \mathcal{H}$. Hence $(Sx, x) \subseteq \mathbb{R}$ for all $x \in \mathcal{H}$ and so $V(S) \subseteq \mathbb{R}$.

(b) $\sigma(S)$ is contained in the closure of $V(S)$ by Lemma 5.42 and so is a subset of $\mathbb{R}$ by part (a).

(c) This is true if $S = 0$, so by working with $\|S\|^{-1}S$ if necessary we may assume that $\|S\| = 1$. From the definition of the norm of S, there exists a sequence $\{x_n\}$ in $\mathcal{H}$ such that $\|x_n\| = 1$ and $\lim_{n\to\infty} \|Sx_n\| = 1$. Then

$$\begin{aligned}\|(I - S^2)x_n\|^2 &= ((I - S^2)x_n, (I - S^2)x_n) \\ &= \|x_n\|^2 + \|S^2x_n\|^2 - 2(S^2x_n, x_n) \\ &\leq 2 - 2(Sx_n, Sx_n) \\ &= 2 - 2\|Sx_n\|^2,\end{aligned}$$

so $\lim_{n\to\infty} \|(I - S^2)x_n\|^2 = 0$ and hence $1 \in \sigma(S^2)$. Thus $1 \in (\sigma(S))^2$ by Theorem 5.39 and hence either 1 or -1 is in $\sigma(S)$ as required.

(d) By (c), Lemma 5.42 and the Cauchy–Schwarz inequality,

$$\|S\| \leq r_\sigma(S) \leq \sup\{|\tau| : \tau \in V(S)\} \leq \|S\|.$$

Hence each of these inequalities is an equality.

(e) Let $\alpha = \inf\{\lambda : \lambda \in \sigma(S)\}$ and $\beta = \sup\{\lambda : \lambda \in \sigma(S)\}$. Let $\lambda \in V(S)$, so that there exists $y \in \mathcal{H}$ with $\|y\| = 1$ and $\lambda = (Sy, y)$.
Suppose that $\lambda < \alpha$. Then $\beta I - S$ has spectrum $\beta - \sigma(S)$ by Theorem 5.39, and so lies in $[0, \beta - \alpha]$. Thus $r_\sigma(\beta I - S) \leq \beta - \alpha$. However,

$$((\beta I - S)y, y) = \beta(y, y) - (Sy, y) = \beta - \lambda > \beta - \alpha.$$

This contradicts part (d) applied to the self-adjoint operator $\beta I - S$.
Suppose that $\lambda > \beta$. Then $S - \alpha I$ has spectrum $\sigma(S) - \alpha$ by Theorem 5.39 and so lies in $[0, \beta - \alpha]$. Thus $r_\sigma(S - \alpha I) \leq \beta - \alpha$. However,

$$((S - \alpha I)y, y) = (Sy, y) - \alpha(y, y) = \lambda - \alpha > \beta - \alpha.$$

This contradicts part (d) applied to the self-adjoint operator $S - \alpha I$.
Therefore $\lambda \in [\alpha, \beta]$. □

It is possible to use Theorem 5.43 to find the norm of a matrix A.

Corollary 5.44

(a) If A is a self-adjoint matrix with eigenvalues $\{\lambda_1, \lambda_2, \ldots, \lambda_n\}$, then

$$\|A\| = \max\{|\lambda_1|, |\lambda_2|, \ldots, |\lambda_n|\}.$$

(b) If B is a square matrix then B^*B is self-adjoint and $\|B\|^2 = \|B^*B\|$.

Proof

(a) $\|A\| = r_\sigma(A) = \max\{|\lambda_1|, |\lambda_2|, \ldots, |\lambda_n|\}$ by Theorem 5.43, since $\sigma(A)$ consists only of eigenvalues.

(b) This follows from Theorem 5.10 and Lemma 5.26. □

Another consequence of Theorem 5.43 is that the spectrum of a self-adjoint operator is non-empty. Using completely different techniques, it can be shown that the the spectrum of any bounded linear operator is non-empty but as we shall not use this result we shall not give further details.

EXERCISES

5.16 Let $T \in B(\ell^2)$ be defined by

$$T(x_1, x_2, x_3, x_4, \ldots) = (x_1, -x_2, x_3, -x_4, \ldots).$$

(a) Show that 1 and -1 are eigenvalues of T with eigenvectors $(1, 0, 0, \ldots)$ and $(0, 1, 0, 0, \ldots)$ respectively.

(b) Find T^2 and hence show that $\sigma(T) = \{-1, 1\}$.

5.17 Let $S \in B(\ell^2)$ be the unilateral shift. Show that $S^*S = I$ but that 0 is an eigenvalue of SS^*.

5.18 Let $c = \{c_n\} \in \ell^\infty$ and let $T_c \in B(\ell^2)$ be the bounded operator defined by $T_c(\{x_n\}) = \{c_n x_n\}$.

(a) Let $c_m \in \{c_n : n \in \mathbb{N}\}$ and let $\{\widetilde{e}_n\}$ be the sequence in ℓ^2 given in Definition 1.59. Show that c_m is an eigenvalue of T_c with eigenvector $\widetilde{e}_m$.

(b) Show that $\{c_n : n \in \mathbb{N}\}^- \subseteq \sigma(T_c)$.

5.19 Let $T \in B(\ell^2)$ be the operator defined in Exercise 5.2.

(a) Find $(T^*)^2$ and show that if $|\mu| < 4$ then μ is an eigenvalue of $(T^*)^2$.

(b) Show that $\sigma(T) = \{\lambda \in \mathbf{C} : |\lambda| \leq 2\}$.

5.20 Find the norms of (a) $A = \begin{bmatrix} 1 & 1 \\ 1 & 2 \end{bmatrix}$; (b) $B = \begin{bmatrix} 1 & 1 \\ 0 & 1 \end{bmatrix}$.

5.21 Let $\mathcal{H}$ be a complex Hilbert space and let $S \in B(\mathcal{H})$ be self-adjoint. Show that S^n is self-adjoint for any $n \in \mathbb{N}$ and deduce that $\|S^n\| = \|S\|^n$.

5.22 Let $\mathcal{H}$ be a complex Hilbert space and let $S \in B(\mathcal{H})$ be self-adjoint. If $\sigma(S)$ contains exactly one point λ, show that $S = \lambda I$.

5.23 Give an example of a operator $T \in B(\ell^2)$ such that $T \neq 0$ but $\sigma(T) = \{0\}$.

5.4 Positive Operators and Projections

If $\mathcal{H}$ is a complex Hilbert space and $S \in B(\mathcal{H})$ is self-adjoint the conditions

(a) $\sigma(S) \subseteq [0, \infty)$,

(b) $(Sx, x) \geq 0$ for all $x \in \mathcal{H}$,

are equivalent, by Theorem 5.43. In the final section of this chapter we investigate the self-adjoint operators which satisfy either of these two conditions in more detail.

Definition 5.45

(a) Let $\mathcal{H}$ be a complex Hilbert space and let $S \in B(\mathcal{H})$. S is *positive* if it is self-adjoint and

$$(Sx, x) \geq 0 \quad \text{for all } x \in \mathcal{H}.$$

(b) If A is a self-adjoint, $n \times n$ matrix then A is *positive* if

$$(Ax, x) \geq 0 \quad \text{for all } x \in \mathbb{C}^n.$$

By the remarks before Definition 5.45 there is an equivalent characterization of positive operators and matrices.

Lemma 5.46

(a) If $\mathcal{H}$ is a complex Hilbert space and $S \in B(\mathcal{H})$ is self-adjoint then S is positive if and only if $\sigma(S) \subseteq [0, \infty)$.

(b) If A is a self-adjoint, $n \times n$ matrix then A is positive if and only if $\sigma(A) \subseteq [0, \infty)$.

The characterization of positivity for matrices given in Lemma 5.46 is usually easier to check than the definition since, for matrices, the spectrum just consists of eigenvalues. Therefore it is quite easy to check whether a self-adjoint matrix is positive. It is not much harder to find examples of positive operators.

Example 5.47

Let $\mathcal{H}$ be a complex Hilbert space, let $R, S \in B(\mathcal{H})$ be positive, let $T \in B(\mathcal{H})$ and let α be a positive real number.

(a) 0 and I are positive operators.

(b) T^*T is positive.

(c) $R + S$ and αS are positive.

Solution

(a) I is self-adjoint by Example 5.23 and it is easy to show that 0 is self-adjoint. If $x \in \mathcal{H}$ then

$$(Ix, x) = (x, x) \geq 0 \text{ and } (0x, x) = (0, x) = 0.$$

Hence 0 and I are positive.

(b) TT^* is self-adjoint by Lemma 5.26. If $x \in \mathcal{H}$ then

$$(T^*Tx, x) = (Tx, Tx) \geq 0.$$

Hence T^*T is positive.

(c) $R + S$ and αS are self-adjoint by Lemma 5.25. If $x \in \mathcal{H}$ then

$$((R + S)x, x) = (Rx, x) + (Sx, x) \geq 0$$

and

$$((\alpha S)x, x) = \alpha(Sx, x) \geq 0.$$

Hence $R + S$ and αS are positive. □

Associated with the positive real numbers there is the idea of an ordering of the real numbers. From the definition of positive operators there is a corresponding idea of ordering of self-adjoint operators.

Notation

Let $\mathcal{H}$ be a complex Hilbert space, let $R, S, T \in B(\mathcal{H})$ be self-adjoint.

(a) If S is positive we write $S \geq 0$ or $0 \leq S$.

(b) More generally, if $T - R$ is positive we write $T \geq R$ or $R \leq T$.

Unlike the order in the real numbers, the ordering of self-adjoint operators is only a partial ordering, that is, there are non-zero self-adjoint operators which are neither positive nor negative. We give a matrix example of this.

Example 5.48

If $A = \begin{bmatrix} 1 & 0 \\ 0 & -1 \end{bmatrix}$ then neither A nor $-A$ is positive.

Solution

A is self-adjoint with eigenvalues ± 1. Therefore $\sigma(A) = \{1, -1\}$ and so neither A nor $-A$ is positive. □

One of the simplest types of positive operator is described in the following definition.

Definition 5.49

Let $\mathcal{H}$ be a complex Hilbert space and let $P \in B(\mathcal{H})$. P is an *orthogonal projection* if

$$P = P^* = P^2.$$

The reason for the terminology "orthogonal projection" for the operators in Definition 5.49 should be rather more apparent after Example 5.50 and Theorem 5.51. More generally, bounded linear operators P for which $P = P^2$ are called projections, and these are useful in Banach spaces, but in the Hilbert space context orthogonal projections are the most useful projections, and are the only projections that will be considered in this book. In fact, in the Hilbert space context, "projection" is often used to mean "orthogonal projection", but to avoid ambiguity we will continue to use the latter terminology.

We note that if P is an orthogonal projection then it is in fact positive, since it is self-adjoint, by definition, and $(Px, x) = (P^2x, x) = (Px, Px) \geq 0$ for all $x \in \mathcal{H}$.

At first sight there may appear to be only two orthogonal projections on any given Hilbert space $\mathcal{H}$, namely 0 and I. However, there are others.

Example 5.50

Let $P : \mathbb{C}^3 \to \mathbb{C}^3$ be the linear transformation defined by $P(x, y, z) = (x, y, 0)$, for all $(x, y, z) \in \mathbb{C}^3$. Then P is an orthogonal projection.

Solution

Since $\mathbb{C}^3$ is finite-dimensional $P \in B(\mathbb{C}^3)$, and clearly $P^2 = P$. It follows from

$$(P(x, y, z), (u, v, w)) = x\overline{u} + y\overline{v} = ((x, y, z), P(u, v, w)),$$

that P is also self-adjoint. Hence P is an orthogonal projection. □

The orthogonal projection P in Example 5.50 has $\mathrm{Im}\,(P) = \{(x, y, 0) : x, y \in \mathbb{C}\}$, and P "projects" vectors "vertically downwards", or "orthogonally", onto this subspace, as shown in Figure 5.1 (where we only draw the action of P on $\mathbb{R}^3$, as we cannot draw $\mathbb{C}^3$).

The orthogonal projection P in Example 5.50 has the matrix representation

$$\begin{bmatrix} 1 & 0 & 0 \\ 0 & 1 & 0 \\ 0 & 0 & 0 \end{bmatrix}.$$

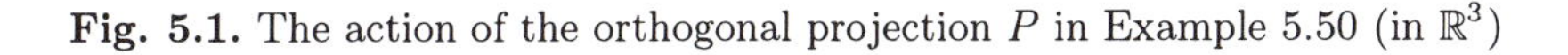

Fig. 5.1. The action of the orthogonal projection P in Example 5.50 (in $\mathbb{R}^3$)

More generally, any $n \times n$ diagonal matrix whose diagonal elements are either 0 or 1 is the matrix of an orthogonal projection in $B(\mathbb{C}^n)$.

One of the reasons why orthogonal projections are important is the link between closed linear subspaces of a Hilbert space $\mathcal{H}$ and orthogonal projections in $B(\mathcal{H})$ which we prove in the following theorem. In the course of the proof it may be helpful to keep in mind the action of the orthogonal projection P in Example 5.50.

Theorem 5.51

Let $\mathcal{H}$ be a complex Hilbert space.

(a) If $\mathcal{M}$ is a closed linear subspace of $\mathcal{H}$ there is an orthogonal projection $P_{\mathcal{M}} \in B(\mathcal{H})$ with range $\mathcal{M}$, kernel $\mathcal{M}^\perp$ and $\|P_{\mathcal{M}}\| \le 1$.

(b) If Q is an orthogonal projection in $B(\mathcal{H})$ then Im Q is a closed linear subspace and $Q = P_{\mathrm{Im}\ Q}$.

Proof

(a) If $x \in \mathcal{H}$ and $x = y + z$ where $y \in \mathcal{M}$ and $z \in \mathcal{M}^\perp$ is the orthogonal decomposition, let $P_{\mathcal{M}} : \mathcal{H} \to \mathcal{H}$ be defined by $P_{\mathcal{M}}(x) = y$. We aim to show that $P_{\mathcal{M}}$ is an orthogonal projection and the first step is to show that $P_{\mathcal{M}}$ is a linear transformation. Let $x_1, x_2 \in \mathcal{H}$, with orthogonal decompositions $x_1 = y_1 + z_1$ and $x_2 = y_2 + z_2$, where y_1, $y_2 \in \mathcal{M}$ and z_1, $z_2 \in \mathcal{M}^\perp$ and let $\lambda, \mu \in \mathbb{C}$. Then as $\mathcal{M}$ and $\mathcal{M}^\perp$ are linear subspaces, $\lambda y_1 + \mu y_2 \in \mathcal{M}$ and $\lambda z_1 + \mu z_2 \in \mathcal{M}^\perp$ so, by uniqueness, the orthogonal decomposition of $\lambda x_1 + \mu x_2$ is $(\lambda y_1 + \mu y_2) + (\lambda z_1 + \mu z_2)$. Hence

$$P_{\mathcal{M}}(\lambda x_1 + \mu x_2) = \lambda y_1 + \mu y_2 = \lambda P_{\mathcal{M}} x_1 + \mu P_{\mathcal{M}} x_2.$$

Therefore $P_{\mathcal{M}}$ is a linear transformation.

Next we show that $P_{\mathcal{M}}$ is bounded and self adjoint. As

$$\|P_{\mathcal{M}} x\|^2 = \|y\|^2 \le \|x\|^2,$$

$P_{\mathcal{M}}$ is bounded and $\|P_{\mathcal{M}}\| \le 1$. Also

$$(P_{\mathcal{M}} x_1, x_2) = (y_1, y_2 + z_2) = (y_1, y_2),$$

since $z_2 \in \mathcal{M}^\perp$ and $y_1 \in \mathcal{M}$, and

$$(x_1, P_{\mathcal{M}} x_2) = (y_1 + z_1, y_2) = (y_1, y_2),$$

since $z_1 \in \mathcal{M}^\perp$ and $y_2 \in \mathcal{M}$. Hence

$$(P_{\mathcal{M}}x_1, x_2) = (x_1, P_{\mathcal{M}}x_2)$$

and so $P_{\mathcal{M}}$ is self-adjoint.

Finally, we check that $P_{\mathcal{M}}$ is an orthogonal projection with range $\mathcal{M}$ and kernel $\mathcal{M}^\perp$. If $w \in \mathcal{M}$, the orthogonal decomposition of w is $w = w + 0$, so $P_{\mathcal{M}}w = w$. Hence $\mathcal{M} \subseteq \text{Im } P_{\mathcal{M}}$. However, $\text{Im } P_{\mathcal{M}} \subseteq \mathcal{M}$ by the definition of $P_{\mathcal{M}}$. Thus $\text{Im } P_{\mathcal{M}} = \mathcal{M}$. Also, for all $x \in \mathcal{H}$,

$$(P_{\mathcal{M}})^2(x) = P_{\mathcal{M}}(P_{\mathcal{M}}x) = P_{\mathcal{M}}y = y = P_{\mathcal{M}}(x)$$

since $y \in \mathcal{M}$ and so $(P_{\mathcal{M}})^2 = P_{\mathcal{M}}$. Therefore $P_{\mathcal{M}}$ is an orthogonal projection. Also, by Lemma 5.11 we have

$$\text{Ker } P_{\mathcal{M}} = (\text{Im } P_{\mathcal{M}}^*)^\perp = (\text{Im } P_{\mathcal{M}})^\perp = \mathcal{M}^\perp.$$

(b) Let $\mathcal{L} = \text{Im } Q$. As Q is a linear transformation, $\mathcal{L}$ is a linear subspace. To show that $\mathcal{L}$ is closed, let $\{y_n\}$ be a sequence in $\mathcal{L}$ which converges to $y \in \mathcal{H}$. As $y_n \in \text{Im } Q$ there exists $x_n \in \mathcal{H}$ such that $y_n = Q(x_n)$ for all $n \in \mathbb{N}$. Hence

$$\begin{aligned} y &= \lim_{n\to\infty} Qx_n \\ &= \lim_{n\to\infty} Q^2 x_n && \text{since } Q^2 = Q \\ &= Q(\lim_{n\to\infty} Qx_n) && \text{since } Q \text{ is continuous} \\ &= Qy \in \text{Im } Q, \end{aligned}$$

and so $\mathcal{L}$ is closed.

If $v \in \mathcal{L}$ then $v = Qx$ for some $x \in \mathcal{H}$ so $Qv = Q^2x = Qx = v$ as $Q^2 = Q$. If $w \in \mathcal{L}^\perp$ then as Q is self-adjoint and $Q^2 w \in \mathcal{L}$,

$$\|Qw\|^2 = (Qw, Qw) = (w, Q^2 w) = 0$$

so $Qw = 0$. Hence, if $x \in \mathcal{H}$ and $x = v + w$, where $v \in \mathcal{L}$ and $w \in \mathcal{L}^\perp$, we have $x = Qv + w$ so

$$P_{\mathcal{L}}x = Qv = Qx$$

as $Qw = 0$. Hence $Q = P_{\text{Im } Q}$. □

To emphasize the manner in which the orthogonal projection is constructed in Theorem 5.51 the following terminology is sometimes used.

Notation

Let $\mathcal{H}$ be a complex Hilbert space and let $\mathcal{M}$ be a closed linear subspace of $\mathcal{H}$. The orthogonal projection $P_{\mathcal{M}} \in B(\mathcal{H})$ with range $\mathcal{M}$ and kernel $\mathcal{M}^\perp$ constructed in Theorem 5.51 is called the orthogonal projection of $\mathcal{H}$ *onto* $\mathcal{M}$.

The orthogonal projection P considered in Example 5.50 is the orthogonal projection onto the subspace $\{(x, y, 0) : x,\ y \in \mathbb{C}\}$.

In the proof of Theorem 5.51, we showed that if $\mathcal{H}$ is a complex Hilbert space, $\mathcal{M}$ is a closed linear subspace of $\mathcal{H}$ and $P_{\mathcal{M}}$ is the orthogonal projection of $\mathcal{H}$ onto $\mathcal{M}$, then $P_{\mathcal{M}} y = y$ for all $y \in \mathcal{M}$ while $P_{\mathcal{M}} z = 0$ for all $z \in \mathcal{M}^{\perp}$.

Lemma 5.52

If $\mathcal{H}$ is a complex Hilbert space, $\mathcal{M}$ is a closed linear subspace of $\mathcal{H}$ and P is the orthogonal projection of $\mathcal{H}$ onto $\mathcal{M}$, then $I - P$ is the orthogonal projection of $\mathcal{H}$ onto $\mathcal{M}^{\perp}$.

Proof

As I and P are self-adjoint so is $I - P$. Also, as $P^2 = P$,

$$(I-P)^2 = I - 2P + P^2 = I - 2P + P = I - P$$

and so $I - P$ is an orthogonal projection. If $x \in \mathcal{H}$ and $x = y + z$ is the orthogonal decomposition of x, where $y \in \mathcal{M}$ and $z \in \mathcal{M}^{\perp}$, then $P(x) = y$ and so $(I - P)(x) = x - y = z$. Hence $I - P$ is the orthogonal projection of $\mathcal{H}$ onto $\mathcal{M}^{\perp}$ by Theorem 5.51. □

For the orthogonal projection P considered in Example 5.50, the operator $I - P$ is given by $(I - P)(x, y, z) = (0, 0, z)$ and is the orthogonal projection onto the subspace $\{(0, 0, z) : z \in \mathbb{C}\}$, see Figure 5.1.

If the closed linear subspace $\mathcal{M}$ has an orthonormal basis then it is possible to give a formula for the orthogonal projection onto $\mathcal{M}$ in terms of this orthonormal basis. This formula is given in Corollary 5.53 whose proof is left as an exercise.

Corollary 5.53

If $\mathcal{H}$ is a complex Hilbert space, $\mathcal{M}$ is a closed linear subspace of $\mathcal{H}$, $\{e_n\}_{n=1}^{J}$ is an orthonormal basis for $\mathcal{M}$, where J is a positive integer or ∞, and P is the orthogonal projection of $\mathcal{H}$ onto $\mathcal{M}$, then

$$Px = \sum_{n=1}^{J} (x, e_n) e_n.$$

By definition, if P is an orthogonal projection then $P = P^2$. If we take any other positive operator T defined on some Hilbert space it raises the question whether there exists an operator R (on the same space) such that $R^2 = T$.

Definition 5.54

(a) Let $\mathcal{H}$ be a complex Hilbert space and let $T \in B(\mathcal{H})$. A *square root* of T is an operator $R \in B(\mathcal{H})$ such that $R^2 = T$.

(b) Let A be a $n \times n$ matrix. A *square root* of A is a matrix B such that $B^2 = A$.

Example 5.55

If λ_1, λ_2 are positive real numbers and $A = \begin{bmatrix} \lambda_1 & 0 \\ 0 & \lambda_2 \end{bmatrix}$, $B = \begin{bmatrix} \sqrt{\lambda_1} & 0 \\ 0 & \sqrt{\lambda_2} \end{bmatrix}$ then $B^2 = A$ so B is a square root of A.

Since all complex numbers have square roots, it might be expected that all complex matrices would have square roots. However, this is not true.

Example 5.56

If $A = \begin{bmatrix} 0 & 1 \\ 0 & 0 \end{bmatrix}$ then there is no 2×2 matrix B such that $B^2 = A$.

Solution

Let $B = \begin{bmatrix} a & b \\ c & d \end{bmatrix}$ and suppose that $B^2 = A$. Then multiplying out we obtain

$$a^2 + bc = 0, \quad b(a+d) = 1, \quad c(a+d) = 0 \text{ and } d^2 + bc = 0.$$

As $b(a+d) \neq 0$ and $c(a+d) = 0$ we deduce that $c = 0$. Then $a^2 = d^2 = 0$, so $a = d = 0$ which contradicts $b(a+d) \neq 0$. Hence, there is no such matrix B. □

Even though all positive real numbers have positive square roots, in view of Example 5.56 it is perhaps surprising that all positive operators have positive square roots. Lemma 5.57 will be the key step in showing that positive operators have square roots.

Lemma 5.57

Let $\mathcal{H}$ be a complex Hilbert space, and let $\mathcal{S}$ be the real Banach space of all self-adjoint operators in $B(\mathcal{H})$. If $S \in \mathcal{S}$ then there exists $\Phi \in B(C_{\mathbb{R}}(\sigma(S)), \mathcal{S})$ such that:

(a) $\Phi(p) = p(S)$ whenever p is a polynomial in $C_{\mathbb{R}}(\sigma(S))$;

(b) $\Phi(fg) = \Phi(f)\Phi(g)$ for all f, $g \in C_{\mathbb{R}}(\sigma(S))$.

Proof

Let $\mathcal{P}$ be the linear subspace of $C_{\mathbb{R}}(\sigma(S))$ consisting of all polynomials. Define $\phi : \mathcal{P} \to \mathcal{S}$ by $\phi(p) = p(S)$. Then ϕ is a linear transformation such that $\phi(pq) = \phi(p)\phi(q)$ for all $p \in \mathcal{P}$ by Lemma 4.36. In addition, by Theorem 5.39,

$$\begin{aligned} \|\phi(p)\| &= \|p(S)\| \\ &= r_\sigma(p(S)) \qquad\qquad \text{since } p(S) \text{ is self-adjoint} \\ &= \sup\{|\mu| : \mu \in \sigma(p(S))\} \\ &= \sup\{|p(\lambda)| : \lambda \in \sigma(S)\} \\ &= \|p\|. \end{aligned}$$

Thus ϕ is an isometry. As $\mathcal{S}$ is a real Banach space and $\mathcal{P}$ is dense in $C_{\mathbb{R}}(\sigma(S))$ by Theorem 1.39, there exists $\Phi \in B(C_{\mathbb{R}}(\sigma(S)), \mathcal{S})$ such that $\Phi(p) = \phi(p)$ by Theorem 4.19. Moreover, as $\phi(pq) = \phi(p)\phi(q)$ for all $p \in \mathcal{P}$, it follows that $\Phi(fg) = \Phi(f)\Phi(g)$ for all f, $g \in C_{\mathbb{R}}(\sigma(S))$ by the density of $\mathcal{P}$ in $C_{\mathbb{R}}(\sigma(S))$ and the continuity of Φ. □

Lemma 5.57 perhaps looks rather technical, so we introduce the following notation to help understand what it means.

Notation

Let $\mathcal{H}$, $\mathcal{S}$, S and Φ be as in Lemma 5.57. For any $f \in C_{\mathbb{R}}(\sigma(S))$ we now denote $\Phi(f)$ by $f(S)$.

In other words, Lemma 5.57 allows us to construct "functions" of a self-adjoint operator S. We had previously defined $p(S)$ when p is a polynomial. Lemma 5.57 extends this to $f(S)$ when $f \in C_{\mathbb{R}}(\sigma(S))$. Suppose now that $\sigma(S) \subseteq [0, \infty)$ and $g : \sigma(S) \to \mathbb{R}$ is defined by $g(x) = x^{1/2}$. Then $g \in C_{\mathbb{R}}(\sigma(S))$ so $g(S)$ makes sense. The notation is intended to suggest that $g(S)$ is a square root of S and we show that this is true in Theorem 5.58.

Theorem 5.58

Let $\mathcal{H}$ be a complex Hilbert space, let $\mathcal{S}$ be the Banach space of all self-adjoint operators in $B(\mathcal{H})$ and let $S \in \mathcal{S}$ be positive.

(a) There exists a positive square root R of S which is a limit of a sequence of polynomials in S.

(b) If Q is any positive square root of S then $R = Q$.

Proof

(a) Let $\mathcal{P}$ be the linear subspace of $C_{\mathbb{R}}(\sigma(S))$ consisting of all polynomials. As S is positive, $\sigma(S) \subseteq [0, \infty)$. Hence $f : \sigma(S) \to \mathbb{R}$ and $g : \sigma(S) \to \mathbb{R}$ and $j : \sigma(S) \to \mathbb{R}$ defined by

$$f(x) = x^{1/4}, \qquad g(x) = x^{1/2} \quad \text{and} \quad j(x) = x,$$

are in $C_{\mathbb{R}}(\sigma(S))$. Let $R = g(S)$ and $T = f(S)$ so R and T are self-adjoint. The set $\mathcal{P}$ is dense in $C_{\mathbb{R}}(\sigma(S))$, by Theorem 1.39. In particular, g is a limit of a sequence of polynomials, and so R is a limit of a sequence of polynomials in S. Also, by Lemma 5.57,

$$R^2 = (g(S))^2 = g^2(S) = j(S) = S,$$

so R is a square root of S and

$$T^2 = (f(S))^2 = f^2(S) = g(S) = R,$$

so R is positive.

(b) As

$$QS = QQ^2 = Q^2Q = SQ,$$

if p is any polynomial then $Qp(S) = p(S)Q$ and so, as R is a limit of a sequence of polynomials in S,

$$QR = RQ.$$

As Q is positive, Q has a positive square root P by part (a). Let $x \in \mathcal{H}$ and let $y = (R - Q)x$. As $R^2 = Q^2 = S$,

$$\begin{aligned} \|Ty\|^2 + \|Py\|^2 &= (T^2y, y) + (P^2y, y) \\ &= ((R + Q)y, y) \\ &= ((R + Q)(R - Q)x, y) \\ &= ((R^2 - Q^2)x, y) \\ &= 0. \end{aligned}$$

Hence $Ty = Py = 0$ and so $T^2y = P^2y = 0$. Therefore $Ry = Qy = 0$ and so

$$\|(R - Q)x\|^2 = ((R - Q)^2x, x) = ((R - Q)y, x) = 0.$$

Thus $R = Q$. □

Notation

Let $\mathcal{H}$ be a complex Hilbert space and let $S \in B(\mathcal{H})$ be positive. The unique positive square root of S constructed in Theorem 5.58 will be denoted by $S^{1/2}$. Similarly, the unique positive square root of a positive matrix A will be denoted by $A^{1/2}$.

In Example 5.55 we showed how to find the square root of a diagonal 2×2 matrix and the same method extends to any diagonal $n \times n$ matrix. Hence, if P is any positive matrix and U is a unitary matrix such that U^*PU is a diagonal matrix D then $P^{1/2} = UD^{1/2}U^*$.

There is an alternative method for computing square roots which follows the construction of the square root more closely. We illustrate this for 2×2 positive matrices with distinct eigenvalues. Let A be any positive matrix which has distinct eigenvalues λ_1 and λ_2 and let

$$p(x) = \frac{x + \sqrt{\lambda_1 \lambda_2}}{\sqrt{\lambda_1} + \sqrt{\lambda_2}}.$$

Then p is a polynomial of degree one such that $p(\lambda_1) = \sqrt{\lambda_1}$ and $p(\lambda_2) = \sqrt{\lambda_2}$. Therefore $p(x) = \sqrt{x}$ for all $x \in \sigma(S)$. By the construction of the square root $A^{1/2} = p(A)$ and it is possible to check this by computing $p(A)^2$. By the Cayley–Hamilton theorem, $A^2 = (\lambda_1 + \lambda_2)A - (\lambda_1\lambda_2)I$ so

$$(p(A))^2 = \left(\frac{A + \sqrt{\lambda_1\lambda_2}I}{\sqrt{\lambda_1} + \sqrt{\lambda_2}} \right)^2 = \frac{A^2 + 2\sqrt{\lambda_1\lambda_2}A + \lambda_1\lambda_2 I}{(\sqrt{\lambda_1} + \sqrt{\lambda_2})^2} = A.$$

It is perhaps fitting that our final result in this chapter should illustrate the analogy between operators on a Hilbert space and the complex numbers so well that it even has the same name as the corresponding result for complex numbers. If $z \in \mathbb{C}$ is invertible then $(\overline{z}z)^{1/2}$, the modulus of z, is positive and $|z((\overline{z}z)^{1/2})^{-1}| = 1$ so $z((\overline{z}z)^{1/2})^{-1} = e^{i\theta}$ for some $\theta \in \mathbb{R}$ with $-\pi < \theta \leq \pi$. The polar form of the complex number z is $e^{i\theta}(\overline{z}z)^{1/2}$. A similar decomposition occurs for invertible operators on a Hilbert space, but this time the factors are a unitary and a positive operator.

Theorem 5.59

Let $\mathcal{H}$ be a complex Hilbert space and let $T \in B(\mathcal{H})$ be invertible. Then $T = UR$, where U is unitary and R is positive.

Proof

As T is invertible so is T^* and T^*T. Now, T^*T is positive by Example 5.47 so T^*T has a positive square root $R = (T^*T)^{1/2}$ by Theorem 5.58. As T^*T is invertible so is R by Theorem 5.39. Let $U = TR^{-1}$. Then U is invertible and so the range of U is $\mathcal{H}$. Also,

$$U^*U = (R^{-1})^*T^*TR^{-1} = R^{-1}R^2R^{-1} = I,$$

and so U is a unitary by Theorem 5.30. □

Notation

(a) Let $\mathcal{H}$ be a complex Hilbert space, let $T \in B(\mathcal{H})$ be invertible. The decomposition $T = UR$, where U is unitary and R is positive, given in Theorem 5.59 is called the *polar decomposition* of T.

(b) If A is an invertible matrix, the corresponding decomposition $A = BC$, where B is a unitary matrix and C is a positive matrix, is called the *polar decomposition* of A.

The proof of Theorem 5.59 indicates the steps required to produce the polar decomposition of an invertible operator. An example to find the polar decomposition of an invertible matrix is given in the exercises.

EXERCISES

5.24 Let $\mathcal{H}$ be a complex Hilbert space and $A \in B(\mathcal{H})$ be invertible and positive. Show that A^{-1} is positive.

5.25 Prove Corollary 5.53.

5.26 Let $\mathcal{H}$ be a complex Hilbert space and let $P \in B(\mathcal{H})$ be such that $P^2 = P$. By considering S^2 where $S = 2P - I$, show that $\sigma(P) \subseteq \{0, 1\}$.

5.27 Let $\mathcal{H}$ be a complex Hilbert space and let P, $Q \in B(\mathcal{H})$ be orthogonal projections.

(a) If $PQ = QP$, show that PQ is an orthogonal projection.

(b) Show that Im P is orthogonal to Im Q if and only if $PQ = 0$.

5.28 Let $\mathcal{H}$ be a complex Hilbert space and let P, $Q \in B(\mathcal{H})$ be orthogonal projections. Show that the following are equivalent.

(a) Im $P \subseteq$ Im Q;

(b) $QP = P$;

(c) $PQ = P$;

(d) $\|Px\| \leq \|Qx\|$ for all $x \in \mathcal{H}$;

(e) $P \leq Q$.

5.29 Let $\mathcal{H}$ be a complex Hilbert space and let $S \in B(\mathcal{H})$ be self-adjoint. If $\sigma(S)$ is the finite set $\{\lambda_1, \lambda_2, \ldots, \lambda_n\}$, show that there exist orthogonal projections $P_1, P_2, \ldots, P_n \in B(\mathcal{H})$ such that $P_j P_k = 0$ if

$j \neq k$, $\sum_{j=1}^{n} P_j = I$ and

$$S = \sum_{j=1}^{n} \lambda_j P_j.$$

5.30 Let $\mathcal{H}$ be a complex Hilbert space and let $S \in B(\mathcal{H})$ be self-adjoint with $\|S\| \leq 1$.

(a) Show that $I - S^2$ is positive.

(b) Show that the operators $S \pm i(I - S^2)^{1/2}$ are unitary.

5.31 (a) Find the positive square root of $A = \begin{bmatrix} 5 & -4 \\ -4 & 5 \end{bmatrix}$.

(b) Find the polar decomposition of $B = \frac{1}{\sqrt{2}} \begin{bmatrix} 2-i & 2i-1 \\ 2+i & -1-2i \end{bmatrix}$.

6

Compact Operators

6.1 Compact Operators

Linear algebra tells us a great deal about the properties of operators between finite-dimensional spaces, and about their spectrum. In general, the situation is considerably more complicated in infinite-dimensional spaces, as we have already seen. However, there is a class of operators in infinite dimensions for which a great deal of the finite-dimensional theory remains valid. This is the class of compact operators. In this chapter we will describe the principal spectral properties of general compact operators on Hilbert spaces and also the more precise results which hold for self-adjoint compact operators.

Compact operators are important not only for the well-developed theory which is available for them, but also because compact operators are encountered in very many important applications. Some of these applications will be considered in more detail in Chapter 7.

As in Chapter 5, the results in Section 6.1 have analogues for real spaces, but in Sections 6.2 and 6.3 it is necessary to use complex spaces in order to discuss the spectral theory of compact operators. Thus for simplicity we will only consider complex spaces in this chapter.

Definition 6.1

Let X and Y be normed spaces. A linear transformation $T \in L(X, Y)$ is *compact* if, for any bounded sequence $\{x_n\}$ in X, the sequence $\{Tx_n\}$ in Y contains a convergent subsequence. The set of compact transformations in $L(X,Y)$ will be denoted by $K(X,Y)$.

Theorem 6.2

Let X and Y be normed spaces and let $T \in K(X,Y)$. Then T is bounded. Thus, $K(X,Y) \subset B(X,Y)$.

Proof

Suppose that T is not bounded. Then for each integer $n \geq 1$ there exists a unit vector x_n such that $\|Tx_n\| \geq n$. Since the sequence $\{x_n\}$ is bounded, by the compactness of T there exists a subsequence $\{Tx_{n(r)}\}$ which converges. But this contradicts $\|Tx_{n(r)}\| \geq n(r)$, so T must be bounded. □

We now prove some simple algebraic properties of compact operators.

Theorem 6.3

Let X, Y, Z be normed spaces.

(a) If S, $T \in K(X,Y)$ and α, $\beta \in \mathbb{C}$ then $\alpha S + \beta T$ is compact. Thus $K(X,Y)$ is a linear subspace of $B(X,Y)$.

(b) If $S \in B(X,Y)$, $T \in B(Y,Z)$ and at least one of the operators S, T is compact, then $TS \in B(X,Z)$ is compact.

Proof

(a) Let $\{x_n\}$ be a bounded sequence in X. Since S is compact, there is a subsequence $\{x_{n(r)}\}$ such that $\{Sx_{n(r)}\}$ converges. Then, since $\{x_{n(r)}\}$ is bounded and T is compact, there is a subsequence $\{x_{n(r(s))}\}$ of the sequence $\{x_{n(r)}\}$ such that $\{Tx_{n(r(s))}\}$ converges. It follows that the sequence $\{\alpha Sx_{n(r(s))} + \beta Tx_{n(r(s))}\}$ converges. Thus $\alpha S + \beta T$ is compact.

(b) Let $\{x_n\}$ be a bounded sequence in X. If S is compact then there is a subsequence $\{x_{n(r)}\}$ such that $\{Sx_{n(r)}\}$ converges. Since T is bounded (and so is continuous), the sequence $\{TSx_{n(r)}\}$ converges. Thus TS is compact. If S is bounded but not compact then the sequence $\{Sx_n\}$ is bounded. Then since T must be compact, there is a subsequence $\{Sx_{n(r)}\}$ such that $\{TSx_{n(r)}\}$ converges, and again TS is compact. □

Remark 6.4

It is clear from the definition of a compact operator, and the above proofs, that when dealing with compact operators we will continually be looking at subsequences $\{x_{n(r)}\}$, or even $\{x_{n(r(s))}\}$, of a sequence $\{x_n\}$. For notational simplicity we will often assume that the subsequence has been relabelled as $\{x_n\}$, and so we can omit the r. Note, however, that this may not be permissible if the original sequence is specified at the beginning of an argument with some particular properties, e.g., if we start with an orthonormal basis $\{e_n\}$ we cannot just discard elements – it would no longer be a basis!

The following theorem shows, as a particular case, that all linear operators on finite-dimensional spaces are compact (recall that if the domain of a linear operator is finite-dimensional then the operator must be bounded, see Theorem 4.9, but linear transformations with finite-dimensional range may be unbounded if their domain is infinite dimensional, see Example 4.10). Thus compact operators are a generalization of operators on finite dimensional spaces.

Theorem 6.5

Let X, Y be normed spaces and $T \in B(X, Y)$.

(a) If T has finite rank then T is compact.

(b) If either $\dim X$ or $\dim Y$ is finite then T is compact.

Proof

(a) Since T has finite rank, the space $Z = \operatorname{Im} T$ is a finite-dimensional normed space. Furthermore, for any bounded sequence $\{x_n\}$ in X, the sequence $\{Tx_n\}$ is bounded in Z, so by the Bolzano–Weierstrass theorem this sequence must contain a convergent subsequence. Hence T is compact.

(b) If $\dim X$ is finite then $r(T) \leq \dim X$, so $r(T)$ is finite, while if $\dim Y$ is finite then clearly the dimension of $\operatorname{Im} T \subset Y$ must be finite. Thus, in either case the result follows from part (a). □

The following theorem, and its corollary, shows that in infinite dimensions there are many operators which are not compact. In fact, compactness is a significantly stronger property than boundedness. This is illustrated further in parts (a) and (b) of Exercise 6.11.

Theorem 6.6

If X is an infinite-dimensional normed space then the identity operator I on X is not compact.

Proof

Since X is an infinite-dimensional normed space the proof of Theorem 2.26 shows there exists a sequence of unit vectors $\{x_n\}$ in X which does not have any convergent subsequence. Hence the sequence $\{Ix_n\} = \{x_n\}$ cannot have a convergent subsequence, and so the operator I is not compact. □

Corollary 6.7

If X is an infinite-dimensional normed space and $T \in K(X)$ then T is not invertible.

Proof

Suppose that T is invertible. Then, by Theorem 6.3, the identity operator $I = T^{-1}T$ on X must be compact. But since X is infinite-dimensional this contradicts Theorem 6.6. □

We now introduce an equivalent characterization of compact operators and an important property of the range of such operators.

Theorem 6.8

Let X, Y be normed spaces and let $T \in L(X, Y)$.

(a) T is compact if and only if, for every bounded subset $A \subset X$, the set $T(A) \subset Y$ is relatively compact.

(b) If T is compact then $\operatorname{Im} T$ and $\overline{\operatorname{Im} T}$ are separable.

Proof

(a) Suppose that T is compact. Let $A \subset X$ be bounded and suppose that $\{y_n\}$ is an arbitrary sequence in $\overline{T(A)}$. Then for each $n \in \mathbb{N}$, there exists $x_n \in A$ such that $\|y_n - Tx_n\| < n^{-1}$, and the sequence $\{x_n\}$ is bounded since A is bounded. Thus, by compactness of T, the sequence $\{Tx_n\}$ contains a convergent subsequence, and hence $\{y_n\}$ contains a convergent subsequence with limit in $\overline{T(A)}$. Since $\{y_n\}$ is arbitrary, this shows that $\overline{T(A)}$ is compact.

Now suppose that for every bounded subset $A \subset X$ the set $T(A) \subset Y$ is relatively compact. Then for any bounded sequence $\{x_n\}$ in X the sequence $\{Tx_n\}$ lies in a compact set, and hence contains a convergent subsequence. Thus T is compact.

(b) For any $r \geq \mathbb{N}$, let $R_r = T(B_r(0)) \subset Y$ be the image of the ball $B_r(0) \subset X$. Since T is compact, the set R_r is relatively compact and so is separable, by Theorem 1.42. Furthermore, since $\operatorname{Im} T$ equals the countable union $\bigcup_{r=1}^{\infty} R_r$, it must also be separable. Finally, if a subset of $\operatorname{Im} T$ is dense in $\operatorname{Im} T$ then it is also dense in $\overline{\operatorname{Im} T}$ (see Exercise 6.14), so $\overline{\operatorname{Im} T}$ is separable.

□

Part (b) of Theorem 6.8 implies that if T is compact then even if the space X is "big" (not separable) the range of T is "small" (separable). In a sense, this is the reason why the theory of compact operators has many similarities with that of operators on finite-dimensional spaces.

We now consider how to prove that a given operator is compact. The following theorem, which shows that the limit of a sequence of compact operators in $B(X, Y)$ is compact, will provide us with a very powerful method of doing this.

Theorem 6.9

If X is a normed space, Y is a Banach space and $\{T_k\}$ is a sequence in $K(X, Y)$ which converges to an operator $T \in B(X, Y)$, then T is compact. Thus $K(X, Y)$ is closed in $B(X, Y)$.

Proof

Let $\{x_n\}$ be a bounded sequence in X. By compactness, there exists a subsequence of $\{x_n\}$, which we will label $\{x_{n(1,r)}\}$ $(= \{x_{n(1,r)}\}_{r=1}^{\infty})$, such that the sequence $\{T_1 x_{n(1,r)}\}$ converges. Similarly, there exists a subsequence $\{x_{n(2,r)}\}$ of $\{x_{n(1,r)}\}$ such that $\{T_2 x_{n(2,r)}\}$ converges. Also, $\{T_1 x_{n(2,r)}\}$ converges since it is a subsequence of $\{T_1 x_{n(1,r)}\}$. Repeating this process inductively, we see that for each $j \in \mathbb{N}$ there is a subsequence $\{x_{n(j,r)}\}$ with the property: for any $k \leq j$ the sequence $\{T_k x_{n(j,r)}\}$ converges. Letting $n(r) = n(r, r)$, for $r \in \mathbb{N}$, we obtain a single subsequence $\{x_{n(r)}\}$ with the property that, for each fixed $k \in \mathbb{N}$, the sequence $\{T_k x_{n(r)}\}$ converges as $r \to \infty$ (this so-called "Cantor diagonalization" type argument is necessary to obtain a single sequence which works simultaneously for all the operators T_k, $k \in \mathbb{N}$; see [7] for other examples of such arguments). We will now show that the sequence $\{Tx_{n(r)}\}$ converges. We do this by showing that $\{Tx_{n(r)}\}$ is a Cauchy sequence, and hence is convergent since Y is a Banach space.

Let $\epsilon > 0$ be given. Since the subsequence $\{x_{n(r)}\}$ is bounded there exists $M > 0$ such that $\|x_{n(r)}\| \leq M$, for all $r \in \mathbb{N}$. Also, since $\|T_k - T\| \to 0$, as $k \to \infty$, there exists an integer $K \geq 1$ such that $\|T_K - T\| < \epsilon/3M$. Next, since $\{T_K x_{n(r)}\}$ converges there exists an integer $R \geq 1$ such that if r, $s \geq R$ then $\|T_K x_{n(r)} - T_K x_{n(s)}\| < \epsilon/3$. But now we have, for r, $s \geq R$,

$$\begin{aligned}\|Tx_{n(r)} - Tx_{n(s)}\| < \|Tx_{n(r)} - T_K x_{n(r)}\| + \|T_K x_{n(r)} - T_K x_{n(s)}\| \\ + \|T_K x_{n(s)} - Tx_{n(s)}\| < \epsilon,\end{aligned}$$

which proves that $\{Tx_{n(r)}\}$ is a Cauchy sequence. □

In applications of Theorem 6.9 it is often the case that the operators T_k are bounded and have finite rank, so are compact by Theorem 6.5. For reference we state this as the following corollary.

Corollary 6.10

If X is a normed space, Y is a Banach space and $\{T_k\}$ is a sequence of bounded, finite rank operators which converges to $T \in B(X, Y)$, then T is compact.

We now give a simple example of how to construct a sequence of finite rank operators which converge to a given operator T. This process is one of the most common ways of proving that an operator is compact.

Example 6.11

The operator $T \in B(\ell^2)$ defined by $T\{a_n\} = \{n^{-1}a_n\}$ is compact (Example 4.5 shows that $T \in B(\ell^2)$).

Solution

For each $k \in \mathbb{N}$ define the operator $T_k \in B(\ell^2)$ by

$$T_k\{a_n\} = \{b_n^k\}, \quad \text{where} \quad \begin{cases} b_n^k = n^{-1}a_n, & n \leq k, \\ b_n^k = 0, & n > k. \end{cases}$$

The operators T_k are bounded and linear, and have finite rank. Furthermore, for any $a \in \ell^2$ we have

$$\|(T_k - T)a\|^2 = \sum_{n=k+1}^{\infty} |a_n|^2/n^2 \leq (k+1)^{-2} \sum_{n=k+1}^{\infty} |a_n|^2 \leq (k+1)^{-2}\|a\|^2.$$

It follows that

$$\|T_k - T\| \leq (k+1)^{-1},$$

and so $\|T_k - T\| \to 0$. Thus T is compact by Corollary 6.10. □

The converse of Corollary 6.10 is not true, in general, when Y is a Banach space, but it is true when Y is a Hilbert space.

Theorem 6.12

If X is a normed space, $\mathcal{H}$ is a Hilbert space and $T \in K(X, \mathcal{H})$, then there is a sequence of finite rank operators $\{T_k\}$ which converges to T in $B(X, \mathcal{H})$.

Proof

If T itself had finite rank the result would be trivial, so we assume that it does not. By Lemma 3.25 and Theorem 6.8 the set $\overline{\operatorname{Im} T}$ is an infinite-dimensional, separable Hilbert space, so by Theorem 3.52 it has an orthonormal basis $\{e_n\}$. For each integer $k \geq 1$, let P_k be the orthogonal projection from $\overline{\operatorname{Im} T}$ onto the linear subspace $\mathcal{M}_k = \operatorname{Sp}\{e_1, \ldots, e_k\}$, and let $T_k = P_k T$. Since $\operatorname{Im} T_k \subset \mathcal{M}_k$, the operator T_k has finite rank. We will show that $\|T_k - T\| \to 0$ as $k \to \infty$.

Suppose that this is not true. Then, after taking a subsequence of the sequence $\{T_k\}$ if necessary, there is an $\epsilon > 0$ such that $\|T_k - T\| \geq \epsilon$ for all k. Thus there exists a sequence of unit vectors $x_k \in X$ such that $\|(T_k - T)x_k\| \geq \epsilon/2$ for all k. Since T is compact, we may suppose that $Tx_k \to y$, for some $y \in \mathcal{H}$ (after again taking a subsequence, if necessary). Now, using the representation of P_m in Corollary 5.53, we have,

$$\begin{aligned}(T_k - T)x_k &= (P_k - I)Tx_k = (P_k - I)y + (P_k - I)(Tx_k - y) \\ &= -\sum_{n=k+1}^{\infty} (y, e_n)e_n + (P_k - I)(Tx_k - y).\end{aligned}$$

Hence, by taking norms we deduce that

$$\epsilon/2 \leq \|(T_k - T)x_k\| \leq \Big(\sum_{n=k+1}^{\infty} (y, e_n)^2 \Big)^{1/2} + 2\|Tx_k - y\|$$

(since $\|P_k\| = 1$, by Theorem 5.51). The right-hand side of this inequality tends to zero as $k \to \infty$, which is a contradiction, and so proves the theorem. □

Using these results we can now show that the adjoint of a compact operator is compact. We first deal with finite rank operators.

Lemma 6.13

If $\mathcal{H}$ is a Hilbert space and $T \in B(\mathcal{H})$, then $r(T) = r(T^*)$ (either as finite numbers or as ∞). In particular, T has finite rank if and only if T^* has finite rank.

Proof

Suppose first that $r(T) < \infty$. For any $x \in \mathcal{H}$, we write the orthogonal decomposition of x with respect to $\operatorname{Ker} T^*$ as $x = u + v$, with $u \in \operatorname{Ker} T^*$ and $v \in (\operatorname{Ker} T^*)^\perp = \overline{\operatorname{Im} T} = \operatorname{Im} T$ (since $r(T) < \infty$). Thus $T^*x = T^*(u+v) = T^*v$, and hence $\operatorname{Im} T^* = T^*(\operatorname{Im} T)$, which implies that $r(T^*) \leq r(T)$. Thus, $r(T^*) \leq r(T)$ when $r(T) < \infty$.

Applying this result to T^*, and using $(T^*)^* = T$, we also see that $r(T) \leq r(T^*)$ when $r(T^*) < \infty$. This proves the lemma when both the ranks are finite, and also shows that it is impossible for one rank to be finite and the other infinite, and so also proves the infinite rank case. □

Theorem 6.14

If $\mathcal{H}$ is a Hilbert space and $T \in B(\mathcal{H})$, then T is compact if and only if T^* is compact.

Proof

Suppose that T is compact. Then by Theorem 6.12 there is a sequence of finite rank operators $\{T_n\}$, such that $\|T_n - T\| \to 0$. By Lemma 6.13, each operator T_n^* has finite rank and, by Theorem 5.10, $\|T_n^* - T^*\| = \|T_n - T\| \to 0$. Hence it follows from Corollary 6.10 that T^* is compact. Thus, if T is compact then T^* is compact. It now follows from this result and $(T^*)^* = T$ that if T^* is compact then T is compact, which completes the proof. □

We end this section by introducing a class of operators which have many interesting properties and applications.

Definition 6.15

Let $\mathcal{H}$ be an infinite-dimensional Hilbert space with an orthonormal basis $\{e_n\}$ and let $T \in B(\mathcal{H})$. If the condition

$$\sum_{n=1}^{\infty} \|Te_n\|^2 < \infty$$

holds then T is a *Hilbert–Schmidt* operator.

At first sight it seems that this definition might depend on the choice of the orthonormal basis of $\mathcal{H}$. The following theorem shows that this is not so. The proof will be left until Exercise 6.8.

Theorem 6.16

Let $\mathcal{H}$ be an infinite-dimensional Hilbert space and let $\{e_n\}$ and $\{f_n\}$ be orthonormal bases for $\mathcal{H}$. Let $T \in B(\mathcal{H})$.

(a) $\sum_{n=1}^{\infty} \|Te_n\|^2 = \sum_{n=1}^{\infty} \|T^* f_n\|^2 = \sum_{n=1}^{\infty} \|Tf_n\|^2$

(where the values of these sums may be either finite or ∞).
Thus the condition for an operator to be Hilbert–Schmidt does not depend on the choice of the orthonormal basis of $\mathcal{H}$.

(b) T is Hilbert–Schmidt if and only if T^* is Hilbert–Schmidt.

(c) If T is Hilbert–Schmidt then it is compact.

(d) The set of Hilbert–Schmidt operators is a linear subspace of $B(\mathcal{H})$.

It will be shown in Exercise 6.11 that finite rank operators are Hilbert–Schmidt, but not all compact operators are Hilbert–Schmidt.

EXERCISES

6.1 Show that for any Banach space X the zero operator $T_0 : X \to X$, defined by $T_0 x = 0$, for all $x \in X$, is compact.

6.2 Let $\mathcal{H}$ be a Hilbert space and let $y, z \in \mathcal{H}$. Define $T \in B(\mathcal{H})$ by $Tx = (x, y)z$ (see Exercise 4.11). Show that T is compact.

6.3 Let X, Y be normed vector spaces. Show that $T \in L(X, Y)$ is compact if and only if for any sequence of vectors $\{x_n\}$ in the closed unit ball $\overline{B_1(0)} \subset X$, the sequence $\{Tx_n\}$ has a convergent subsequence.

6.4 Show that an orthonormal sequence $\{e_n\}$ in a Hilbert space $\mathcal{H}$ cannot have a convergent subsequence.

6.5 Show that for any Banach spaces X and Y the set of compact operators $K(X, Y)$ is a Banach space (with the usual operator norm obtained from $B(X, Y)$).

6.6 Show that for any $T \in B(\mathcal{H})$, $r(T) = r(T^*T)$.
[Hint: see Exercise 5.5(b).]

6.7 Let $\mathcal{H}$ be an infinite-dimensional Hilbert space with an orthonormal basis $\{e_n\}$ and let $T \in B(\mathcal{H})$. Show that if T is compact then $\lim_{n\to\infty} \|Te_n\| = 0$.

6.8 Prove Theorem 6.16.

6.9 Let $\mathcal{H}$ be an infinite-dimensional Hilbert space and let $S, T \in B(\mathcal{H})$. Prove the following results.

(a) If either S or T is Hilbert–Schmidt, then ST is Hilbert–Schmidt.

(b) If T has finite rank then it is Hilbert–Schmidt.
[Hint: use Exercise 3.26 to construct an orthonormal basis of $\mathcal{H}$ consisting of elements of $\operatorname{Im} T$ and $(\operatorname{Im} T)^\perp$, and use part (a) of Theorem 6.16, with $\{e_n\} = \{f_n\}$.]

6.10 Let k be a non-zero continuous function on $[-\pi, \pi]$ and define the operator $T_k \in B(L^2[-\pi, \pi])$ by $(T_k g)(t) = k(t)g(t)$. Show that T_k is not compact.
[Hint: use Corollary 3.57 and Exercise 6.7.]

6.11 Let $\mathcal{H}$ be an infinite-dimensional Hilbert space and let $\{e_n\}$, $\{f_n\}$, be orthonormal sequences in $\mathcal{H}$. Let $\{\alpha_n\}$ be a sequence in $\mathbb{C}$ and define a linear operator $T : \mathcal{H} \to \mathcal{H}$ by

$$Tx = \sum_{n=1}^{\infty} \alpha_n (x, e_n) f_n.$$

Show that:

(a) T is bounded if and only if the sequence $\{\alpha_n\}$ is bounded;

(b) T is compact if and only if $\lim_{n\to\infty} \alpha_n = 0$;

(c) T is Hilbert–Schmidt if and only if $\sum_{n=1}^{\infty} |\alpha_n|^2 < \infty$;

(d) T has finite rank if and only if there exists $N \in \mathbb{N}$ such that $\alpha_n = 0$ for $n \geq N$.

It follows that each of these classes of operators is strictly contained in the preceding class. In particular, not all compact operators are Hilbert–Schmidt.

6.12 Suppose that $\{\alpha_n\}$ is a bounded sequence in $\mathbb{C}$ and consider the operator $T \in B(\mathcal{H})$ defined in Exercise 6.11.

(a) Show that $y \in \operatorname{Im} T$ if and only if

$$y = \sum_{n=1}^{\infty} \alpha_n \xi_n f_n, \quad \text{for some } \{\xi_n\} \in \ell^2.$$

Deduce that if infinitely many of the numbers α_n are non-zero and $\lim_{n\to\infty} \alpha_n = 0$ then $\operatorname{Im} T$ is not closed.

(b) Show that the adjoint operator $T^* : \mathcal{H} \to \mathcal{H}$ is given by

$$T^* x = \sum_{n=1}^{\infty} \overline{\alpha}_n (x, f_n) e_n.$$

Show also that $\operatorname{Im} T$ is dense in $\mathcal{H}$ if and only if $\operatorname{Ker} T^* = \{0\}$. Deduce that if $\mathcal{H}$ is separable then there exists a compact operator on $\mathcal{H}$ whose range is dense in $\mathcal{H}$ but not equal to $\mathcal{H}$.

(c) Show that if the orthonormal sequences $\{e_n\}$ and $\{f_n\}$ are the same and all the numbers α_n are real, then T is self-adjoint.

6.13 It was shown in Exercise 6.12 that if $\mathcal{H}$ is separable then there exists a compact operator on $\mathcal{H}$ whose range is dense in $\mathcal{H}$. Is this possible if $\mathcal{H}$ is not separable?

6.14 Suppose that (M, d) is a metric space and $A \subset M$. Prove the following results.

(a) A is relatively compact if and only if any sequence $\{a_n\}$ in A has a subsequence which converges in M.
[Hint: for the "if" part, consider an arbitrary sequence $\{x_n\}$ in $\overline{A}$ and construct a "nearby" sequence $\{a_n\}$ in A.]

(b) If $B \subset A$ is dense in A then it is dense in $\overline{A}$.

(c) If A is compact then, for each $r \in \mathbb{N}$, there exists a finite set $B_r \subset A$ with the property: for any $a \in A$ there exists a point $b \in B_r$ such that $d(a, b) < r^{-1}$.

(d) Deduce from part (c) that if A is compact then it is separable.

6.15 Let X be a separable, infinite-dimensional, normed space and suppose that $0 < \alpha < 1$. Show that there exists a sequence of unit vectors $\{x_n\}$ in X such that $\|x_m - x_n\| \geq \alpha$ for any $m, n \in \mathbb{N}$ with $m \neq n$ and $\overline{\operatorname{Sp}}\{x_n\} = X$.
[Hint: follow the proofs of Theorems 3.40 and 3.52, using Riesz' lemma (Theorem 2.25) to find the required vector in the inductive step.]

6.2 Spectral Theory of Compact Operators

From now on in this chapter we will suppose that $\mathcal{H}$ is a complex Hilbert space and $T \in K(\mathcal{H})$ (we require $\mathcal{H}$ to be complex in order that we may discuss spectral theory). If $\mathcal{H}$ is finite-dimensional then we know that the spectrum $\sigma(T)$ consists of a non-empty, finite collection of eigenvalues, each having finite multiplicity (see Definition 1.13). For general operators in infinite-dimensional spaces the spectrum can be very different, but for compact operators the spectrum has many similarities with the finite-dimensional case. Specifically, we will show that if $\mathcal{H}$ is infinite-dimensional then $\sigma(T)$ consists of a countable (possibly empty or finite) collection of non-zero eigenvalues, each having finite multiplicity, together with the point $\lambda = 0$, which necessarily belongs to $\sigma(T)$ but need not be an eigenvalue or, if it is, it need not have finite multiplicity. To describe this structure of $\sigma(T)$ the following notation will be convenient.

Definition 6.17

Let $\mathcal{K}$ be a Hilbert space and let $S \in B(\mathcal{K})$. We define the sets

$$\begin{aligned} \sigma_{\mathrm{p}}(S) &= \{\lambda : \lambda \text{ is an eigenvalue of } S\}, \\ \rho(S) &= \mathbb{C} \setminus \sigma(S). \end{aligned}$$

The set $\sigma_{\mathrm{p}}(S)$ is the *point spectrum* of S, while $\rho(S)$ is the *resolvent set* of S.

We begin our discussion of $\sigma(T)$ by dealing with the point $\lambda = 0$.

Theorem 6.18

If $\mathcal{H}$ is infinite-dimensional then $0 \in \sigma(T)$. If $\mathcal{H}$ is separable then either $0 \in \sigma_{\mathrm{p}}(T)$ or $0 \in \sigma(T) \setminus \sigma_{\mathrm{p}}(T)$ may occur. If $\mathcal{H}$ is not separable then $0 \in \sigma_{\mathrm{p}}(T)$.

Proof

If we had $0 \in \rho(T)$, then T would be invertible. However, since $\mathcal{H}$ is infinite-dimensional this contradicts Corollary 6.7, so we must have $0 \in \sigma(T)$. We leave the remainder of the proof to the exercises (see Exercises 6.16 and 6.17). □

We now consider the case $\lambda \neq 0$ and we first prove some preliminary results.

Theorem 6.19

If $\lambda \neq 0$ then $\mathrm{Ker}\,(T - \lambda I)$ has finite dimension.

Proof

Suppose that $\mathcal{M} = \operatorname{Ker}(T - \lambda I)$ is infinite-dimensional. Since the kernel of a bounded operator is closed (by Lemma 4.11), the space $\mathcal{M}$ is an infinite-dimensional Hilbert space, and there is an orthonormal sequence $\{e_n\}$ in $\mathcal{M}$ (by Theorem 3.40). Since $e_n \in \operatorname{Ker}(T-\lambda I)$ we have $Te_n = \lambda e_n$ for each $n \in \mathbb{N}$, and since $\lambda \neq 0$ the sequence $\{\lambda e_n\}$ cannot have a convergent subsequence, since $\{e_n\}$ is orthonormal (see Exercise 6.4). This contradicts the compactness of T, which proves the theorem. □

Theorem 6.20

If $\lambda \neq 0$ then $\operatorname{Im}(T - \lambda I)$ is closed.

Proof

Let $\{y_n\}$ be a sequence in $\operatorname{Im}(T - \lambda I)$, with $\lim_{n\to\infty} y_n = y$. Then for each n we have $y_n = (T-\lambda I)x_n$, for some x_n, and since $\operatorname{Ker}(T-\lambda I)$ is closed, x_n has an orthogonal decomposition of the form $x_n = u_n + v_n$, with $u_n \in \operatorname{Ker}(T-\lambda I)$ and $v_n \in \operatorname{Ker}(T - \lambda I)^\perp$. We will show that the sequence $\{v_n\}$ is bounded.

Suppose not. Then, after taking a subsequence if necessary, we may suppose that $\|v_n\| \neq 0$, for all n, and $\lim_{n\to\infty} \|v_n\| = \infty$. Putting $w_n = v_n/\|v_n\|$, $n = 1, 2, \ldots$, we have $w_n \in \operatorname{Ker}(T - \lambda I)^\perp$, $\|w_n\| = 1$ (so the sequence $\{w_n\}$ is bounded) and

$$(T - \lambda I)w_n = y_n/\|v_n\| \to 0,$$

since $\{y_n\}$ is bounded (because it is convergent). Also, by the compactness of T we may suppose that $\{Tw_n\}$ converges (after taking a subsequence if necessary). By combining these results it follows that the sequence $\{w_n\}$ converges (since $\lambda \neq 0$). Letting $w = \lim_{n\to\infty} w_n$, we see that $\|w\| = 1$ and

$$(T - \lambda I)w = \lim_{n\to\infty} (T - \lambda I)w_n = 0,$$

so $w \in \operatorname{Ker}(T - \lambda I)$. However, $w_n \in \operatorname{Ker}(T - \lambda I)^\perp$ so

$$\|w - w_n\|^2 = (w - w_n, w - w_n) = 1 + 1 = 2,$$

which contradicts $w_n \to w$. Hence the sequence $\{v_n\}$ is bounded.

Now, by the compactness of T we may suppose that $\{Tv_n\}$ converges. Then $v_n = \lambda^{-1}(Tv_n - (T-\lambda I)v_n) = \lambda^{-1}(Tv_n - y_n)$, for $n \in \mathbb{N}$, so the sequence $\{v_n\}$ converges. Let its limit be v. Then

$$y = \lim_{n\to\infty} y_n = \lim_{n\to\infty} (T - \lambda I)v_n = (T - \lambda I)v,$$

and so $y \in \operatorname{Im}(T - \lambda I)$. This proves that $\operatorname{Im}(T - \lambda I)$ is closed. □

Since T^* is also compact, Theorems 6.19 and 6.20 also apply to T^*, and in particular, the set $\mathrm{Im}\,(T^* - \overline{\lambda}I)$ is closed when $\lambda \neq 0$. Thus, from Corollary 3.36 and Lemma 5.11 we have the following result.

Corollary 6.21

If $\lambda \neq 0$ then

$$\mathrm{Im}\,(T - \lambda I) = \mathrm{Ker}\,(T^* - \overline{\lambda}I)^\perp, \quad \mathrm{Im}\,(T^* - \overline{\lambda}I) = \mathrm{Ker}\,(T - \lambda I)^\perp.$$

We can now begin to discuss the structure of the non-zero part of $\sigma(T)$ and $\sigma(T^*)$ (again, the following results apply to T^* as well as to T).

Theorem 6.22

For any real $t > 0$, the set of all distinct eigenvalues λ of T with $|\lambda| \geq t$ is finite.

Proof

Suppose instead that for some $t_0 > 0$ there is a sequence of distinct eigenvalues $\{\lambda_n\}$ with $|\lambda_n| \geq t_0$ for all n, and let $\{e_n\}$ be a sequence of corresponding unit eigenvectors. We will now construct, inductively, a particular sequence of unit vectors $\{y_n\}$. Let $y_1 = e_1$. Now consider any integer $k \geq 1$. By Lemma 1.14 the set $\{e_1, \ldots, e_k\}$ is linearly independent, thus the set $\mathcal{M}_k = \mathrm{Sp}\,\{e_1, \ldots, e_k\}$ is k-dimensional and so is closed by Corollary 2.20. Any $e \in \mathcal{M}_k$ can be written as $e = \alpha_1 e_1 + \ldots + \alpha_k e_k$, and we have

$$(T - \lambda_k I)e = \alpha_1(\lambda_1 - \lambda_k)e_1 + \ldots + \alpha_{k-1}(\lambda_{k-1} - \lambda_k)e_{k-1},$$

and so if $e \in \mathcal{M}_k$,

$$(T - \lambda_k I)e \in \mathcal{M}_{k-1}.$$

Similarly, if $e \in \mathcal{M}_k$,

$$Te \subset \mathcal{M}_k.$$

Next, $\mathcal{M}_k$ is a closed subspace of $\mathcal{M}_{k+1}$ and not equal to $\mathcal{M}_{k+1}$, so the orthogonal complement of $\mathcal{M}_k$ in $\mathcal{M}_{k+1}$ is a non-trivial linear subspace of $\mathcal{M}_{k+1}$. Hence there is a unit vector $y_{k+1} \in \mathcal{M}_{k+1}$ such that $(y_{k+1}, e) = 0$ for all $e \in \mathcal{M}_k$, and $\|y_{k+1} - e\| \geq 1$. Repeating this process inductively, we construct a sequence $\{y_n\}$.

It now follows from the construction of the sequence $\{y_n\}$ that for any integers m, n with $n > m$,

$$\|Ty_n - Ty_m\| = |\lambda_n|\,\|y_n - \lambda_n^{-1}[-(T - \lambda_n)y_n + Ty_m]\| \geq |\lambda_n| \geq t_0,$$

since, by the above results, $-(T-\lambda_n)y_n + Ty_m \in \mathcal{M}_{n-1}$. This shows that the sequence $\{Ty_n\}$ cannot have a convergent subsequence. This contradicts the compactness of T, and so proves the theorem. □

By taking the union of the finite sets of eigenvalues λ with $|\lambda| \geq r^{-1}$, $r = 1, 2, \ldots$, we obtain the following corollary of Theorem 6.22.

Corollary 6.23

The set $\sigma_{\mathrm{p}}(T)$ is at most countably infinite. If $\{\lambda_n\}$ is any sequence of distinct eigenvalues of T then $\lim_{n\to\infty} \lambda_n = 0$.

We note that it is possible for a compact operator T on an infinite-dimensional space to have no eigenvalues at all, see Exercise 6.17. In that case, by Theorem 6.18 and Theorem 6.25 below, $\sigma(T) = \{0\}$.

We will now show that for any compact operator T, all the non-zero points of $\sigma(T)$ must be eigenvalues. Since T^* is also compact, it follows from this and Lemma 5.37 that if $\lambda \neq 0$ is an eigenvalue of T then $\overline{\lambda}$ is an eigenvalue of T^*. We will also prove that these eigenvalues have equal and finite multiplicity. These results are standard in the finite-dimensional setting. We will prove them in the infinite-dimensional case in two steps:

(a) we consider finite rank operators and reduce the problem to the finite-dimensional case;

(b) we consider general compact operators and reduce the problem to the finite rank case.

The following notation will be helpful in the proof of the next lemma. Suppose that X, Y are normed spaces and $C \in B(X)$, $D \in B(X,Y)$, $E \in B(Y,X)$ and $F \in B(Y)$. We can define an operator $M \in B(X \times Y)$ by

$$M(x,y) = (Ax + By, Cx + Dy),$$

see Exercise 6.18. This operator may be written in "matrix" form as

$$M \begin{bmatrix} x \\ y \end{bmatrix} = \begin{bmatrix} C & D \\ E & F \end{bmatrix} \begin{bmatrix} x \\ y \end{bmatrix},$$

where, formally, we use the standard matrix multiplication rules to evaluate the matrix product, even though the elements in the matrices are operators or vectors – this is valid so long as we keep the correct order of the operators and vectors.

Lemma 6.24

If T has finite rank and $\lambda \neq 0$, then either: (a) $\lambda \in \rho(T)$ *and* $\overline{\lambda} \in \rho(T^*)$; or (b) $\lambda \in \sigma_{\mathrm{p}}(T)$ *and* $\overline{\lambda} \in \sigma_{\mathrm{p}}(T^*)$. Furthermore, $n(T - \lambda I) = n(T^* - \overline{\lambda} I) < \infty$.

Proof

Let $\mathcal{M} = \operatorname{Im} T$ and $\mathcal{N} = \operatorname{Ker} T^* = \mathcal{M}^\perp$ (by Lemma 5.11). Since $\mathcal{M}$ is finite-dimensional it is closed, so any $x \in \mathcal{H}$ has an orthogonal decomposition $x = u + v$, with $u \in \mathcal{M}$, $v \in \mathcal{N}$. Using this decomposition, we can identify any $x \in \mathcal{H}$ with a unique element $(u, v) \in \mathcal{M} \times \mathcal{N}$, and vice versa (alternatively, this shows that the space $\mathcal{H}$ is isometrically isomorphic to the space $\mathcal{M} \times \mathcal{N}$). Also,

$$(T - \lambda I)(u + v) = Tu - \lambda u + Tv - \lambda v,$$

and we have $Tu - \lambda u \in \mathcal{M}$, $Tv \in \mathcal{M}$ and $-\lambda v \in \mathcal{N}$. It follows from this that we can express the action of the operator $(T - \lambda I)$ in matrix form by

$$(T - \lambda I) \begin{bmatrix} u \\ v \end{bmatrix} = \begin{bmatrix} (T - \lambda I)|_{\mathcal{M}} & T|_{\mathcal{N}} \\ 0 & -\lambda I|_{\mathcal{N}} \end{bmatrix} \begin{bmatrix} u \\ v \end{bmatrix},$$

where $(T - \lambda I)|_{\mathcal{M}} \in B(\mathcal{M})$, $T|_{\mathcal{N}} \in B(\mathcal{N}, \mathcal{M})$ and $I|_{\mathcal{N}} \in B(\mathcal{N})$ denote the restrictions of the operators $T - \lambda I$, T and I to the spaces $\mathcal{M}$ and N. We now write $C = (T - \lambda I)|_{\mathcal{M}}$. It follows from Lemma 1.12 and Corollary 4.45 that either C is invertible ($n(C) = 0$) or $n(C) > 0$, and so, from Exercise 6.18, either $T - \lambda I$ is invertible or $n(T - \lambda I) = n(C) > 0$, that is, either $\lambda \in \rho(T)$ or $\lambda \in \sigma_{\mathrm{p}}(T)$.

Now let $P_{\mathcal{M}}$, $P_{\mathcal{N}}$ denote the orthogonal projections of $\mathcal{H}$ onto $\mathcal{M}$, $\mathcal{N}$. Using $I = P_{\mathcal{M}} + P_{\mathcal{N}}$ and $\mathcal{N} = \operatorname{Ker} T^*$, we have

$$(T^* - \overline{\lambda} I)(u + v) = (T^* - \overline{\lambda} I)u - \overline{\lambda} v = P_{\mathcal{M}}(T^* - \overline{\lambda} I)u + P_{\mathcal{N}} T^* u - \overline{\lambda} v.$$

Hence $T^* - \overline{\lambda} I$ can be represented in matrix form by

$$(T^* - \overline{\lambda} I) \begin{bmatrix} u \\ v \end{bmatrix} = \begin{bmatrix} P_{\mathcal{M}}(T^* - \overline{\lambda} I)|_{\mathcal{M}} & 0 \\ P_{\mathcal{N}}(T^*)|_{\mathcal{M}} & -\overline{\lambda} I|_{\mathcal{N}} \end{bmatrix} \begin{bmatrix} u \\ v \end{bmatrix}.$$

Also, $C^* = P_{\mathcal{M}}(T^* - \overline{\lambda} I)|_{\mathcal{M}} \in B(\mathcal{M})$ (see Exercise 6.20). Again by finite-dimensional linear algebra, $n(C^*) = n(C)$. It now follows from Exercise 6.18 that if $n(C) = 0$ then $T - \lambda I$ and $T^* - \overline{\lambda} I$ are invertible, while if $n(C) > 0$ then $n(T - \lambda I) = n(T^* - \overline{\lambda} I) = n(C) > 0$, so $\lambda \in \sigma_{\mathrm{p}}(T)$ and $\overline{\lambda} \in \sigma_{\mathrm{p}}(T^*)$. □

We now extend the results of Lemma 6.24 to the case of a general compact operator T.

Theorem 6.25

If T is compact and $\lambda \neq 0$, then either: (a) $\lambda \in \rho(T)$ *and* $\overline{\lambda} \in \rho(T^*)$; or (b) $\lambda \in \sigma_{\mathrm{p}}(T)$ *and* $\overline{\lambda} \in \sigma_{\mathrm{p}}(T^*)$. Furthermore, $n(T - \lambda I) = n(T^* - \overline{\lambda} I) < \infty$.

Proof

We first reduce the problem to the case of a finite rank operator. By Theorem 6.12 there is a finite rank operator T_F on $\mathcal{H}$ with $\|\lambda^{-1}(T - T_F)\| < \frac{1}{2}$, so by Theorem 4.40 and Lemma 5.14, the operators $S = I - \lambda^{-1}(T - T_F)$ and S^* are invertible. Now, letting $G = T_F S^{-1}$ we see that

$$T - \lambda I = (G - \lambda I)S, \quad \text{and so} \quad T^* - \overline{\lambda} I = S^*(G^* - \overline{\lambda} I).$$

Since S and S^* are invertible it follows that $T - \lambda I$ and $T^* - \overline{\lambda} I$ are invertible if and only if $G - \lambda I$ and $G^* - \overline{\lambda} I$ are invertible, and $n(T - \lambda I) = n(G - \lambda I)$, $n(T^* - \overline{\lambda} I) = n(G^* - \overline{\lambda} I)$ (see Exercise 6.21). Now, since $\operatorname{Im} G \subset \operatorname{Im} T_F$ the operator G has finite rank, so the first results of the theorem follow from Lemma 6.24. □

We now consider the following equations:

$$(T - \lambda I)x = 0, \quad (T^* - \overline{\lambda} I)y = 0, \tag{6.1}$$

$$(T - \lambda I)x = p, \quad (T^* - \overline{\lambda} I)y = q \tag{6.2}$$

(equations of the form (6.1), with zero right-hand sides, are called *homogeneous* while equations of the form (6.2), with non-zero right-hand sides, are called *inhomogeneous*). The results of Theorem 6.25 (together with Corollary 6.21) can be restated in terms of the solvability of these equations.

Theorem 6.26 (The Fredholm alternative)

If $\lambda \neq 0$ then one or other of the following alternatives holds.

(a) Each of the homogeneous equations (6.1) has only the solution $x = 0$, $y = 0$, respectively, while the corresponding inhomogeneous equations (6.2) have unique solutions x, y for any given p, $q \in \mathcal{H}$.

(b) There is a finite number $m_\lambda > 0$ such that each of the homogeneous equations (6.1) has exactly m_λ linearly independent solutions, say x_n, y_n, $n = 1, \ldots, m_\lambda$, respectively, while the corresponding inhomogeneous equations (6.2) have solutions if and only if p, $q \in \mathcal{H}$ satisfy the conditions

$$(p, y_n) = 0, \quad (q, x_n) = 0, \quad n = 1, \ldots, m_\lambda. \tag{6.3}$$

Proof

The result follows immediately from Theorem 6.25. Alternative (a) corresponds to the case $\lambda \in \rho(T)$, while alternative (b) corresponds to the case $\lambda \in \sigma_{\rm p}(T)$. In this case, $m_\lambda = n(T - \lambda I)$. It follows from Corollary 6.21 that the conditions on p, q in (b) ensure that $p \in \operatorname{Im}(T - \lambda I)$, $q \in \operatorname{Im}(T^* - \overline{\lambda} I)$, respectively, so solutions of (6.2) exist. □

The dichotomy expressed in Theorem 6.26 between unique solvability of the equations and solvability if and only if a finite set of conditions holds is often called the *Fredholm alternative*; this dichotomy was discovered by Fredholm in his investigation of certain integral equations (which give rise to equations of the above form with compact integral operators, see Chapter 7). More generally, if the operator $T - \lambda I$ in (6.1) and (6.2) is replaced by a bounded linear operator S then S is said to satisfy the Fredholm alternative if the corresponding equations again satisfy the alternatives in Theorem 6.26. A particularly important feature of the Fredholm alternative is the following restatement of alternative (a) in Theorem 6.26.

Corollary 6.27

If $\lambda \neq 0$ and the equation

$$(T - \lambda I)x = 0 \tag{6.4}$$

has only the solution $x = 0$ then $T - \lambda I$ is invertible, and the equation

$$(T - \lambda I)x = p \tag{6.5}$$

has the unique solution $x = (T - \lambda I)^{-1}p$ for any $p \in \mathcal{H}$. This solution depends continuously on p.

Proof

The hypothesis ensures that λ is not an eigenvalue of T, so by alternative (a) of Theorem 6.26, $\lambda \in \rho(T)$ and hence $T - \lambda I$ is invertible. The rest of the corollary follows immediately from this. □

In essence, Corollary 6.27 states that "uniqueness of solutions of equation (6.5) implies existence of solutions". This is an extremely useful result. In many applications it is relatively easy to prove uniqueness of solutions of a given equation. If the equation has the form (6.5) and we know that the operator T is compact then we can immediately deduce the existence of a solution.

Many problems in applied mathematics can be reduced to solving an equation of the form

$$Ru = f, \tag{6.6}$$

for some linear operator R and some given function (or "data") f. In order for this equation to be a reasonable model of a physical situation it should have certain properties. Hadamard proposed the following definition.

Definition 6.28

Equation (6.6) (or the corresponding physical model) is said to be *well-posed* if the following properties hold:

(a) A solution u exists for every f.

(b) The solution u is unique for each f.

(c) The solution u depends continuously on f in a suitable sense.

The motivation for properties (a) and (b) is fairly clear – the model will not be very useful if solutions do not exist or there are several solutions. The third property is motivated by the fact that in any physical situation the data f will not be known precisely, so it is desirable that small variations in the data should not produce large variations in the predicted solution. However, the statements of properties (a)–(c) in Definition 6.28 are rather vague mathematically. For instance, what does "for every f" mean, and what is a "suitable sense" for continuous dependence of the solution. These properties are usually made more precise by choosing, for instance, suitable normed or Banach spaces X, Y and a suitable operator $R \in B(X, Y)$ with which to represent the problem. The space Y usually incorporates desirable features of the data being modelled, while X incorporates corresponding desirable features of the solution being sought. In such a set-up it is clear that equation (6.6) being well-posed is equivalent to the operator R being invertible, and this is often proved using Corollary 6.27.

We now investigate rather more fully the nature of the set of solutions of equation (6.5) and the dependence of these solutions on p in the case where alternative (b) holds in Theorem 6.26.

Theorem 6.29

Suppose that $\lambda \neq 0$ is an eigenvalue of T. If $p \in \mathrm{Im}\,(T - \lambda I)$ (that is, p satisfies (6.3)) then equation (6.5) has a unique solution $S_\lambda(p) \in \mathrm{Ker}\,(T - \lambda I)^\perp$. The function $S_\lambda : \mathrm{Im}\,(T - \lambda I) \to \mathrm{Ker}\,(T - \lambda I)^\perp$ is linear and bounded, and the set of solutions of (6.5) has the form

$$S_\lambda p + \mathrm{Ker}\,(T - \lambda I). \tag{6.7}$$

Proof

Since $p \in \operatorname{Im}(T-\lambda I)$ there exists a solution x_0 of (6.5). Let P be the orthogonal projection of $\mathcal{H}$ onto $\operatorname{Ker}(T-\lambda I)^\perp$, and let $u_0 = Px_0$. Then $x_0 - u_0 \in \operatorname{Ker}(T-\lambda I)$, and so $(T-\lambda I)u_0 = (T-\lambda I)x_0 = p$. Therefore, u_0 is also a solution of (6.5), and any vector of the form $u_0 + z$, with $z \in \operatorname{Ker}(T-\lambda I)$, is a solution of (6.5). On the other hand, if x is a solution of (6.5) then $(T-\lambda I)(u_0 - x) = p - p = 0$, so $u_0 - x \in \operatorname{Ker}(T-\lambda I)$, and hence x has the form $x = u_0 + z$, $z \in \operatorname{Ker}(T-\lambda I)$. Thus the set of solutions of (6.5) has the form (6.7).

Next, it can be shown (see Exercise 6.23) that $u_0 \in \operatorname{Ker}(T-\lambda I)^\perp$ is uniquely determined by p so we may define a function $S_\lambda : \operatorname{Im}(T-\lambda I) \to \operatorname{Ker}(T-\lambda I)^\perp$ by $S_\lambda(p) = u_0$, for $p \in \operatorname{Im}(T-\lambda I)$. Using uniqueness, it can now be shown that the function S_λ is linear (see Exercise 6.23).

Finally, suppose that S_λ is not bounded. Then there exists a sequence of unit vectors $\{p_n\}$, such that $\|S_\lambda p_n\| \neq 0$, for all $n \in \mathbb{N}$, and $\lim_{n\to\infty} \|S_\lambda p_n\| = \infty$. Putting $w_n = \|S_\lambda p_n\|^{-1} S_\lambda p_n$, we see that $w_n \in \operatorname{Ker}(T-\lambda I)^\perp$, $\|w_n\| = 1$ and $(T-\lambda I)w_n = \|S_\lambda p_n\|^{-1} p_n \to 0$ as $n \to \infty$. Now, exactly as in the second paragraph of the proof of Theorem 6.20, we can show that these properties lead to a contradiction, which proves the result. □

Theorem 6.29 shows that the solution $S_\lambda p$ satisfies $\|S_\lambda p\| \le C\|p\|$, for some constant $C > 0$. However, such an inequality cannot hold for all solutions x of (6.5) since there are solutions x of the form $S_\lambda p + z$, with $z \in \operatorname{Ker}(T-\lambda I)$ having arbitrarily large $\|z\|$.

EXERCISES

6.16 Show that if $\mathcal{H}$ is not separable then $0 \in \sigma_\mathrm{p}(T)$ for any compact operator T on $\mathcal{H}$.
[Hint: use Exercise 3.19 and Theorem 6.8.]

6.17 Define operators $S, T \in B(\ell^2)$ by

$$Sx = \left(0, \frac{x_1}{1}, \frac{x_2}{2}, \frac{x_3}{3}, \dots\right), \quad Tx = \left(\frac{x_2}{1}, \frac{x_3}{2}, \frac{x_4}{3}, \dots\right).$$

Show that these operators are compact, $\sigma(S) = \{0\}$, $\sigma_\mathrm{p}(S) = \emptyset$, and $\sigma(T) = \sigma_\mathrm{p}(T) = \{0\}$. Show also that $\operatorname{Im} S$ is not dense in ℓ^2, but $\operatorname{Im} T$ is dense.
[Hint: see Example 5.35.]

6.18 Suppose that X, Y are normed spaces and $C \in B(X)$, $D \in B(X,Y)$, $E \in B(Y,X)$ and $F \in B(Y)$. Let $(x,y) \in X \times Y$ (recall that the normed space $X \times Y$ was defined in Example 2.8), and define $M(x, y) = (Ax + By, Cx + Dy)$. Show that $M \in B(X \times Y)$.

Now suppose that $F = I_Y$ is the identity on Y, and consider the operators on $X \times Y$ represented by the matrices

$$M_1 = \begin{bmatrix} C & D \\ 0 & I_Y \end{bmatrix}, \quad M_2 = \begin{bmatrix} C & 0 \\ E & I_Y \end{bmatrix}$$

(here, 0 denotes the zero operator on the appropriate spaces). Show that if C is invertible then M_1 and M_2 are invertible.
[Hint: use the operator C^{-1} to explicitly construct matrix inverses for M_1 and M_2.]

Show also that if C is not invertible then $\operatorname{Ker} M_1 = (\operatorname{Ker} C) \times \{0\}$. Find a similar representation for $\operatorname{Ker} M_2$.

6.19 Suppose that $\mathcal{M}$ and $\mathcal{N}$ are Hilbert spaces, so that $\mathcal{M} \times \mathcal{N}$ is a Hilbert space with the inner product defined in Example 3.10 (see Exercise 3.11). Show that if $M \in B(\mathcal{M} \times \mathcal{N})$ is an operator of the form considered in Exercise 6.18, then the adjoint operator M^* has the matrix form

$$M^* = \begin{bmatrix} A^* & C^* \\ B^* & D^* \end{bmatrix}.$$

6.20 Prove that $C^* = P_{\mathcal{M}}(T^* - \overline{\lambda} I)_{\mathcal{M}}$ in the proof of Lemma 6.24.

6.21 Let X be a normed space and S, $B \in B(X)$, with B invertible. Let $T = SB$. Show that:

(a) T is invertible if and only if S is;

(b) $n(T) = n(S)$.

Show that these results also hold if $T = BS$.

6.22 Suppose that T is a compact operator on a Hilbert space $\mathcal{H}$. Show that if $r(T)$ is finite then $\sigma(T)$ is a finite set.

6.23 Show that for $p \in \operatorname{Im}(T - \lambda I)$ the solution $u_0 \in (\operatorname{Ker}(T - \lambda I))^\perp$ of (6.5) constructed in the proof of Theorem 6.29 is the unique solution lying in the subspace $(\operatorname{Ker}(T - \lambda I))^\perp$. Hence show that the function S_λ constructed in the same proof is linear.

6.3 Self-adjoint Compact Operators

Throughout this section we will suppose that $\mathcal{H}$ is a complex Hilbert space and $T \in B(\mathcal{H})$ is self-adjoint and compact. In this case the previous results regarding the spectrum of T can be considerably improved. In a sense, the main reason for this is the following rather trivial-seeming lemma. We first need a definition.

Definition 6.30

Let X be a vector space and let $S \in L(X)$. A linear subspace $W \subset X$ is said to be *invariant* under S if $S(W) \subset W$.

Lemma 6.31

Let $\mathcal{K}$ be a Hilbert space and let $S \in B(\mathcal{K})$ be self-adjoint. If $\mathcal{M}$ is a closed linear subspace of $\mathcal{K}$ which is invariant under S then $\mathcal{M}^\perp$ is also invariant under S.

Proof

For any $u \in \mathcal{M}$ and $v \in \mathcal{M}^\perp$ we have $(Sv, u) = (v, Su) = 0$ (since S is self-adjoint and $Su \in \mathcal{M}$), so $Sv \in \mathcal{M}^\perp$, and hence $S(\mathcal{M}^\perp) \subset \mathcal{M}^\perp$, which proves the lemma. □

This lemma will enable us to "split up" or decompose a self-adjoint operator and look at its action on various linear subspaces $\mathcal{M} \subset \mathcal{H}$, and also on the orthogonal complements $\mathcal{M}^\perp$. For a general operator $S \in B(\mathcal{H})$, even if S is invariant on $\mathcal{M}$ it need not be invariant on $\mathcal{M}^\perp$, so this strategy fails in general. However, Lemma 6.31 ensures that this works for self-adjoint T. The main subspaces that we use to decompose T will be $\operatorname{Ker} T$ and $\overline{\operatorname{Im} T}$ (since $0 \in \operatorname{Ker} T$ and $\operatorname{Im} T \subset \overline{\operatorname{Im} T}$, it is clear that both these spaces are invariant under T). Since T is self-adjoint it follows from Corollary 3.36 and Lemma 5.11 that

$$\overline{\operatorname{Im} T} = (\operatorname{Ker} T)^\perp. \tag{6.8}$$

From now on, P will denote the orthogonal projection of $\mathcal{H}$ onto $\overline{\operatorname{Im} T}$. It then follows from (6.8) that $I - P$ is the orthogonal projection onto $\operatorname{Ker} T$. Also, the space $\overline{\operatorname{Im} T}$ is a separable Hilbert space (separability follows from Theorem 6.8). We will see that we can construct an orthonormal basis of $\overline{\operatorname{Im} T}$ consisting of eigenvectors of T (regardless of whether $\mathcal{H}$ is separable). Since the restriction of T to the space $\operatorname{Ker} T$ is trivial, this will give a complete representation of the action of T on $\mathcal{H}$.

Note that equation (6.8) certainly need not hold if T is not self-adjoint – consider the operator on $\mathbb{C}^2$ whose matrix is

$$\begin{bmatrix} 0 & 1 \\ 0 & 0 \end{bmatrix}. \tag{6.9}$$

We saw above that a general compact operator need have no non-zero eigenvalue. This cannot happen when T is a non-zero, self-adjoint, compact operator.

Theorem 6.32

At least one of the numbers $\|T\|$, $-\|T\|$, is an eigenvalue of T.

Proof

If T is the zero operator the result is trivial so we may suppose that T is non-zero. By Theorem 5.43 at least one of $\|T\|$ or $-\|T\|$ is in $\sigma(T)$, so by Theorem 6.25 this point must belong to $\sigma_{\mathrm{p}}(T)$. □

We can summarize what we know at present about the spectrum of T in the following theorem.

Theorem 6.33

The set of non-zero eigenvalues of T is non-empty and is either finite or consists of a sequence which tends to zero. Each non-zero eigenvalue is real and has finite multiplicity. Eigenvectors corresponding to different eigenvalues are orthogonal.

Proof

Most of the theorem follows from Corollary 6.23 and Theorems 5.43, 6.19 and 6.32. To prove the final result, suppose that λ_1, $\lambda_2 \in \mathbb{R}$, are distinct eigenvalues with corresponding eigenvectors e_1, e_2. Then, since T is self-adjoint

$$\lambda_1(e_1, e_2) = (Te_1, e_2) = (e_1, Te_2) = \lambda_2(e_1, e_2),$$

which, since $\lambda_1 \neq \lambda_2$, implies that $(e_1, e_2) = 0$. □

In view of Theorem 6.33 we can now order the eigenvalues of T in the form of a non-empty, finite list $\lambda_1, \ldots, \lambda_J$, or a countably infinite list λ_1, $\lambda_2, \ldots$, in such a way that $|\lambda_n|$ decreases as n increases and each eigenvalue λ_n is repeated in the list according to its multiplicity (more precisely, if λ is an eigenvalue of T with multiplicity $m_\lambda > 0$, then λ is repeated exactly m_λ times in the list). Furthermore, for each n we can use the by the Gram–Schmidt algorithm to construct an orthonormal basis of each space $\operatorname{Ker}(T - \lambda_n I)$ consisting of exactly m_{λ_n} eigenvectors. Thus, listing the eigenvectors so constructed in the same order as the eigenvalues, we obtain a list of corresponding eigenvectors of the form $e_1, \ldots, e_J$ or $e_1, e_2, \ldots$. By the construction, eigenvectors in this list corresponding to the same eigenvalue are orthogonal, while by Theorem 6.33, eigenvectors corresponding to different eigenvalues are orthogonal. Hence the complete list is an orthonormal set.

At present we do not know how many non-zero eigenvalues there are. To deal with both the finite and infinite case we will, for now, denote this number by J, where J may be a finite integer or "$J = \infty$", and we will write the above lists in the form $\{\lambda_n\}_{n=1}^J$, $\{e_n\}_{n=1}^J$. We will show that J is, in fact, equal to $r(T)$, the rank of T (which may be finite or ∞ here). We will also show that $\{e_n\}_{n=1}^J$ is an orthonormal basis for the Hilbert space $\overline{\operatorname{Im} T}$.

Theorem 6.34

The number of non-zero eigenvalues of T (repeated according to multiplicity) is equal to $r(T)$. The set of eigenvectors $\{e_n\}_{n=1}^{r(T)}$ constructed above is an orthonormal basis for $\overline{\operatorname{Im} T}$ and the operator T has the representation,

$$Tx = \sum_{n=1}^{r(T)} \lambda_n (x, e_n) e_n, \qquad (6.10)$$

where $\{\lambda_n\}_{n=1}^{r(T)}$ is the set of non-zero eigenvalues of T.

Proof

Let $\mathcal{M} = \overline{\operatorname{Sp}}\,\{e_n\}_{n=1}^J$, so that $\{e_n\}_{n=1}^J$ is an orthonormal basis for $\mathcal{M}$ (by Theorem 3.47). We will show that $\mathcal{M} = \overline{\operatorname{Im} T}$, and hence we must have $J = r(T)$ (in either the finite or infinite case). Recall that if $r(T) < \infty$ then $\operatorname{Im} T$ is closed, so $\operatorname{Im} T = \overline{\operatorname{Im} T}$.

By Theorem 3.47, for any $u \in \mathcal{M}$ we have $u = \sum_{n=1}^J \alpha_n e_n$, where $\alpha_n = (u, e_n)$, $n = 1, \ldots, J$. Thus, if $J = \infty$,

$$u = \lim_{k \to \infty} \sum_{n=1}^k \alpha_n \lambda_n^{-1} T e_n = \lim_{k \to \infty} T\Big(\sum_{n=1}^k \alpha_n \lambda_n^{-1} e_n\Big) \in \overline{\operatorname{Im} T},$$

and so $\mathcal{M} \subset \overline{\operatorname{Im} T}$; a similar argument holds when J is finite (without the limits). From this we obtain $\operatorname{Ker} T = \overline{\operatorname{Im} T}^\perp \subset \mathcal{M}^\perp$ (by (6.8) and Lemma 3.29).

We will now show that $\mathcal{M}^\perp \subset \operatorname{Ker} T$, which implies that $\mathcal{M}^\perp = \operatorname{Ker} T$, and hence $\mathcal{M} = \mathcal{M}^{\perp\perp} = \overline{\operatorname{Im} T}$ (by Corollary 3.35 and (6.8)), which is the desired result.

If $J = \infty$ and $u \in \mathcal{M}$, we have

$$\begin{aligned} Tu = T\Big(\lim_{k\to\infty}\sum_{n=1}^{k}\alpha_n e_n\Big) &= \lim_{k\to\infty}\sum_{n=1}^{k}\alpha_n T e_n = \lim_{k\to\infty}\sum_{n=1}^{k}\lambda_n\alpha_n e_n \\ &= \sum_{n=1}^{\infty}\lambda_n\alpha_n e_n \in \mathcal{M}, \end{aligned}$$

and again a similar calculation holds (without the limits) if $J < \infty$. Thus $\mathcal{M}$ is invariant under T. Lemma 6.31 now implies that $\mathcal{N} = \mathcal{M}^\perp$ is invariant under T. Let $T_\mathcal{N}$ denote the restriction of T to $\mathcal{N}$. It can easily be checked that $T_\mathcal{N}$ is a self-adjoint, compact operator on the Hilbert space $\mathcal{N}$, see Exercise 6.24. Now suppose that $T_\mathcal{N}$ is not the zero operator on $\mathcal{N}$. Then by Theorem 6.32, $T_\mathcal{N}$ must have a non-zero eigenvalue, say $\widetilde{\lambda}$, with corresponding non-zero eigenvector $\widetilde{e} \in \mathcal{N}$, so by definition, $T\widetilde{e} = T_\mathcal{N}\widetilde{e} = \widetilde{\lambda}\widetilde{e}$. However, this implies that $\widetilde{\lambda}$ is a non-zero eigenvalue of T, so we must have $\widetilde{\lambda} = \lambda_n$, for some n, and $\widetilde{e}$ must belong to the subspace spanned by the eigenvectors corresponding to λ_n. But this subspace lies in $\mathcal{M}$, so $\widetilde{e} \in \mathcal{M}$ which contradicts $\widetilde{e} \in \mathcal{N} = \mathcal{M}^\perp$ (since $\widetilde{e} \neq 0$). Thus $T_\mathcal{N}$ must be the zero operator. In other words, $Tv = T_\mathcal{N}v = 0$ for all $v \in$N, or $\mathcal{M}^\perp = \mathcal{N} \subset \operatorname{Ker} T$, which is what we asserted above, and so completes the proof that $\mathcal{M} = \overline{\operatorname{Im} T}$.

Finally, for any $x \in \mathcal{H}$ we have $(I - P)x \in \mathcal{M}^\perp$ so

$$(x, e_n) = (Px + (I-P)x, e_n) = (Px, e_n), \tag{6.11}$$

for all n (since $e_n \in \mathcal{M}$), and hence

$$Tx = T(Px + (I-P)x) = TPx = \sum_{n=1}^{J}\lambda_n(Px, e_n)e_n = \sum_{n=1}^{J}\lambda_n(x, e_n)e_n,$$

by the above calculation. □

The representation (6.10) of the self-adjoint operator T is an infinite-dimensional version of the well-known result in finite dimensional linear algebra that a self-adjoint matrix can be diagonalized by choosing a basis consisting of eigenvectors of the matrix.

The orthonormal set of eigenvectors $\{e_n\}_{n=1}^{r(T)}$ constructed above is an orthonormal basis for the space $\overline{\operatorname{Im} T}$ but not for the whole space $\mathcal{H}$, unless $\overline{\operatorname{Im} T} = \mathcal{H}$. By (6.8) and Lemma 3.29, this holds when $\operatorname{Ker} T = \{0\}$, that is, when T is one-to-one, so we have the following result.

Corollary 6.35

If $\operatorname{Ker} T = \{0\}$ then the set of eigenvectors $\{e_n\}_{n=1}^{r(T)}$ is an orthonormal basis for $\mathcal{H}$. In particular, if $\mathcal{H}$ is infinite-dimensional and $\operatorname{Ker} T = \{0\}$ then T has infinitely many distinct eigenvalues.

If $\mathcal{H}$ is separable we can also obtain a basis of $\mathcal{H}$ consisting of eigenvectors of T, even when $\operatorname{Ker} T \neq \{0\}$.

Corollary 6.36

Suppose that $\mathcal{H}$ is separable. Then there exists an orthonormal basis of $\mathcal{H}$ consisting entirely of eigenvectors of T. This basis has the form $\{e_n\}_{n=1}^{r(T)} \cup \{z_m\}_{m=1}^{n(T)}$, where $\{e_n\}_{n=1}^{r(T)}$ is an orthonormal basis of $\overline{\operatorname{Im} T}$ and $\{z_m\}_{m=1}^{n(T)}$ is an orthonormal basis of $\operatorname{Ker} T$.

Proof

Since $\mathcal{H}$ is separable, $\operatorname{Ker} T$ is a separable Hilbert space (by Lemma 3.25 and Exercise 3.25), so by Theorem 3.52 there is an orthonormal basis for $\operatorname{Ker} T$, which we write in the form $\{z_m\}_{m=1}^{n(T)}$ (where $n(T)$ may be finite or infinite). By definition, for each m we have $Tz_m = 0$, so z_m is an eigenvector of T corresponding to the eigenvalue $\lambda = 0$. Now, the union $E = \{e_n\}_{n=1}^{r(T)} \cup \{z_m\}_{m=1}^{n(T)}$ is a countable orthonormal set in $\mathcal{H}$. In fact, it is a basis for $\mathcal{H}$. To see this, notice that by Theorem 3.47 we have

$$\begin{aligned} x = Px + (I-P)x &= \sum_{n=1}^{r(T)} (Px, e_n)e_n + \sum_{m=1}^{n(T)} ((I-P)x, z_m)z_m \\ &= \sum_{n=1}^{r(T)} (x, e_n)e_n + \sum_{m=1}^{n(T)} (x, z_m)z_m, \end{aligned}$$

using (6.11) and, similarly, $((I-P)x, z_m) = (x, z_m)$ for each m. Hence, by Theorem 3.47 again, E is an orthonormal basis for $\mathcal{H}$. □

In Theorem 6.26 we discussed the existence of solutions of the equations (6.5) for the case of a general compact operator T. When T is self-adjoint we can use the representation of T in (6.10) to give a corresponding representation of the solutions.

Theorem 6.37

Let $\{\lambda_n\}_{n=1}^{r(T)}$ and $\{e_n\}_{n=1}^{r(T)}$ be the set of non-zero eigenvalues of T and the corresponding orthonormal set of eigenvectors constructed above. Then for any $\lambda \neq 0$, one of the following alternatives holds for the equation

$$(T - \lambda I)x = p. \tag{6.12}$$

(a) If λ is not an eigenvalue then equation (6.12) has a unique solution, and this solution has the form

$$x = \sum_{n=1}^{r(T)} \frac{(p, e_n)}{\lambda_n - \lambda} e_n - \frac{1}{\lambda}(I - P)p. \tag{6.13}$$

(b) If λ is an eigenvalue then, letting E denote the set of integers n for which $\lambda_n = \lambda$, equation (6.12) has a solution if and only if

$$(p, e_n) = 0, \quad n \in E. \tag{6.14}$$

If (6.14) holds then the set of solutions of (6.12) has the form

$$x = \sum_{\substack{n=1 \\ n \notin E}}^{r(T)} \frac{(p, e_n)}{\lambda_n - \lambda} e_n - \frac{1}{\lambda}(I - P)p + z, \tag{6.15}$$

where $z = \sum_{n \in E} \alpha_n e_n$ is an arbitrary element of $\operatorname{Ker}(T - \lambda I)$.

Proof

The existence of solutions of (6.12) under the stated conditions follows from Theorem 6.26. To show that the solutions have the stated form we note that since $\{e_n\}_{n=1}^{r(T)}$ is an orthonormal basis for $\overline{\operatorname{Im} T} = (\operatorname{Ker} T)^\perp$, we have

$$x = \sum_{n=1}^{r(T)} (x, e_n)e_n + (I - P)x, \quad p = \sum_{n=1}^{r(T)} (p, e_n)e_n + (I - P)p,$$

(using (6.11)) and hence, from (6.12),

$$(T - \lambda I)x = \sum_{n=1}^{r(T)} (x, e_n)(\lambda_n - \lambda)e_n - \lambda(I - P)x = \sum_{n=1}^{r(T)} (p, e_n)e_n + (I - P)p.$$

Taking the inner product of both sides of this formula with e_k, for any $1 \leq k \leq r(T)$, and assuming that λ is not an eigenvalue we have

$$(x, e_k)(\lambda_k - \lambda) = (p, e_k) \quad \text{and so} \quad (x, e_k) = \frac{(p, e_k)}{\lambda_k - \lambda}. \tag{6.16}$$

Also, taking the orthogonal projection of both sides of the above formula onto $\operatorname{Ker} T$ yields

$$-\lambda(I-P)x=(I-P)p.$$

The formula (6.13) now follows immediately from these two results. This proves alternative (a). The proof of alternative (b) is similar. Notice that when $k \in E$, conditions (6.14) ensure that the first equation in (6.16) is satisfied by arbitrary coefficients $(x, e_k) = \alpha_k$ say (and we avoid the difficulty caused by the term $\lambda_n - \lambda$ in the denominator). The corresponding term $\alpha_k e_k$ contributes to the arbitrary element in $\operatorname{Ker}(T-\lambda I)$ in the expression for the solution x. □

EXERCISES

6.24 Suppose that $\mathcal{H}$ is a Hilbert space, $T \in B(\mathcal{H})$ is a self-adjoint, compact operator on $\mathcal{H}$ and $\mathcal{N}$ is a closed linear subspace of $\mathcal{H}$ which is invariant under T. Let $T_{\mathcal{N}}$ denote the restriction of T to $\mathcal{N}$. Show that $T_{\mathcal{N}}$ is a self-adjoint, compact operator on the Hilbert space $\mathcal{N}$.

6.25 Let $\{\lambda_n\}$ be a sequence of non-zero real numbers which tend to zero. Show that if $\mathcal{H}$ is an infinite-dimensional Hilbert space then there exists a self-adjoint, compact operator on $\mathcal{H}$ whose set of non-zero eigenvalues is the set $\{\lambda_n\}$.
[Hint: see Exercise 6.11.]

6.26 If $S \in B(\mathcal{H})$ is a positive, compact operator then Lemma 5.46 shows that any eigenvalue of S is positive, while Theorem 5.58 constructs a positive square root R of S. Using the notation of Theorem 6.34, define $\widetilde{R} \in L(\mathcal{H})$ by

$$\widetilde{R}x=\sum_{n=1}^{r(S)}\sqrt{\lambda_n}(x,e_n)e_n.$$

Show that $\widetilde{R}$ is a positive, compact square root of S (and so equals R by the uniqueness result in Theorem 5.58). Are there other square roots of S? If so, are they positive?

6.27 (Singular value decomposition.) Recall that if $S \in B(\mathcal{H})$ is a compact operator (S need not be self-adjoint) then the operator S^*S is positive, $\operatorname{Ker} S^*S = \operatorname{Ker} S$ and $r(S) = r(S^*S)$, by Example 5.47 and Exercises 5.5 and 6.6. Let $\{\lambda_n\}_{n=1}^{r(S)}$ be the non-zero (hence positive) eigenvalues of the positive operator S^*S, and let $\{e_n\}_{n=1}^{r(S)}$ be the corresponding orthonormal set of eigenvectors. For $n = 1, \ldots, r(S)$, let $\mu_n = \sqrt{\lambda_n}$ (taking the positive square roots). Show that there exists an orthonormal set $\{f_n\}_{n=1}^{r(S)}$, in $\mathcal{H}$, such that

$$Se_n = \mu_n f_n, \quad S^* f_n = \mu_n e_n, \quad n = 1, \ldots, r(S) \tag{6.17}$$

and

$$Sx = \sum_{n=1}^{r(S)} \mu_n (x, e_n) f_n, \tag{6.18}$$

for any $x \in \mathcal{H}$. The real numbers μ_n, $n = 1, \ldots, r(S)$, are called the *singular values* of S and the formula (6.18) is called the *singular value decomposition* of S (if 0 is an eigenvalue of S^*S then 0 is also a singular value of S, but this does not contribute to the singular value decomposition (6.18)).

Show that if S is self-adjoint then the non-zero singular values of S are the absolute values of the eigenvalues of S.

6.28 It follows from Exercise 6.27 that any compact operator S on a Hilbert space $\mathcal{H}$ has the form discussed in Exercises 6.11 and 6.12. In fact, the operators discussed there seem to be more general since the numbers α_n may be complex, whereas the numbers α_n in Exercise 6.27 are real and positive. Are these operators really more general?

6.29 Obtain the singular value decomposition of the operator on $\mathbb{C}^2$ represented by the matrix (6.9).

6.30 Let $\mathcal{H}$ be an infinite-dimensional Hilbert space with an orthonormal basis $\{g_n\}$, and let S be a Hilbert–Schmidt operator on $\mathcal{H}$ (see Definition 6.15), with singular values $\{\mu_n\}_{n=1}^{r(S)}$ (where $r(S)$ may be finite or infinite). Show that

$$\sum_{n=1}^{r(S)} \|Sg_n\|^2 = \sum_{n=1}^{r(S)} \mu_n^2.$$

6.31 Show that if S is a compact operator on a Hilbert space $\mathcal{H}$ with infinite rank then $\operatorname{Im} S$ is not closed.
[Hint: use Exercises 6.12 and 6.27.]

7 Integral and Differential Equations

In this chapter we consider two of the principal areas of application of the theory of compact operators from Chapter 6. These are the study of integral and differential equations. Integral equations give rise very naturally to compact operators and so the theory can be applied almost immediately to such equations. On the other hand, as we have seen before, differential equations tend to give rise to unbounded linear transformations, so the theory of compact operators cannot be applied directly. However, with a bit of effort the differential equations can be transformed into certain integral equations whose corresponding linear operators are compact. In effect, we construct compact integral operators by "inverting" the unbounded differential linear transformations and we apply the theory to these integral operators. Thus, in a sense, the theory of differential equations which we will consider is a consequence of the theory of integral equations. We therefore consider integral equations first.

7.1 Fredholm Integral Equations

We have already looked briefly at integral operators acting in the Banach space $X = C[a,b]$, a, $b \in \mathbb{R}$, in Examples 4.7 and 4.41. However, we will now consider such operators more fully, using also the space $\mathcal{H} = L^2[a,b]$. For many purposes here the space $\mathcal{H}$ is more convenient than X because $\mathcal{H}$ is a Hilbert space rather than a Banach space, so all the results of Chapter 6 are available in this setting. Throughout this chapter $(\cdot\,,\cdot)$ will denote the usual inner product on

$L^2[a,b]$. However, to distinguish between the norms on $\mathcal{H}$ and X, these will be denoted by $\|\cdot\|_{\mathcal{H}}$ and $\|\cdot\|_X$ respectively. We use a similar notation for norms of operators on these spaces. We will also need the following linear subspaces of X. For $k = 1, 2, \ldots,$ let

$$C^k[a,b] = \{u \in C[a,b] : u^{(n)} \in C[a,b],\ n = 1,\ldots,k\},$$

where $u^{(n)}$ denotes the nth derivative of u. These subspaces can be given norms which turn them into Banach spaces in their own right, but we will merely consider them as subspaces of X.

We note that, since we will be using the spectral theory from Chapter 6, it will be assumed throughout this chapter, unless otherwise stated, that all the spaces used are complex. Of course, in many situations one has "real" equations and real-valued solutions are desired. In many cases it can easily be deduced (often by simply taking a complex conjugate) that the solutions we obtain in complex spaces are in fact real-valued. An example of such an argument is given in Exercise 7.1. Similar arguments work for the equations discussed in the following sections, but for brevity we will not consider this further.

We define the sets

$$\begin{aligned} R_{a,b} &= [a,b] \times [a,b] \subset \mathbb{R}^2, \\ \Delta_{a,b} &= \{(s,t) \in R_{a,b} : t \le s\} \end{aligned}$$

(geometrically, $R_{a,b}$ is a rectangle in the plane $\mathbb{R}^2$ and $\Delta_{a,b}$ is a triangle in $R_{a,b}$). Suppose that $k : R_{a,b} \to \mathbb{C}$ is continuous. For each $s \in [a,b]$ we define the function $k_s \in X$ by $k_s(t) = k(s,t)$, $t \in [a,b]$ (see the solution of Example 4.7). Also, we define numbers M, N by

$$\begin{aligned} M &= \max\{|k(s,t)| : (s,t) \in R_{a,b}\}, \\ N^2 &= \int_a^b \int_a^b |k(s,t)|^2\, ds\, dt = \int_a^b \Big\{ \int_a^b |k_s(t)|^2\, dt \Big\}\, ds = \int_a^b \|k_s\|_{\mathcal{H}}^2\, dt. \end{aligned}$$

Now, for any $u \in \mathcal{H}$, define a function $f : [a,b] \to \mathbb{C}$ by

$$f(s) = \int_a^b k(s,t)u(t)\, dt. \tag{7.1}$$

Lemma 7.1

For any $u \in \mathcal{H}$ the function f defined by (7.1) belongs to $X \subset \mathcal{H}$. Furthermore,

$$\|f\|_X \le M(b-a)^{1/2}\|u\|_{\mathcal{H}}, \tag{7.2}$$

$$\|f\|_{\mathcal{H}} \le N\|u\|_{\mathcal{H}}. \tag{7.3}$$

Proof

Suppose that $\epsilon > 0$ and $s \in [a,b]$. Choosing $\delta > 0$ as in the solution of Example 4.7 and following that solution, we find that for any $s' \in [a,b]$, with $|s - s'| < \delta$,

$$\begin{aligned}|f(s) - f(s')| &\leq \int_a^b |k(s,t) - k(s',t)||u(t)|\,dt \leq \|k_s - k_{s'}\|_{\mathcal{H}}\|u\|_{\mathcal{H}} \\ &\leq \epsilon(b-a)^{1/2}\|u\|_{\mathcal{H}},\end{aligned}$$

by the Cauchy–Schwarz inequality (note that we are using the $L^2[a,b]$ norm here rather than the $C[a,b]$ norm used in Example 4.7). This shows that f is continuous. A similar calculation shows that

$$|f(s)| \leq \int_a^b |k(s,t)||u(t)|\,dt \leq \|k_s\|_{\mathcal{H}}\|u\|_{\mathcal{H}},$$

from which we can derive (7.2), and also

$$\int_a^b |f(s)|^2\,ds \leq \|u\|_{\mathcal{H}}^2 \int_a^b \|k_s\|_{\mathcal{H}}^2\,ds,$$

from which we obtain (7.3). □

By Lemma 7.1 we can now define an operator $K : \mathcal{H} \to \mathcal{H}$ by putting $Ku = f$, for any $u \in \mathcal{H}$, where f is defined by (7.1). It can readily be verified that K is linear and, by Lemma 7.1, K is bounded with

$$\|K\|_{\mathcal{H}} \leq N.$$

The operator K is called a *Fredholm integral operator* (or simply an *integral operator*), and the function k is called the *kernel* of the operator K. This is a different usage of the term "kernel" to its previous use to denote the kernel (null-space) of a linear operator. Unfortunately, both these terminologies are well established. To avoid confusion, from now on we will use the term "kernel" to mean the above kernel function of an integral operator and we will avoid using it in the other sense. However, we will continue to use the notation $\operatorname{Ker} T$ to denote the null-space of an operator T.

If we regard f as known and u as unknown then (7.1) is called an *integral equation.* In fact, this type of equation is known as a *first kind Fredholm integral equation.* A *second kind Fredholm integral equation* is an equation of the form

$$f(s) = u(s) - \mu \int_a^b k(s,t)u(t)\,dt, \quad s \in [a,b], \tag{7.4}$$

where $0 \neq \mu \in \mathbb{C}$. Equations (7.1) and (7.4) can be written in operator form as

$$Ku = f, \tag{7.5}$$

$$(I - \mu K)u = f. \tag{7.6}$$

It will be seen below that there is an extensive and satisfactory theory of solvability of second kind equations, based on the results of Chapter 6, while the theory for first kind equations is considerably less satisfactory.

Remark 7.2

In the $L^2[a,b]$ setting it is not necessary for the kernel k to be a continuous function. Assuming that k is square integrable on the region $R_{a,b}$ would suffice for most of the results below, and many such operators arise in practice. However, Lemma 7.1 would then be untrue, and rather more Lebesgue measure and integration theory than we have outlined in Chapter 1 would be required even to show that the formula (7.1) defines a reasonable (measurable) function f in this case. It would also be more difficult to justify the various integral manipulations which will be performed below. Since all our applications will have continuous kernels we will avoid these purely measure-theoretic difficulties by assuming throughout that k is continuous. Standard Riemann integration theory will then (normally) suffice. This assumption also has a positive benefit (in addition to avoiding Lebesgue integration theory) in that it will enable us to prove Theorem 7.12 below, which would be untrue for general L^2 kernels k, and which will strengthen some results in the following sections on differential equations.

One of the most important properties of K is that it is compact. Once we have shown this we can apply the theory from Chapter 6 to the operator K and to equations (7.5) and (7.6).

Theorem 7.3

The integral operator $K : \mathcal{H} \to \mathcal{H}$ is compact.

Proof

We will show that K is a Hilbert–Schmidt operator, and hence is compact (see Exercise 6.7). Let $\{e_n\}$ be an orthonormal basis for $\mathcal{H}$ (such a basis exists, see Theorem 3.54 for instance). For each $s \in [a,b]$ and $n \in \mathbb{N}$,

$$(Ke_n)(s) = \int_a^b k(s,t)e_n(t)\,dt = (k_s, \overline{e}_n),$$

where $\overline{e}_n$ is the complex conjugate of e_n. By Theorem 3.47, the sequence $\{\overline{e}_n\}$ is also an orthonormal basis and

$$\sum_{n=1}^{\infty} \|Ke_n\|_{\mathcal{H}}^2 = \sum_{n=1}^{\infty} \int_a^b |(k_s, \overline{e}_n)|^2 \, ds = \int_a^b \sum_{n=1}^{\infty} |(k_s, \overline{e}_n)|^2 \, ds$$
$$= \int_a^b \|k_s\|_{\mathcal{H}}^2 \, ds < \infty,$$

which completes the proof. We note that, by Lemma 7.1, each term $|(k_s, \overline{e}_n)|^2$ is continuous in s and non-negative, so the above analytic manipulations can be justified using Riemann integration. □

In Chapter 6 the adjoint of an operator played an important role. We now describe the adjoint, K^*, of the operator K.

Theorem 7.4

The adjoint operator $K^* : \mathcal{H} \to \mathcal{H}$ of K is given by the formula

$$(K^*v)(t) = \int_a^b \overline{k(s,t)} v(s) \, ds,$$

for any $v \in \mathcal{H}$.

Proof

For any $u, v \in \mathcal{H}$,

$$(Ku, v) = \int_a^b \Big\{ \int_a^b k(s,t)u(t) \, dt \Big\} \overline{v(s)} \, ds = \int_a^b u(t) \overline{\Big\{ \int_a^b \overline{k(s,t)} v(s) \, ds \Big\}} \, dt$$
$$= (u, K^*v),$$

by the definition of the adjoint (see also the remark below). Since u and v are arbitrary this proves the result. □

Remark 7.5

The change of order of the integrations in the above proof is not trivial for general functions $u, v \in L^2[a, b]$. It follows from a general theorem in Lebesgue integration called Fubini's theorem, see Theorem 6.7 in [4] or Theorem 2.4.17 in [8]. However, we can avoid using Fubini's theorem by the following method, which is commonly used in proving results in L^2 and other spaces. Let X be a normed space and suppose that we can choose a dense subset $Y \subset X$ consisting of "nice" elements y for which it is relatively easy to prove the required formula. We then extend this to the whole space by letting $x \in X$ be arbitrary and choosing a sequence $\{y_n\}$ in Y with $y_n \to x$ and taking limits in the formula. Assuming that the formula has suitable continuity properties this procedure will be valid (see Corollary 1.29). For instance, in the above case the dense set we choose is the set of continuous functions (by Theorem 1.61 this set is dense in $L^2[a,b]$). For continuous functions u, v the above proof only requires the theory of Riemann integration. Now, for any u, $v \in \mathcal{H}$, choose sequences of continuous functions $\{u_n\}$, $\{v_n\}$ such that $u_n \to u$, $v_n \to v$. The desired result holds for all u_n, v_n, so taking the limit and using the continuity of the integral operators and the inner product then gives the result for u, $v \in \mathcal{H}$. Of course we have not entirely avoided the use of Lebesgue integration theory by this method since we needed to know that the set of continuous functions is dense in $\mathcal{H}$ (Theorem 1.61).

Definition 7.6

If $k(s,t) = \overline{k(t,s)}$, for all s, $t \in [a,b]$, then k is *Hermitian*. If k is real-valued and Hermitian then k is *symmetric*.

Corollary 7.7

If k is Hermitian (or symmetric) then the integral operator K is self-adjoint.

We can now apply the general results of Chapter 6 to equations (7.5) and (7.6). We begin by considering the first kind equation (7.5). Since K is compact it follows from Corollary 6.7 that K is not invertible in $B(\mathcal{H})$. Thus, in the terminology of Chapter 6, equation (7.5) is not well-posed, that is, for a given $f \in \mathcal{H}$, the equation may not have a solution, or, if it does, this solution may not depend continuously on f. First kind equations are not completely devoid of interest and they do arise in certain applications. However, their theory is considerably more delicate than that of second kind equations and we will not consider them further here.

Next we consider the second kind equation (7.6). We first observe that, although it is conventional to write equation (7.6) in the above form, the position of the parameter μ in (7.6) is slightly different to that of λ in the equations considered in Chapter 6, see (6.1) and (6.2) in particular. We thus make the following definition.

Definition 7.8

Let V be a vector space and $T \in L(V)$. A scalar $\mu \in \mathbb{F}$ is a *characteristic value* of T if the equation $v - \mu Tv = 0$ has a non-zero solution $v \in V$. Characteristic values of matrices are defined similarly.

Note that, for any T, the point $\mu = 0$ cannot be a characteristic value of T, and $\mu \neq 0$ is a characteristic value of T if and only if $\lambda = \mu^{-1}$ is an eigenvalue of T. Thus there is a one-to-one correspondence between characteristic values and non-zero eigenvalues of T.

Now, the homogeneous version of (7.6) can be written as $(I - \mu K)u = 0$ so μ is a characteristic value of K if this equation has a non-zero solution u (note that here, "non-zero" u means non-zero as an element of X or $\mathcal{H}$, that is, $u \not\equiv 0$ – it does not mean that $u(s) \neq 0$ for all $s \in [a, b]$). We can therefore explain the distinction between first and second kind equations by the observation that the first kind equation (7.5) corresponds to the ("bad") case $\lambda = 0$ in Chapter 6, while the second kind equation (7.6) corresponds to the ("good") case $\lambda \neq 0$.

With these remarks in mind, the results of Chapter 6 can easily be applied to equation (7.6). In particular, Theorem 6.26 yields the following result.

Theorem 7.9

For any fixed $\mu \in \mathbb{C}$ the Fredholm alternative holds for the second kind Fredholm integral equation (7.6). That is, either (a) μ is not a characteristic value of K and the equation has a unique solution $u \in \mathcal{H}$ for any given $f \in \mathcal{H}$; or (b) μ is a characteristic value of K and the corresponding homogeneous equation has non-zero solutions, while the inhomogeneous equation has (non-unique) solutions if and only if f is orthogonal to the subspace $\operatorname{Ker}(I - \overline{\mu} K^*)$. Furthermore, if the kernel k is Hermitian then the results of Section 6.3 also apply to (7.6).

Remark 7.10

The dichotomy between first and second kind equations (that is, between $\lambda = 0$ and $\lambda \neq 0$) may still seem somewhat mysterious. Another way of seeing why there is such a distinction is to note that, by Lemma 7.1, the range of K consists of continuous functions. Clearly, therefore, equation (7.5) cannot have a solution if f is discontinuous. Now, although the set of continuous functions is dense in $L^2[a, b]$, there is a well-defined sense in which there are more discontinuous than continuous functions in $L^2[a, b]$ (it is very easy to turn a continuous function into a discontinuous function – just change the value at a single point; it is not so easy to go the other way in general). Thus (7.5) is only solvable for a "small", but dense, set of functions f. This is reflected in the lack of well-posedness of this equation. On the other hand, rearranging (7.6) merely leads to the necessary condition that the difference $f-u$ must be continuous (even though f and u need not be continuous individually). This clearly does not prevent solutions existing, it merely tells us something about the solution, and so is consistent with the above solvability theory.

Example 7.11

We illustrate the above results by considering the second kind Fredholm equation

$$u(s) - \mu \int_0^1 e^{s-t} u(t)\, dt = f(s), \tag{7.7}$$

for $s \in [0,1]$ and some constant $\mu \neq 0$. By rearranging this equation it is clear that any solution u has the form

$$u(s) = f(s) + \Big(\mu \int_0^1 e^{-t} u(t)\, dt \Big) e^s = f(s) + ce^s, \tag{7.8}$$

where c is an unknown (at present) constant. Substituting this into (7.7) we find that

$$c(1-\mu) = \mu \int_0^1 e^{-t} f(t)\, dt. \tag{7.9}$$

Now suppose that $\mu \neq 1$. Then c is uniquely determined by (7.9), and so (7.7) has the unique solution

$$u(s) = f(s) + \frac{\mu \int_0^1 e^{-t} f(t)\, dt}{1-\mu} e^s.$$

On the other hand, if $\mu = 1$, then (7.9) has no solution unless

$$\int_0^1 e^{-t} f(t)\, dt = 0, \tag{7.10}$$

in which case any $c \in \mathbb{C}$ satisfies (7.9), and the formula (7.8), with arbitrary c, provides the complete set of solutions of (7.7). In this case, the homogeneous equation corresponding to the adjoint operator is

$$v(t) - \int_0^1 e^{s-t} v(s)\, ds = 0,$$

for $s \in [0,1]$, and it can be shown, by similar arguments, that any solution of this equation is a scalar multiple of the function $v(t) = e^{-t}$, that is, the set of solutions of this equation is a 1-dimensional subspace spanned by this function. Thus the condition (7.10) coincides with the solvability condition given by the Fredholm alternative.

So far we have only considered solutions of equation (7.6) in the space $\mathcal{H}$, primarily because this space is a Hilbert space and the general theory works better in Hilbert spaces. However, in particular problems the Banach space $X = C[a, b]$ may be more natural. We will show that, when k is continuous, solvability results in X can be deduced from those in $\mathcal{H}$. We first note that by Lemma 7.1, $Ku \in X$, for all $u \in X$, so we can also regard K as an operator from X to X. When it is necessary to emphasize the distinction, we will let $K_{\mathcal{H}}$, K_X, denote the operator K considered in the spaces $\mathcal{H}$ and X, respectively. This distinction is not trivial. It is conceivable that the equation $u = \mu K u$ could have a non-zero solution $u \in \mathcal{H}$, but no non-zero solution in X, that is, $K_{\mathcal{H}}$ might have more characteristic values than K_X (any non-zero solution $u \in X$ is automatically in $\mathcal{H}$, so any characteristic value of K_X is a characteristic value of $K_{\mathcal{H}}$). Also, equation (7.6) might only have a solution $u \in \mathcal{H} \backslash X$, that is, (7.6) could be solvable in $\mathcal{H}$ but not in X, even when $f \in X$. These possibilities can certainly occur for general L^2 kernels, but the following theorem shows that they cannot occur when k is continuous. We leave the proof to Exercise 7.4.

Theorem 7.12

Suppose that k is continuous, $\mu \neq 0$, $f \in X$ and $u \in \mathcal{H}$ is a solution of (7.6). Then:

(a) $u \in X$;

(b) μ is a characteristic value of $K_{\mathcal{H}}$ if and only if it is a characteristic value of K_X (thus it is unnecessary to distinguish between $K_{\mathcal{H}}$ and K_X when discussing the characteristic values);

(c) if μ is not a characteristic value of K then, for any $f \in X$, equation (7.6) has a unique solution $u \in X$, and there exists a constant $C > 0$ (independent of f) such that $\|u\|_X \leq C\|f\|_X$. Thus the operator $I - \mu K_X : X \to X$ is invertible.

EXERCISES

7.1 Suppose that k is a real-valued kernel and $\mu \in \mathbb{R}$ is not a characteristic value of K. Show that if the function f in equation (7.6) is real-valued then the unique solution u given by Theorem 7.9 is real-valued. This can be interpreted as a solvability result in the spaces $L^2_{\mathbb{R}}[a, b]$ or $C_{\mathbb{R}}[a, b]$ (with Theorem 7.12).

7.2 (Degenerate kernels) A kernel k having the form

$$k(s,t) = \sum_{j=1}^{n} p_j(s)q_j(t),$$

is said to be a *degenerate kernel*. We assume that p_j, $q_j \in X$, for $1 \leq j \leq n$ (so k is continuous), and the sets of functions $\{p_1, \ldots, p_n\}$, $\{q_1, \ldots, q_n\}$, are linearly independent (otherwise the number of terms in the sum for k could simply be reduced). If K is the corresponding operator on X, show that if (7.6) has a solution u then it must have the form

$$u = f + \mu \sum_{j=1}^{n} \alpha_j p_j, \qquad (7.11)$$

for some $\alpha_1, \ldots, \alpha_n$ in $\mathbb{C}$. Now, letting $\alpha = (\alpha_1, \ldots, \alpha_n)$, show that α must satisfy the matrix equation

$$(I - \mu W)\alpha = \beta, \qquad (7.12)$$

where the elements of the matrix $W = [w_{ij}]$ and the vector $\beta = (\beta_1, \ldots, \beta_n)$ are given by

$$w_{ij} = \int_a^b q_i(t)p_j(t)\,dt, \quad \beta_j = \int_a^b q_j(t)f(t)\,dt.$$

Deduce that the set of characteristic values of K is equal to the set of characteristic values of the matrix W, and if μ is not a characteristic value of W then equation (7.6) is uniquely solvable and the solution u can be constructed by solving the matrix equation (7.12) and using the formula (7.11).

7.3 Consider the equation

$$u(s) - \mu \int_0^1 (s+t)u(t)\,dt = f(s), \quad s \in [0,1].$$

Use the results of Exercise 7.2 to show that the characteristic values of the integral operator in this equation are the roots of the equation $\mu^2 + 12\mu - 12 = 0$. Find a formula for the solution of the equation, for general f, when $\mu = 2$.

7.4 Prove Theorem 7.12.
[Hint: for (a), rearrange equation (7.6) and use Lemma 7.1, and for (c), the Fredholm alternative results show that $u \in \mathcal{H}$ exists and $\|u\|_{\mathcal{H}} \leq C\|f\|_{\mathcal{H}}$, for some constant $C > 0$. Use this and (7.2) to obtain the required inequality.]

7.2 Volterra Integral Equations

In this section we consider a special class of integral operators K having the form

$$(Ku)(s) = \int_a^s k(s,t)u(t)\,dt, \quad s \in [a,b], \tag{7.13}$$

where the upper limit of the integral in the definition of K is variable and the kernel $k : \Delta_{a,b} \to \mathbb{C}$ is continuous. When K has the form (7.13) it is said to be a *Volterra integral operator*, and the corresponding equations (7.5) and (7.6) are said to be first and second kind *Volterra integral equations* (rather than Fredholm equations).

Volterra operators can be regarded as a particular type of Fredholm operator by extending the definition of the kernel k from the set $\Delta_{a,b}$ to the set $R_{a,b}$ by defining $k(s,t) = 0$ when $t > s$. Unfortunately, this extended kernel will, in general, be discontinuous along the line $s = t$ in $R_{a,b}$. Thus, since the proofs in Section 7.1 relied on the continuity of k on the set $R_{a,b}$, we cannot simply assert that these results hold for Volterra operators. However, the proofs of all the results in Section 7.1 can be repeated, using the Volterra form of operator (7.13), and we find that these results remain valid for this case. This is because the integrations that arise in this case only involve the values of k on the set $\Delta_{a,b}$, where we have assumed that k is continuous, so in effect the kernel has the continuity properties required in Section 7.1. In fact, as was mentioned in Remark 7.2, most of the results in Section 7.1 are actually valid for L^2 kernels, and the extended Volterra type kernel is certainly in L^2, so in this more general setting the Volterra operators are special cases of the Fredholm operators. The numbers M and N are defined here as in Section 7.1, using the extended kernel. We will say that K has *continuous kernel* when k is continuous on the set $\Delta_{a,b}$ (irrespective of the continuity of the extended kernel).

The solvability theory for second kind Volterra equations is even simpler than for the corresponding Fredholm equations since, as will be shown below, such operators have no non-zero eigenvalues, that is, only one of the alternatives of the Fredholm alternative can hold and, for any μ, equation (7.6) is solvable for all $f \in \mathcal{H}$. We consider solutions in $\mathcal{H}$ here, and use the $\mathcal{H}$ norm – solutions in X will be considered in the exercises.

Lemma 7.13

If K is a Volterra integral operator on $\mathcal{H}$ then there exists a constant $C > 0$ such that $\|K^n\|_{\mathcal{H}} \le C^n/n!$, for any integer $n \ge 1$.

Proof

For any $u \in \mathcal{H}$ we have, using the Cauchy–Schwarz inequality,

$$\begin{aligned}|(Ku)(s)| &\leq \int_a^s |k_s(t)||u(t)|\,dt \leq \Big(\int_a^s |k_s(t)|^2\,dt\Big)^{1/2}\|u\|_{\mathcal{H}}\\ &\leq M(b-a)^{1/2}\|u\|_{\mathcal{H}},\end{aligned}$$

$$|(K^2u)(s)| \leq \int_a^s |k_s(t)||(Ku)(t)|\,dt \leq (s-a)M^2(b-a)^{1/2}\|u\|_{\mathcal{H}}.$$

Now by induction, using a similar calculation to that for the second inequality above, we can show that for each $n \geq 2$,

$$|(K^nu)(s)| \leq \int_a^s |k_s(t)||(K^{n-1}u)(t)|\,dt \leq \frac{(s-a)^{n-1}}{(n-1)!}M^n(b-a)^{1/2}\|u\|_{\mathcal{H}}$$

(the second inequality above yields the case $n = 2$). By integration we obtain

$$\|K^nu\|_{\mathcal{H}} \leq \frac{M^n(b-a)^n}{(n-1)!\sqrt{2n-1}}\|u\|_{\mathcal{H}} \leq \frac{C^n}{n!}\|u\|_{\mathcal{H}},$$

for some constant C, from which the result follows immediately. □

Theorem 7.14

A Volterra integral operator K on $\mathcal{H}$ has no non-zero eigenvalues. Hence, for such a K, equation (7.6) has a unique solution $u \in \mathcal{H}$ for all $\mu \in \mathbb{C}$ and $f \in \mathcal{H}$.

Proof

Suppose that λ is an eigenvalue of K, that is, $Ku = \lambda u$, for some $u \neq 0$. Then, by Lemma 7.13, for all integers $n \geq 1$,

$$|\lambda|^n\|u\|_{\mathcal{H}} = \|K^nu\|_{\mathcal{H}} \leq \|K^n\|_{\mathcal{H}}\|u\|_{\mathcal{H}} \leq \frac{C^n}{n!}\|u\|_{\mathcal{H}},$$

which implies that $|\lambda| = 0$. □

Remark 7.15

A bounded operator T on a general Banach space Y is said to be *nilpotent* if $T^n = 0$ for some $n \geq 1$, and *quasi-nilpotent* if $\|T^n\|^{1/n} \to 0$ as $n \to \infty$. Lemma 7.13 shows that any Volterra integral operator is quasi-nilpotent. The proof of Theorem 7.14 actually shows that any quasi-nilpotent operator cannot have non-zero eigenvalues so, in particular, a compact, quasi-nilpotent operator such as a Volterra integral operator cannot have any non-zero points in its spectrum (since all non-zero points in the spectrum are eigenvalues). In fact, with some more effort it can be shown that all quasi-nilpotent operators have no non-zero points in their spectrum.

EXERCISES

7.5 Consider the Volterra integral operator K on X defined by

$$(Ku)(s) = \int_a^s u(t)\, dt, \quad s \in [a, b],$$

for $u \in X$. Show that for this operator the first kind equation (7.5) has a solution $u = u(f) \in X$ if and only if $f \in C^1[a, b]$. Show also that the solution does not depend continuously on f, in other words, for any $C > 0$, the inequality $\|u(f)\|_X \leq C\|f\|_X$ does not hold for all $f \in C^1[a, b]$.

7.6 Let S be a quasi-nilpotent operator on a Banach space Y (S need not be compact). Show that the operator $I - S$ is invertible. [Hint: see the proof of Theorem 4.40.]

Deduce that there is no non-zero point in the spectrum of S.

Show that any Volterra integral operator K with continuous kernel k is quasi-nilpotent on the space X. Deduce that, for this K, equation (7.6) has a unique solution $u \in X$ for all $f \in X$ (these results are the analogue, in X, of Lemma 7.13 and Corollary 7.14 in $\mathcal{H}$).

7.3 Differential Equations

We now consider the application of the preceding theory to differential equation problems. The primary difficulty with applying the above operator theory to differential equations is that the differential operators which arise are generally unbounded. There are two main ways to overcome this difficulty. One way would be to develop a theory of unbounded linear transformations – this can be done but is rather complicated and too long for this book. The other way is to reformulate the differential equation problems, which lead to unbounded transformations, as integral equation problems, which lead to bounded integral operators. This method will be described here. We will only consider second order, ordinary differential equations, but the methods can be extended to much more general problems.

The first type of problem we consider is the *initial value problem.* Consider the differential equation

$$u''(s) + q_1(s)u'(s) + q_0(s)u(s) = f(s), \tag{7.14}$$

on an interval $[a, b]$, where q_0, q_1, g are (known) continuous functions on $[a, b]$, together with the *initial conditions*

$$u(a) = \alpha_0, \quad u'(a) = \alpha_1, \tag{7.15}$$

for some constants α_0, $\alpha_1 \in \mathbb{C}$. A *solution* of the initial value problem is an element $u \in C^2[a, b]$ satisfying (7.14) and (7.15). In the following two lemmas we will show that this problem is equivalent, in a certain sense, to a Volterra integral equation.

Let $X = C[a, b]$ and let Y_i be the set of functions $u \in C^2[a, b]$ satisfying (7.15). For any $u \in Y_i$, let $T_i u = u''$ (since $u \in C^2[a, b]$ this definition makes sense and $T_i u \in X$). It can easily be verified that T_i is a linear transformation, that is $T_i \in L(Y_i, X)$, but T_i is not bounded (see Example 4.8 for a similar example of a linear transformation which is not bounded).

Lemma 7.16

The transformation $T_i \in L(Y_i, X)$ is bijective.

Proof

Suppose that $u \in Y_i$ and let $w = T_i u \in X$. Integrating w and using the conditions (7.15) we obtain

$$u'(s) = \alpha_1 + \int_a^s w(t)\,dt, \tag{7.16}$$

$$\begin{aligned} u(s) &= \alpha_0 + \alpha_1(s-a) + \int_a^s \Big\{ \int_a^r w(t)\,dt \Big\}\,dr \\ &= \alpha_0 + \alpha_1(s-a) + \int_a^s \Big\{ \int_t^s dr \Big\}\,w(t)\,dt \\ &= \alpha_0 + \alpha_1(s-a) + \int_a^s (s-t)w(t)\,dt. \end{aligned} \tag{7.17}$$

Thus, for any $u \in Y_i$ we can express u in terms of $w = T_i u$ by the final line of (7.17). Now, suppose that u_1, $u_2 \in Y_i$ satisfy $T_i u_1 = T_i u_2 = w$, for some $w \in X$. Then (7.17) shows that $u_1 = u_2$, so T_i is injective.

Next, suppose that $w \in X$ is arbitrary and define u by the final line of (7.17). Then, by reversing the above calculations, we see that u' is given by (7.16) and $u'' = w$, so $u \in C^2[a,b]$, and also u satisfies (7.15). Thus, starting from arbitrary $w \in X$ we have constructed $u \in Y_i$ such that $T_i u = w$. It follows that T_i is surjective, and hence by the previous result, T_i must be bijective. □

Lemma 7.17

The function $u \in Y_i$ is a solution of the initial value problem (7.14), (7.15) if and only if $w = T_i u \in X$ satisfies the Volterra equation.

$$w(s) - \int_a^s k(s,t)w(t)\,dt = h(s), \tag{7.18}$$

where

$$\begin{aligned} k(s,t) &= -q_1(s) - q_0(s)(s-t), \\ h(s) &= f(s) - (\alpha_0 + \alpha_1(s-a))q_0(s) - \alpha_1 q_1(s). \end{aligned}$$

Proof

See Exercise 7.7. □

Lemma 7.17 shows that the initial value problem (7.14), (7.15), is equivalent to the second kind Volterra integral equation (7.18), via the linear transformation T_i, in the sense that if either problem has a solution then the other problem also has a solution, and these solutions are related by the bijective transformation T_i. Using this result we can now show that the initial value problem is uniquely solvable for all $f \in X$.

Theorem 7.18

The initial value problem (7.14), (7.15) has a unique solution $u \in X$ for any α_0, $\alpha_1 \in \mathbb{C}$ and any $f \in X$.

Proof

It follows from Exercise 7.6 that equation (7.18) has a unique solution $w \in X$ for any $f \in X$. Hence, the theorem follows from the equivalence between the initial value problem and (7.18) proved in Lemma 7.17. □

The initial value problem imposed two conditions on the solution u at the single point a. Another important type of problem arises when a single condition is imposed at each of the two points a, b. We consider the conditions

$$u(a) = u(b) = 0. \tag{7.19}$$

The problem (7.14), (7.19) is called a *boundary value problem*, and the conditions (7.19) are called *boundary conditions.* A *solution* of the boundary value problem is an element $u \in C^2[a,b]$ satisfying (7.14) and (7.19). We will treat this boundary value problem by showing that it is equivalent to a second kind Fredholm equation. To simplify the discussion we suppose that the function $q_1 = 0$, and we write $q_0 = q$, that is, we consider the equation

$$u''(s) + q(s)u(s) = f(s), \tag{7.20}$$

together with (7.19).

Let Y_b be the set of functions $u \in C^2[a,b]$ satisfying (7.19), and define $T_b \in L(Y_b, X)$ by $T_b u = u''$.

Lemma 7.19

The transformation $T_b \in L(Y_b, X)$ is bijective. If $w \in X$ then the solution $u \in Y_b$ of the equation $T_b u = w$ is given by

$$u(s) = \int_a^b g_0(s,t)w(t)\,dt, \tag{7.21}$$

where

$$g_0(s,t) = \begin{cases} -\dfrac{(s-a)(b-t)}{b-a}, & \text{if } a \le s \le t \le b, \\[2ex] -\dfrac{(b-s)(t-a)}{b-a}, & \text{if } a \le t \le s \le b. \end{cases}$$

Proof

Suppose that $u \in Y_b$ and let $w = T_b u$. Then from (7.17) we see that

$$u(s) = \gamma(s-a) + \int_a^s (s-t)w(t)\,dt,$$

for some constant $\gamma \in \mathbb{C}$ (unknown at present). This formula satisfies the boundary condition at a. By substituting this into the boundary condition at b we obtain

$$0 = \gamma(b-a) + \int_a^b (b-t)w(t)\,dt,$$

and by solving this equation for γ we obtain

$$\begin{aligned}
u(s) &= -\frac{s-a}{b-a}\int_a^b (b-t)w(t)\,dt + \int_a^s (s-t)w(t)\,dt \\
&= \int_a^s \left\{ (s-t) - \frac{s-a}{b-a}(b-t) \right\} w(t)\,dt - \int_s^b \frac{(s-a)(b-t)}{b-a} w(t)\,dt \\
&= \int_a^b g_0(s,t)w(t)\,dt,
\end{aligned}$$

which is the formula for u in terms of w in the statement of the lemma. Given this formula the proof that T_b is bijective now proceeds as in the proof of Lemma 7.16. □

It is easy to see that the function g_0 in Lemma 7.19 is continuous on $R_{a,b}$. We let G_0 denote the integral operator with kernel g_0.

Lemma 7.20

If $u \in C^2[a,b]$ is a solution of (7.19), (7.20) then it satisfies the second kind Fredholm integral equation

$$u(s) + \int_a^b g_0(s,t)q(t)u(t)\,dt = h(s), \tag{7.22}$$

where

$$h(s) = \int_a^b g_0(s,t)f(t)\,dt.$$

Conversely, if $u \in C[a,b]$ satisfies (7.22) then $u \in Y_b$ and satisfies (7.19), (7.20).

Proof

See Exercise 7.7. □

The equivalence between the initial value problem and the integral equation described in Lemma 7.20 is somewhat more direct than the equivalence described in Lemma 7.17, in that the same function u satisfies both problems. However, it is important to note that we only need to assume that $u \in C[a,b]$ and satisfies (7.22) to conclude that $u \in C^2[a,b]$ and satisfies (7.19), (7.20). That is, we do not need to assume at the outset that $u \in C^2[a,b]$. This is crucial because when we use the results from Section 7.1 to try to solve the integral equation they will yield, at best, solutions in $C[a,b]$.

We cannot now use the integral equation to deduce that the boundary value problem has a unique solution for all $f \in X$ as we could for the initial value problem. In fact, it follows from the discussion in Section 7.4 below of eigenvalue problems for differential operators that if $q \equiv -\lambda$, for a suitable constant λ, then the homogeneous problem has non-zero solutions, so uniqueness certainly cannot hold in this case. However, we can show that the problem has a unique solution if the function q is sufficiently small.

Theorem 7.21

If $(b-a)^2\|q\|_X < 4$, then the boundary value problem (7.19), (7.20) has a unique solution u for any $f \in X$.

Proof

Let $Q : X \to X$ be the integral operator whose kernel is the function $g_0 q$ as in (7.22). Then it follows from Theorem 4.40 that equation (7.22) has a unique solution $u \in X$, for any $f \in X$, if $\|Q\|_X < 1$, and, from Example 4.7, this will be true if $(b-a)\max\{|g_0(s,t)q(t)|\} < 1$. Now, from the definition of g_0 and elementary calculus it can be shown that the maximum value of $(b-a)|g_0(s,t)|$ is $(b-a)^2/4$, so the required condition will hold if $(b-a)^2\|q\|_X < 4$. The theorem now follows from the above equivalence between the boundary value problem and (7.22). □

EXERCISES

7.7 Prove: (a) Lemma 7.17; (b) Lemma 7.20.

7.4 Eigenvalue Problems and Green's Functions

In Lemma 7.20 we showed that the boundary value problem (7.19), (7.20) is equivalent to the integral equation (7.22). Unfortunately, the integral operator in equation (7.22) is not in any sense an inverse of the differential problem, it merely converts a differential equation problem into an integral equation (that is, to find u we still need to solve an equation). However, with a bit more effort we can find an integral operator which will give us the solution u of the boundary value problem directly (that is, by means of an integration, without solving any equation).

To illustrate the ideas involved we first consider the following simple boundary value problem

$$\begin{aligned} u'' &= f, \\ u(a) &= u(b) = 0. \end{aligned} \tag{7.23}$$

As in Section 7.3, we let Y_b denote the set of functions $u \in C^2[a,b]$ satisfying $u(a) = u(b) = 0$, but we now let $T_0 : Y_b \to X$ denote the mapping defined by $T_0 u = u''$. It follows from Lemma 7.19 that for any $f \in X$ the solution of (7.23) is given directly by the formula $u = G_0 f$. Thus, in a sense, T_0 is "invertible", and its "inverse" is the integral operator G_0. Unfortunately, T_0 is not bounded and we have not defined inverses of unbounded linear transformations so we will not attempt to make this line of argument rigorous. We note, however, that G_0 is *not* the inverse of linear transformation from u to u'' alone, the boundary conditions are also crucially involved in the definition of T_0 and in the construction of G_0.

The integral operator G_0 is a *solution operator* (or *Green's operator*) for the boundary value problem (7.23); the kernel g_0 of this solution operator is called a *Green's function* for the problem (7.23).

We can also consider the eigenvalue problem associated with (7.23),

$$\begin{aligned} u'' &= \lambda u, \\ u(a) &= u(b) = 0. \end{aligned} \tag{7.24}$$

As usual, an *eigenvalue* of this problem is a number λ for which there exists a non-zero solution $u \in Y_b$, and any such solution is called an *eigenfunction* (in this context the term eigenfunction is commonly used rather than eigenvector).

We now consider what the integral equation approach can tell us about the problem (7.24). Putting $f = \lambda u$, we see that if (7.24) holds then

$$u = \lambda G_0 u, \tag{7.25}$$

and conversely, if (7.25) holds then (7.24) holds (note that to assert the converse statement from the above discussion we require that $u \in X$; however, it follows from Theorem 7.12 that if (7.25) holds for $u \in \mathcal{H}$ then necessarily $u \in X$ – hence it is sufficient for (7.25) to hold in either X or $\mathcal{H}$). Thus λ is an eigenvalue of (7.24) if and only if λ^{-1} is an eigenvalue of G_0 (the existence of G_0 shows that $\lambda = 0$ cannot be an eigenvalue of (7.24)), and the eigenfunctions of (7.24) coincide with those of G_0. Furthermore, the kernel g_0 is continuous and symmetric, so the operator G_0 is compact and self-adjoint in the space $\mathcal{H}$, so the results of Section 7.1, together with those of Chapter 6, show that there is a sequence of eigenvalues $\{\lambda_n\}$, and a corresponding orthonormal sequence of eigenfunctions $\{e_n\}$, and the set $\{e_n\}$ is an orthonormal basis for $\mathcal{H}$ (it will be proved below that $\operatorname{Ker} G_0 = \{0\}$, so by Corollary 6.35 there must be infinitely many eigenvalues).

In this simple situation the eigenvalues and eigenfunctions of the problem (7.24) can be explicitly calculated.

Example 7.22

Suppose that $a = 0$, $b = \pi$. By solving (7.24) we find that for each $n \geq 1$,

$$\lambda_n = -n^2, \quad e_n(s) = (2/\pi)^{1/2} \sin ns, \ s \in [0, \pi],$$

see Exercise 7.8 for the (well-known) calculations. Thus the eigenfunctions of (7.24) are exactly the functions in the basis considered in Theorem 3.56, and the eigenfunction expansion corresponds to the standard Fourier sine series expansion on the interval $[0, \pi]$.

Thus, qualitatively, the above results agree with the calculations in Example 7.22, although the integral equation does not tell us precisely what the eigenvalues or eigenfunctions are. However, combining the fact that the set of eigenfunctions is an orthonormal basis for $\mathcal{H}$ with the specific form of the eigenfunctions found in Example 7.22 gives us an alternative proof of Theorem 3.56 (at least, once we know that $\mathcal{H}$ is separable, which can be proved in its own right using Exercise 3.27). However, if this was all that could be achieved with the integral equation methods the theory would not seem worthwhile. Fortunately, these methods can be used to extend the above results to a much more general class of problems, for which the eigenvalues and eigenfunctions cannot be explicitly calculated

We will consider the boundary value problem

$$\begin{aligned} u'' + qu &= f, \\ u(a) = u(b) &= 0, \end{aligned} \tag{7.26}$$

where $q \in C_{\mathbb{R}}[a, b]$. This problem may still look rather restricted, but much more general boundary value problems can be transformed to this form by a suitable change of variables (see (7.31) and Lemma 7.30 below) so in fact we are dealing with a broad class of problems. The boundary value problem (7.26) and the more general one (7.31) are called *Sturm–Liouville* boundary value problems.

For this problem we will again construct a kernel g such that the corresponding integral operator G is a solution operator; again the kernel g will be called a *Green's function* for the problem. We will also consider the corresponding eigenvalue problem

$$\begin{aligned} u'' + qu &= \lambda u, \\ u(a) = u(b) &= 0, \end{aligned} \tag{7.27}$$

and we will obtain the spectral properties of this problem from those of G.

We first note that for a solution operator to exist it is necessary that there is no non-zero solution of the homogeneous form of problem (7.26), in other words, $\lambda = 0$ is not an eigenvalue of (7.27). We will now construct the Green's function for (7.26).

Let u_l, u_r be the solutions, on $[a, b]$, of the equation $u'' + qu = 0$, with the conditions

$$\begin{aligned} u_l(a) &= 0, \quad u_l'(a) = 1, \\ u_r(b) &= 0, \quad u_r'(b) = 1 \end{aligned} \tag{7.28}$$

(this is a pair of initial value problems, so by Theorem 7.18 these solutions exist – a trivial modification of the theorem is required in the case of u_r since the "initial values" for u_r are imposed at b rather than at a).

Lemma 7.23

If 0 is not an eigenvalue of (7.27) then there is a constant $\xi_0 \neq 0$ such that $u_l(s)u_r'(s) - u_r(s)u_l'(s) = \xi_0$ for all $s \in [a, b]$.

Proof

Define the function $h = u_l u_r' - u_r u_l'$. Then

$$h' = u_l u_r'' - u_r u_l'' = -q u_l u_r + q u_r u_l = 0$$

(by the definition of u_l and u_r), so h is constant, say $h(s) = \xi_0$ for all $s \in [a, b]$. Furthermore, from (7.28), $\xi_0 = \xi(a) = -u_r(a)$, and $u_r(a) \neq 0$, since otherwise u_r would be a non-zero solution of (7.27) with $\lambda = 0$, which would contradict the assumption that 0 is not an eigenvalue of (7.27). □

Theorem 7.24

If 0 is not an eigenvalue of (7.27) then the function

$$g(s,t) = \begin{cases} \xi_0^{-1} u_l(s) u_r(t), & \text{if } a \leq s \leq t \leq b, \\ \xi_0^{-1} u_r(s) u_l(t), & \text{if } a \leq t \leq s \leq b, \end{cases}$$

is a Green's function for the boundary value problem (7.26), that is, if G is the integral operator with kernel g and $f \in X$, then $u = Gf$ is the unique solution of (7.26).

Proof

Choose an arbitrary function $f \in X$ and let $u = Gf$. Then, using the definitions of g, u_l, u_r, and differentiating, we obtain, for $s \in [a,b]$,

$$\begin{aligned}
\xi_0 u(s) &= u_r(s)\int_a^s u_l(t)f(t)\,dt + u_l(s)\int_s^b u_r(t)f(t)\,dt,\\
\xi_0 u'(s) &= u_r'(s)\int_a^s u_l(t)f(t)\,dt + u_r(s)u_l(s)f(s) + u_l'(s)\int_s^b u_r(t)f(t)\,dt\\
&\qquad - u_l(s)u_r(s)f(s)\\
&= u_r'(s)\int_a^s u_l(t)f(t)\,dt + u_l'(s)\int_s^b u_r(t)f(t)\,dt,\\
\xi_0 u''(s) &= u_r''(s)\int_a^s u_l(t)f(t)\,dt + u_r'(s)u_l(s)f(s) + u_l''(s)\int_s^b u_r(t)f(t)\,dt\\
&\qquad - u_l'(s)u_r(s)f(s)\\
&= -q(s)u_r(s)\int_a^s u_l(t)f(t)\,dt - q(s)u_l(s)\int_s^b u_r(t)f(t)\,dt + \xi_0 f(s)\\
&= -q(s)\xi_0 u(s) + \xi_0 f(s).
\end{aligned}$$

Thus the function $u = Gf$ satisfies the differential equation in (7.26). It can readily be shown, using the above formula for $\xi_0 u(s)$ and the conditions (7.28), that u also satisfies the boundary conditions in (7.26), and so u is a solution of (7.26). Furthermore, it follows from the assumption that 0 is not an eigenvalue of (7.27) that the solution of (7.26) must be unique, so $u = Gf$ is the only solution of (7.26). Thus G is the solution operator for (7.26) and g is a Green's function. □

For the problem (7.23) it is clear that $u_l(s) = s - a$, $u_r(s) = s - b$ and $\xi_0 = -u_r(a) = b - a$, so that the Green's function g_0 has the above form.

We now relate the spectral properties of the problem (7.26) to those of the operator $G_{\mathcal{H}}$.

Lemma 7.25

Suppose that 0 is not an eigenvalue of (7.27). Then:

(a) $G_{\mathcal{H}}$ is compact and self-adjoint;

(b) $\operatorname{Ker} G_{\mathcal{H}} = \{0\}$, that is, 0 is not an eigenvalue of $G_{\mathcal{H}}$;

(c) λ is an eigenvalue of (7.27) if and only if λ^{-1} is an eigenvalue of $G_{\mathcal{H}}$;

(d) the eigenfunctions of (7.27) corresponding to λ coincide with the eigenfunctions of $G_{\mathcal{H}}$ corresponding to λ^{-1}.

Proof

It is clear that the Green's function g is continuous and symmetric, so part (a) follows from Theorem 7.3 and Corollary 7.7. Also, it follows from Theorem 7.24 that $\operatorname{Im} G_X = Y_b$, so $Y_b \subset \operatorname{Im} G_{\mathcal{H}}$. Since Y_b is dense in $\mathcal{H}$ (see Exercise 7.11), part (b) now follows from (6.8) and Lemma 3.29. Next, parts (c) and (d) clearly hold for the operator G_X, by Theorem 7.24, and it follows from Theorem 7.12 that the result also holds for the operator $G_{\mathcal{H}}$. □

The main spectral properties of the boundary value problem (7.27) can now be proved (even without the assumption that 0 is not an eigenvalue of (7.27)).

Theorem 7.26

There exists a sequence of eigenvalues $\{\lambda_n\}$ and a corresponding orthonormal sequence of eigenfunctions $\{e_n\}$ of the boundary value problem (7.27) with the following properties:

(a) each λ_n is real and the corresponding eigenspace is 1-dimensional;

(b) the sequence $\{\lambda_n\}$ can be ordered so that $\lambda_1 > \lambda_2 > \dots$, and $\lambda_n \to -\infty$ as $n \to \infty$;

(c) the sequence $\{e_n\}$ is an orthonormal basis for $\mathcal{H}$.

Proof

Suppose initially that 0 is not an eigenvalue of (7.27). The existence and property (c) of the sequences $\{\lambda_n\}$ and $\{e_n\}$ follows from Lemma 7.25 and the results of Section 6.3 (the fact that $\{e_n\}$ is a basis for $\mathcal{H}$, and hence the sequences $\{\lambda_n\}$ and $\{e_n\}$ have infinitely many terms, follows from part (b) of Lemma 7.25 and Corollary 6.35). The reality of the eigenvalues follows from (a) and (c) of Lemma 7.25 and Theorem 6.33. Now suppose that u_1, u_2 are two non-zero eigenfunctions of (7.27), corresponding to an eigenvalue λ. Then $u_1'(a) \neq 0$ (otherwise uniqueness of the solution of the initial value problem (7.14), (7.15) would imply that $u_1 \equiv 0$), so we have $u_2'(a) = \gamma u_1'(a)$, for some constant γ, and hence, again by uniqueness of the solution of the initial value problem, $u_2 \equiv \gamma u_1$. This shows that the eigenspace corresponding to any eigenvalue is one-dimensional, and so proves part (a). Next, it follows from Exercise 7.10 that the set of eigenvalues is bounded above, and from Corollary 6.23 the eigenvalues λ_n^{-1} of G tend to zero, so we must have $\lambda_n \to -\infty$ as $n \to \infty$. It follows that we can order the eigenvalues in the manner stated in the theorem. Since the eigenspace of any eigenvalue is one-dimensional, each number λ_n only occurs once in the sequence of eigenvalues, so we obtain the strict inequality in the ordering of the eigenvalues (we observe that if the eigenspace associated with an eigenvalue λ_n is m-dimensional then the number $\lambda = \lambda_n$ must occur in the sequence m times in order to obtain the correct correspondence with the sequence of eigenfunctions, all of which must be included in order to obtain an orthonormal basis of $\mathcal{H}$).

Next suppose that 0 is an eigenvalue of (7.27). Choose a real number α which is not an eigenvalue of (7.27) (such a number exists by Exercise 7.10) and rewrite (7.27) in the form

$$\begin{gathered} u'' + \widetilde{q}u = u'' + (q - \alpha)u = (\lambda - \alpha)u = \widetilde{\lambda}u, \\ u(a) = u(b) = 0 \end{gathered}$$

(writing $\widetilde{q} = q - \alpha$, $\widetilde{\lambda} = \lambda - \alpha$). Now, $\widetilde{\lambda} = 0$ is not an eigenvalue of this problem (otherwise $\lambda = \alpha$ would be an eigenvalue of the original problem) so the results just proved apply to this problem, and hence to the original problem (the shift in the parameter $\widetilde{\lambda}$ back to the parameter λ does not affect any of the assertions in the theorem). □

We can also combine (7.26) and (7.27) to obtain the slightly more general problem

$$\begin{gathered} u'' + qu - \lambda u = f, \\ u(a) = u(b) = 0. \end{gathered} \tag{7.29}$$

Solvability results for this problem follow immediately from the above results, but nevertheless we will state them as the following theorem.

Theorem 7.27

If λ is not an eigenvalue of (7.27) then, for any $f \in X$, (7.29) has a unique solution $u \in X$, and there exists a Green's operator $G(\lambda)$ for (7.29) with a corresponding Green's function $g(\lambda)$ (that is, a function $g(\lambda, s, t)$).

Proof

Writing $\widetilde{q} = q - \lambda$ and replacing q in (7.26) and (7.27) with $\widetilde{q}$, the theorem follows immediately from the previous results. □

By Theorem 7.26, any element $u \in \mathcal{H}$ can be written in the form

$$u = \sum_{n=1}^{\infty} (u, e_n) e_n, \tag{7.30}$$

where the convergence is with respect to the norm $\| \cdot \|_{\mathcal{H}}$. This is often a useful expansion, but there are occasions when it is desirable to have convergence with respect to the norm $\| \cdot \|_X$. We can obtain this for suitable functions u.

Theorem 7.28

If $u \in Y_b$ then the expansion (7.30) converges with respect to the norm $\| \cdot \|_X$.

Proof

Define a linear transformation $T : Y_b \to X$ by $Tu = u'' + qu$, for $u \in Y_b$. For any integer $k \geq 1$,

$$\begin{aligned} T\Big(u - \sum_{n=1}^{k} (u, e_n) e_n \Big) &= Tu - \sum_{n=1}^{k} (u, e_n) \lambda_n e_n = Tu - \sum_{n=1}^{k} (u, \lambda_n e_n) e_n \\ &= Tu - \sum_{n=1}^{k} (u, Te_n) e_n = Tu - \sum_{n=1}^{k} (Tu, e_n) e_n \end{aligned}$$

(the final step is a special case of part (c) of Lemma 7.30). Now, since G is the solution operator for (7.26), we have

$$u - \sum_{n=1}^{k} (u, e_n) = G\Big(Tu - \sum_{n=1}^{k} (Tu, e_n) \Big),$$

so by (7.2) (where M is defined for the kernel g as in Section 7.1)

$$\begin{aligned} \Big\| u - \sum_{n=1}^{k} (u, e_n) \Big\|_X &= \Big\| G\Big(Tu - \sum_{n=1}^{k} (Tu, e_n) \Big) \Big\|_X \\ &\leq M(b-a)^{1/2} \Big\| Tu - \sum_{n=1}^{k} (Tu, e_n) \Big\|_{\mathcal{H}} \to 0, \quad \text{as } k \to \infty \end{aligned}$$

since $\{e_n\}$ is an orthonormal basis for $\mathcal{H}$, which proves the theorem. □

We noted above that the eigenfunctions of the problem (7.24) are the functions in the orthonormal basis considered in Theorem 3.56. Thus Theorem 7.28 yields the following result.

Corollary 7.29

If $a = 0$, $b = \pi$ and $S = \{s_n\}$ is the orthonormal basis in Theorem 3.56, then, for any $u \in Y_b$, the series

$$u = \sum_{n=1}^{\infty} (u, s_n) s_n,$$

converges with respect to the norm $\| \cdot \|_X$.

Corollary 7.29 shows that the Fourier sine series converges uniformly on $[0, \pi]$ when $u \in Y_b$, that is, when $u(0) = u(\pi) = 0$ and $u \in C^2[0, \pi]$. In particular, the series converges at the points 0, π. Now, since $s_n(0) = s_n(\pi) = 0$ (this is clear anyway, but also these are the boundary conditions in the problem (7.24)), the condition $u(0) = u(\pi) = 0$ is clearly necessary for this result. However, the condition that $u \in C^2[0, \pi]$ is not necessary, but some smoothness condition on u is required. Thus Corollary 7.29 is not the optimal convergence result for this series, but it has the virtue of being a simple consequence of a broad theory of eigenfunction expansions for second order differential operators. As in Section 3.5, we do not wish to delve further into the specific theory of Fourier series.

We conclude this section by showing that the above results for the problem (7.27) hold for a much more general class of problems. We consider the general eigenvalue problem

$$\begin{aligned} \frac{d}{ds}\left(p(s)\frac{du}{ds}(s)\right) + q(s)u(s) = \lambda u(s), \quad s \in [a, b], \\ u(a) = u(b) = 0, \end{aligned} \tag{7.31}$$

where $p \in C^2_{\mathbb{R}}[a, b]$, $q \in C_{\mathbb{R}}[a, b]$ and $p(s) > 0$, $s \in [a, b]$. It will be shown that this problem can be transformed into a problem having the form (7.27) by using the following changes of variables

$$t = t(s) = \int_a^s p(y)^{-1/2}\, dy, \quad \widetilde{u}(t) = u(s(t))p(s(t))^{1/4}. \tag{7.32}$$

We can make this more precise as follows. Let $c = t(b) = \int_a^b p(y)^{-1/2}\, dy$. Now, $dt/ds = p(s)^{-1/2} > 0$ on $[a, b]$, so the mapping $f : [a, b] \to [0, c]$ defined by $f(s) = t(s)$ (using the above formula) is differentiable and strictly increasing, so is invertible, with a differentiable inverse $g : [0, c] \to [a, b]$. Thus, we may regard t as a function $t = f(s)$ of s, or s as a function $s = g(t)$ of t – the latter interpretation is used in the above definition of the function $\widetilde{u}$. The change of variables (7.32) is known as the *Liouville transform*. The following lemma will be proved in Exercise 7.12.

Lemma 7.30

(a) The change of variables (7.32) transforms the problem (7.31) into the problem

$$\begin{aligned} \frac{d^2\widetilde{u}}{dt^2}(t) + \widetilde{q}(t)\widetilde{u}(t) &= \lambda\widetilde{u}(t), \quad t \in [0,c], \\ \widetilde{u}(0) = \widetilde{u}(c) &= 0, \end{aligned} \tag{7.33}$$

where

$$\widetilde{q}(t) = q(s(t)) + \frac{1}{16}\left(\frac{1}{p(s(t))}\left(\frac{dp}{ds}(s(t))\right)^2 - 4\,\frac{d^2p}{ds^2}(s(t))\right).$$

(b) Suppose that $u,\, v \in L^2[a,b]$ are transformed into $\widetilde{u},\, \widetilde{v} \in L^2(0,c)$ by the change of variables (7.32). Then

$$(u,v) = (\widetilde{u},\widetilde{v}),$$

where $(\cdot\,,\cdot)$ denotes the $L^2[a,b]$ inner product on the left of this formula, and the $L^2(0,c)$ inner product on the right.

(c) For any $u \in Y_b$, define $Tu \in X$ to be the function on the left-hand side of the equation in the problem (7.31). Then, for any $u,\, v \in Y_b$,

$$(Tu,v) = (u,Tv),$$

where $(\cdot\,,\cdot)$ denotes the $L^2[a,b]$ inner product.

As a consequence of parts (a) and (b) of Lemma 7.30 we now have the following theorem.

Theorem 7.31

Theorems 7.26 and 7.28 hold for the general problem (7.31).

Proof

The transformed problem (7.33) has the form of problem (7.31), and part (b) of Lemma 7.30 shows that the Liouville transform does not change the value of the inner product. In particular, by Theorem 7.26, there exists an orthonormal basis of eigenfunctions $\{e_n\}$ of the transformed problem, and this corresponds to a set of orthonormal eigenfunctions $\{f_n\}$ of the original problem. To see that $\{f_n\}$ is also a basis we observe that if it is not then there exists a non-zero function $g \in L^2[a,b]$ orthogonal to the set $\{f_n\}$ (by part (a) of Theorem 3.47), and hence the transformed function $\widetilde{g} \neq 0$ is orthogonal to the set $\{e_n\}$, which contradicts the fact that this set is a basis. □

Part (c) of Lemma 7.30 shows that, in a sense, T is "self-adjoint", and this is the underlying reason why these results hold for the problem (7.31) (particularly the fact that the eigenfunctions form an orthonormal basis, which is certainly related to self-adjointness). This is also why equation (7.31) is written in the form it is, which seems slightly strange at first sight. However, T is not bounded, and we have not defined self-adjointness for unbounded linear transformations – this can be done, but involves a lot of work which we have avoided here by converting the problem to the integral form.

Finally, we remark that much of the theory in this chapter carries over to multi-dimensional integral equations and partial differential equations, with very similar results. However, the technical details are usually considerably more complicated. We will not pursue this any further here.

EXERCISES

7.8 Calculate the eigenvalues and normalized eigenfunctions for the boundary value problem

$$\begin{aligned} u'' &= \lambda u, \\ u(0) &= u(\pi) = 0 \end{aligned}$$

(consider the cases $\lambda > 0$, $\lambda < 0$, $\lambda = 0$ separately).

7.9 Calculate the Green's function for the problem

$$\begin{aligned} u'' - \lambda u &= f, \\ u(0) &= u(\pi) = 0, \end{aligned}$$

when $\lambda \neq 0$ (consider the cases $\lambda > 0$, $\lambda < 0$ separately).

7.10 Show that if λ is an eigenvalue of (7.27) then $\lambda \leq \|q\|_X$.
[Hint: by definition, the equation $u'' + qu = \lambda u$ has a non-zero solution $u \in X$; take the $L^2[a,b]$ inner product of this equation with u and integrate by parts.]

7.11 Prove the following results.

(a) Given an arbitrary function $z \in \mathcal{H}$, and $\epsilon > 0$, show that there exists a function $w \in C^2[a,b]$ such that $\|z - w\|_{\mathcal{H}} < \epsilon$.
[Hint: see the proof of Theorem 3.54.]

(b) Given an arbitrary function $w \in C^2[a,b]$, and $\delta > 0$ such that $a + \delta \in (a,b)$, show that there exists a cubic polynomial p_δ such that, if we define a function v_δ by

$$v_\delta(s) = \begin{cases} p_\delta(s), & \text{if } s \in [a, a+\delta], \\ w(s), & \text{if } s \in [a+\delta, b], \end{cases}$$

then $v_\delta \in C^2[a,b]$ and $v_\delta(a) = 0$.
[Hint: find p_δ by using these conditions on v_δ to write down a set of linear equations for the coefficients of p_δ.]

(c) Use the construction of p_δ in part (b) to show that, for any $\epsilon > 0$, there exists $\delta > 0$ such that $\|w - v_\delta\|_{\mathcal{H}} < \epsilon$.

(d) Deduce that the set Y_b is dense in $\mathcal{H}$.

This result is rather surprising at first sight (draw some pictures), and depends on properties of the $L^2[a,b]$ norm – it is certainly not true in $C[a,b]$.

7.12 Prove Lemma 7.30.
[Hint: (a) use the chain rule to differentiate the formula $u(s) = p(s)^{-1/4}\widetilde{u}(t(s))$, obtained by rearranging (7.32);
(b) use the standard formula for changing the variable of integration from the theory of Riemann integration (since the transformation is differentiable, this rule extends to the Lebesgue integral – you may assume this);
(c) use integration by parts (twice).]

8
Solutions to Exercises

Chapter 2

2.1 Let (x,y), $(u,v) \in Z$ and let $\alpha \in \mathbb{F}$.

(a) $\|(x,y)\| = \|x\|_1 + \|y\|_2 \geq 0$.

(b) If $(x,y) = 0$ then $x = 0$ and $y = 0$. Hence $\|x\|_1 = \|y\|_2 = 0$ and so $\|(x,y)\| = 0$.

Conversely, if $\|(x,y)\| = 0$ then $\|x\|_1 = \|y\|_2 = 0$. Thus $x = 0$ and $y = 0$ and so $(x,y) = 0$.

(c) $$\begin{aligned}\|\alpha(x,y)\| &= \|(\alpha x, \alpha y)\| = \|\alpha x\|_1 + \|\alpha y\|_2 = |\alpha|\|x\|_1 + |\alpha|\|y\|_2 \\ &= |\alpha|\|(x,y)\|.\end{aligned}$$

(d) $$\begin{aligned}\|(x,y) + (u,v)\| &= \|(x+u, y+v)\| = \|x+u\|_1 + \|y+v\|_2 \\ &\leq \|x\|_1 + \|u\|_1 + \|y\|_2 + \|v\|_2 = \|(x,y)\| + \|(u,v)\|.\end{aligned}$$

2.2 Let f, $g \in F_b(S,X)$ and let $\alpha \in \mathbb{F}$.

(a) $\|f\|_b = \sup\{\|f(s)\| : s \in S\} \geq 0$.

(b) If $f = 0$ then $f(s) = 0$ for all $s \in S$ so that $\|f(s)\| = 0$ for all $s \in S$ and hence $\|f\|_b = 0$.

Conversely, if $\|f\|_b = 0$ then $\|f(s)\| = 0$ for all $s \in S$. Thus $f(s) = 0$ for all $s \in S$ and so $f = 0$.

(c) $\|\alpha f\|_b = \sup\{\|\alpha f(s)\| : s \in S\}$
$= \sup\{|\alpha|\|f(s)\| : s \in S\} = |\alpha| \sup\{\|f(s)\| : s \in S\} = |\alpha|\|f\|_b$.

(d) $\|(f+g)(s)\| \leq \|f(s)\| + \|g(s)\| \leq \|f\|_b + \|g\|_b$ for any $s \in S$. Hence

$$\|f+g\|_b = \sup\{\|(f+g)(s)\| : s \in S\} \leq \|f\|_b + \|g\|_b.$$

2.3 We use the standard norms on the spaces.

(a) $\|f_n\| = \sup\{|f_n(x)| : x \in [0,1]\} = 1$.

(b) As f_n is continuous, the Riemann and Lebesgue integrals of f are the same, so that $\|f_n\| = \int_0^1 x^n\, dx = \frac{1}{n}$.

2.4 If $\alpha = \dfrac{r}{\|x\|}$ then $\|\alpha x\| = \alpha\|x\| = r$.

2.5 Let $\epsilon > 0$.

(a) Suppose that $\{(x_n, y_n)\}$ converges to (x, y) in Z. Then there exists $N \in \mathbb{N}$ such that

$$\|(x_n - x, y_n - y)\| = \|(x_n, y_n) - (x, y)\| \leq \epsilon$$

when $n \geq N$. Thus $\|x_n - x\|_1 \leq \|(x_n - x, y_n - y)\| \leq \epsilon$ and $\|y_n - y\|_2 \leq \|(x_n - x, y_n - y)\| \leq \epsilon$ when $n \geq N$. Thus $\{x_n\}$ converges to x in X and $\{y_n\}$ converges to y in Y.

Conversely, suppose that $\{x_n\}$ converges to x in X and $\{y_n\}$ converges to y in Y. Then there exist $N_1, N_2 \in \mathbb{N}$ such that $\|x_n - x\|_1 \leq \dfrac{\epsilon}{2}$ when $n \geq N_1$ and $\|y_n - y\|_2 \leq \dfrac{\epsilon}{2}$ when $n \geq N_2$. Let $N_0 = \max(N_1, N_2)$. Then

$$\|(x_n, y_n) - (x, y)\| = \|(x_n - x, y_n - y)\| = \|x_n - x\|_1 + \|y_n - y\|_2 \leq \epsilon$$

when $n \geq N_0$. Hence $\{(x_n, y_n)\}$ converges to (x, y) in Z.

(b) Suppose that $\{(x_n, y_n)\}$ is Cauchy in Z. Then there exists $N \in \mathbb{N}$ such that

$$\|(x_n - x_m, y_n - y_m)\| = \|(x_n, y_n) - (x_m, y_m)\| \leq \epsilon$$

when $m, n \geq N$. Thus $\|x_n - x_m\|_1 \leq \|(x_n - x_m, y_n - y_m)\| \leq \epsilon$ and $\|y_n - y_m\|_2 \leq \|(x_n - x_m, y_n - y_m)\| \leq \epsilon$ when $m, n \geq N$. Thus $\{x_n\}$ is Cauchy in X and $\{y_n\}$ is Cauchy in Y.

Conversely, suppose that $\{x_n\}$ is Cauchy in X and $\{y_n\}$ is Cauchy in Y. Then there exist $N_1, N_2 \in \mathbb{N}$ such that $\|x_n - x_m\|_1 \leq \dfrac{\epsilon}{2}$ when $m, n \geq$

N_1 and $\|y_n - y_m\|_2 \leq \frac{\epsilon}{2}$ when $m,\, n \geq N_2$. Let $N_0 = \max(N_1, N_2)$. Then

$$\begin{aligned}\|(x_n, y_n) - (x_m, y_m)\| &= \|(x_n - x_m, y_n - y_m)\| \\ &= \|x_n - x_m\|_1 + \|y_n - y_m\|_2 \\ &\leq \epsilon\end{aligned}$$

when $m,\, n \geq N_0$. Hence $\{(x_n, y_n)\}$ is Cauchy in Z.

2.6 Suppose that $\|\cdot\|_1$ and $\|\cdot\|_2$ are equivalent on $\mathcal{P}$. Then there exist $M, m > 0$ such that

$$m\|p\|_1 \leq \|p\|_2 \leq M\|p\|_1$$

for all $p \in \mathcal{P}$. As $m > 0$ there is $n \in \mathbb{N}$ such that $\frac{1}{n} < m$. Let $p_n : [0,1] \to \mathbb{R}$ be defined by $p_n(x) = x^n$. Then $\|p_n\|_1 = 1$ and $\|p_n\|_2 = \frac{1}{n}$ by Exercise 2.3. Hence

$$m = m\|p_n\|_1 \leq \|p_n\|_2 = \frac{1}{n},$$

which is a contradiction. Thus $\|\cdot\|_1$ and $\|\cdot\|_2$ are not equivalent.

2.7 The sets are shown in Figure 8.1.

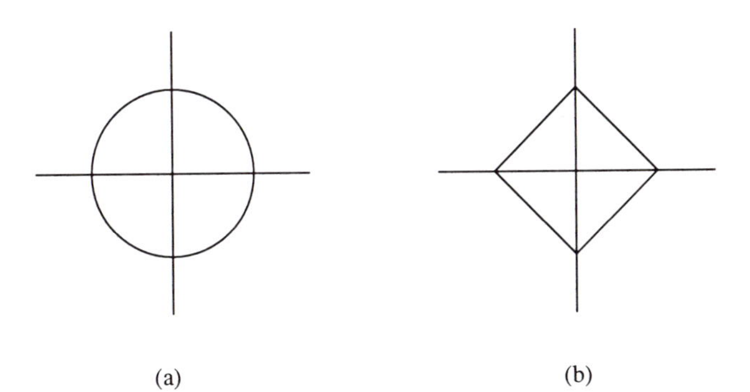

Fig. 8.1. (a) The standard norm; (b) the alternative norm

2.8 Let $\epsilon > 0$. There exists N in $\mathbb{N}$ such that $\|x_n - x_m\|_1 < \frac{\epsilon}{K}$ when $m,\, n \geq \mathbb{N}$. Hence, when $m,\, n \geq \mathbb{N}$,

$$\|x_n - x_m\| \leq K\|x_n - x_m\|_1 < \epsilon.$$

Therefore $\{x_n\}$ is Cauchy in the metric space (X, d).

2.9 As $\|x\| \leq \frac{1}{m}\|x\|_1$ for all $x \in X$, if $\{x_n\}$ is Cauchy in the metric space (X, d_1), then $\{x_n\}$ is Cauchy in the metric space (X, d) by Exercise 2.8.

Conversely, as $\|x\|_1 \leq M\|x\|$ for all $x \in X$, if $\{x_n\}$ is Cauchy in the metric space (X, d), then $\{x_n\}$ is Cauchy in the metric space (X, d_1) by Exercise 2.8.

2.10 If $x = \left(1, \frac{1}{2}, \frac{1}{3}, \dots\right)$ then $x \in \ell^2 \setminus S$. For each $n \in \mathbb{N}$, let $x_n = \left(1, \frac{1}{2}, \frac{1}{3}, \dots, \frac{1}{n}, 0, 0, \dots\right)$. Then $x_n \in S$ and

$$\|x - x_n\|^2 = \|\left(0, 0, \dots, 0, \frac{1}{n+1}, \frac{1}{n+2}, \dots\right)\|^2 = \sum_{j=n+1}^{\infty} \left(\frac{1}{j}\right)^2.$$

Hence $\lim_{n\to\infty} \|x - x_n\| = 0$ and so $\lim_{n\to\infty} x_n = x$. Therefore $x \in \overline{S} \setminus S$ and thus S is not closed.

2.11 (a) $\left\|\frac{\eta x}{2\|x\|}\right\| = \frac{\eta\|x\|}{2\|x\|} = \frac{\eta}{2} < \eta$, so by the given condition $\frac{\eta x}{2\|x\|} \in Y$.

(b) Let $x \in X \setminus \{0\}$. As Y is open there exists $\eta > 0$ such that $\{y \in X : \|y\| < \eta\} \subseteq Y$. Hence $\frac{\eta x}{2\|x\|} \in Y$ by part (a). As a scalar multiple of a vector in Y is also in Y we have $x = \frac{2\|x\|}{\eta}\left(\frac{\eta x}{2\|x\|}\right) \in Y$. So $X \subseteq Y$. As Y is a subset of X by definition, it follows that $Y = X$.

2.12 (a) Let $\{x_n\}$ be a sequence in T which converges to $x \in X$. Then $\|x_n\| \leq 1$ for all $n \in \mathbb{N}$, and so $\|x\| = \lim_{n\to\infty} \|x_n\| \leq 1$, by Theorem 2.11. Thus $x \in T$ so T is closed.

(b) $\|x - x_n\| = \|x - (1 - \frac{1}{n})x\| = \frac{1}{n}\|x\| \leq \frac{1}{n}$ for all $n \in \mathbb{N}$. Therefore $\lim_{n\to\infty} x_n = x$.

As $S \subseteq T$ and T is closed, $\overline{S} \subseteq T$. Conversely, if $x \in T$ and x_n is defined as above then $\|x_n\| = (1 - \frac{1}{n})\|x\| \leq 1 - \frac{1}{n} < 1$ so $x_n \in S$. Thus x is a limit of a sequence of elements of S, so $x \in \overline{S}$. Hence $T \subseteq \overline{S}$ so $T = \overline{S}$.

2.13 Let $\{(x_n, y_n)\}$ be a Cauchy sequence in Z. Then $\{x_n\}$ is Cauchy in X and $\{y_n\}$ is Cauchy in Y by Exercise 2.5. As X and Y are Banach spaces, $\{x_n\}$ converges to $x \in X$ and $\{y_n\}$ converges to $y \in Y$. Hence $\{(x_n, y_n)\}$ converges to (x, y) by Exercise 2.5. Therefore Z is a Banach space.

2.14 Let $\{f_n\}$ be a Cauchy sequence in $F_b(S, X)$ and let $\epsilon > 0$. There exists $N \in \mathbb{N}$ such that $\|f_n - f_m\|_b < \epsilon$ when $n, m > N$. For all $s \in S$

$$\|f_n(s) - f_m(s)\| \leq \|f_n - f_m\|_b < \epsilon$$

when n, $m > N$ and so it follows that $\{f_n(s)\}$ is a Cauchy sequence in X. Since X is complete $\{f_n(s)\}$ converges, so we define a function $f : S \to X$ by

$$f(s) = \lim_{n\to\infty} f_n(s).$$

As $\|f_n(s) - f_m(s)\| < \epsilon$ for all $s \in S$ when n, $m > N$, taking the limit as m tends to infinity we have $\|f_n(s) - f(s)\| \leq \epsilon$ when $n > N$. Thus

$$\|f(s)\| \leq \epsilon + \|f_n(s)\| \leq \epsilon + \|f_n\|_b$$

when $n > N$ for all $s \in S$. Therefore f is a bounded function so $f \in F_b(S, X)$ and $\lim_{n\to\infty} f_n = f$ as $\|f_n - f\|_b \leq \epsilon$ when $n > N$. Thus $F_b(S, X)$ is a Banach space.

Chapter 3

3.1 Rearranging the condition we obtain $(x, u-v) = 0$ for all $x \in V$. If $x = u-v$ then $(u - v, u - v) = 0$, and hence $u - v = 0$, by part (b) in the definition of the inner product. Thus $u = v$.

3.2 Let $x = \sum_{n=1}^k \lambda_n \widehat{e}_n$, $y = \sum_{n=1}^k \mu_n \widehat{e}_n$, $z = \sum_{n=1}^k \sigma_n \widehat{e}_n \in V$, α, $\beta \in \mathbb{F}$. We show that the formula in Example 3.6 defines an inner product on V by verifying that all the properties in Definition 3.1 or 3.3 hold.

(a) $(x, x) = \sum_{n=1}^k \lambda_n \overline{\lambda}_n \geq 0$.

(b) If $(x, x) = 0$ then $\sum_{n=1}^k |\lambda_n|^2 = 0$ and so $\lambda_n = 0$ for all n. Thus $x = 0$. Conversely, if $x = 0$ then $\lambda_n = 0$ for all n, so $(x, x) = 0$.

(c)
$$\begin{aligned}(\alpha x + \beta y, z) &= \textstyle\sum_{n=1}^k (\alpha\lambda_n + \beta\mu_n)\overline{\sigma}_n \\ &= \alpha \textstyle\sum_{n=1}^k \lambda_n \overline{\sigma}_n + \beta \textstyle\sum_{n=1}^k \mu_n \overline{\sigma}_n \\ &= \alpha(x, z) + \beta(y, z).\end{aligned}$$

(d) $(x, y) = \sum_{n=1}^k \lambda_n \overline{\mu}_n = \overline{\sum_{n=1}^n \mu_n \overline{\lambda}_n} = \overline{(y, x)}$.

3.3 (a) The inequality follows from

$$0 \leq (|a| - |b|)^2 = |a|^2 - 2|a||b| + |b|^2.$$

(b)
$$\begin{aligned}|a + b|^2 &= (a + b)(\overline{a} + \overline{b}) = |a|^2 + a\overline{b} + \overline{a}b + |b|^2 \\ &\leq |a|^2 + 2|a||b| + |b|^2 \leq 2(|a|^2 + |b|^2) \qquad \text{(by inequality (a))}.\end{aligned}$$

We now show that ℓ^2 is a vector space. Let $x = \{x_n\}$, $y = \{y_n\} \in \ell^2$ and $\alpha \in \mathbb{F}$. Then $\sum_{n=1}^{\infty} |\alpha x_n|^2 = |\alpha|^2 \sum_{n=1}^{\infty} |x_n|^2 < \infty$, so $\alpha x \in \ell^2$. Also, for any $n \in \mathbb{N}$, $|x_n + y_n|^2 \leq 2(|x_n|^2 + |y_n|^2)$ (by inequality (b)) so

$$\sum_{n=1}^{\infty} |x_n + y_n|^2 \leq 2 \sum_{n=1}^{\infty} (|x_n|^2 + |y_n|^2) < \infty,$$

and hence $x+y \in \ell^2$. Thus, ℓ^2 is a vector space. Moreover, by inequality (a),

$$\sum_{n=1}^{\infty} |x_n \overline{y}_n| \leq \tfrac{1}{2} \sum_{n=1}^{\infty} (|x_n|^2 + |y_n|^2) < \infty,$$

so $x\overline{y} \in \ell^1$ (in the proof of the analogous result in Example 3.7 we used Hölder's inequality, since it was available, but the more elementary form of inequality used here would have sufficed). Since $x\overline{y} \in \ell^1$, it follows that the formula $(x, y) = \sum_{n=1}^{\infty} x_n \overline{y}_n$ is well-defined. We now verify that this formula defines an inner product on ℓ^2.

(a) $(x, x) = \sum_{n=1}^{\infty} |x_n|^2 \geq 0$.

(b) If $(x, x) = 0$ then $\sum_{n=1}^{\infty} |x_n|^2 = 0$ and so $x_n = 0$ for all n. Thus $x = 0$. Conversely, if $x = 0$ then $\lambda_n = 0$ for all n, so $(x, x) = 0$.

(c) $$\begin{aligned}(\alpha x + \beta y, z) &= \textstyle\sum_{n=1}^{\infty} (\alpha x_n + \beta y_n) \overline{z}_n \\ &= \alpha \textstyle\sum_{n=1}^{\infty} x_n \overline{z}_n + \beta \textstyle\sum_{n=1}^{\infty} y \overline{z}_n \\ &= \alpha(x, z) + \beta(y, z)\end{aligned}$$

(d) $(x, y) = \sum_{n=1}^{\infty} x_n \overline{y}_n = \overline{\sum_{n=1}^{\infty} y_n \overline{x}_n} = \overline{(y, x)}$.

3.4 (a) By expanding the inner products we obtain

$$\begin{aligned}(u + v, x + y) &= (u, x) + (v, y) + (u, y) + (v, x) \\ (u - v, x - y) &= (u, x) + (v, y) - (u, y) - (v, x).\end{aligned}$$

Subtracting these gives the result.

(b) In the identity in part (a), replace v by iv and y by iy to obtain

$$\begin{aligned}(u + iv, x + iy) - (u - iv, x - iy) &= 2(u, iy) + 2(iv, x) \\ &= -2i(u, y) + 2i(v, x).\end{aligned}$$

Multiply this equation by i and add to the identity in part (a) to obtain

$$(u + v, x + y) - (u - v, x - y) + i(u + iv, x + iy) - i(u - iv, x - iy)$$

$$= 2(u, y) + 2(v, x) + 2(u, y) - 2(v, x) = 4(u, y).$$

3.5 (a) Putting $u = y$ and $v = x$ in the identity in part (a) of Exercise 3.4 and using the definition of the induced norm gives the result.

(b) Putting $u = x$ and $v = y$ in the identity in part (a) of Exercise 3.4 and using $(x, y) = (y, x)$ gives the result.

(c) Putting $u = x$ and $v = y$ in the identity in part (b) of Exercise 3.4 gives the result.

3.6

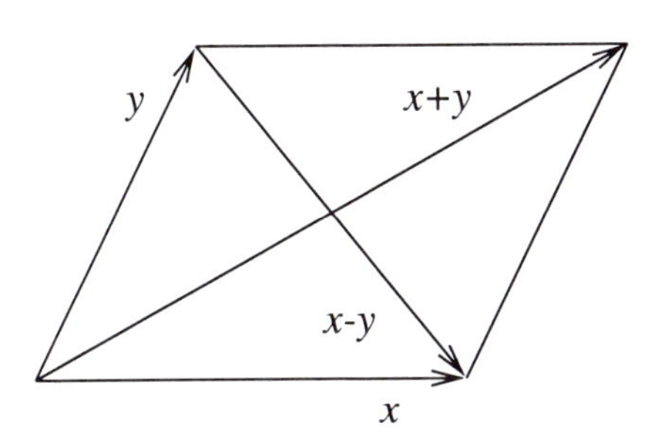

Fig. 8.2. Parallelogram with sides x and y in the plane

Writing out the cosine rule for any two adjacent triangles in the above figure (using the angles θ and $\pi - \theta$ at the centre) and adding these yields the parallelogram rule in this case (using $\cos(\pi - \theta) = -\cos\theta$).

3.7 From the definition of the norm $\|\cdot\|_1$ on $\mathbb{R}^k$ we have

$$\|\widehat{e}_1 + \widehat{e}_2\|_1^2 + \|\widehat{e}_1 - \widehat{e}_2\|_1^2 = 2^2 + 2^2 = 8,$$
$$2(\|\widehat{e}_1\|_1^2 + \|\widehat{e}_2\|_1^2) = 2(1 + 1) = 4.$$

Thus the parallelogram rule does not hold and so the norm cannot be induced by an inner product.

3.8 From

$$(x + \alpha y).(x + \alpha y) = \|x\|^2 + \|y\|^2 + 2\Re e\,(\alpha(y, x)),$$
$$(x - \alpha y).(x - \alpha y) = \|x\|^2 + \|y\|^2 - 2\Re e\,(\alpha(y, x)),$$

it follows that, for any $\alpha \neq 0$,

$$\|x + \alpha y\| = \|x - \alpha y\| \iff \Re e\,(\alpha(y, x)) = 0 \iff (x, y) = 0.$$

3.9 We have $\|a\| = \sqrt{18}$, $\|b\| = \sqrt{2}$, and putting $v_1 = \|a\|^{-1}a$ and $v_2 = \|b\|^{-1}b$ it can be checked that $\{v_1, v_2\}$ is an orthonormal set. This can be extended to an orthonormal basis of $\mathbb{R}^3$ by adding the vector $(1, 0, 0)$ to the set and applying the Gram–Schmidt algorithm, to obtain the vector $v_3 = \frac{1}{3}(2, -1, 2)$ orthogonal to v_1 and v_2.

3.10 The Gram–Schmidt algorithm yields
$e_1 = \frac{1}{\sqrt{2}}$, $e_2 = \sqrt{\frac{3}{2}}x$, $e_3 = \sqrt{\frac{5}{8}}(3x^2 - 1)$.

3.11 By Example 3.10, $\mathcal{H}$ is an inner product space so we only need to show that it is complete with respect to the norm induced by the inner product defined in Example 3.10. Now, Exercise 2.14 shows that $\mathcal{H}$ is complete with respect to the norm defined in Example 2.8, while Remark 3.11 notes that this norm is equivalent to the induced norm, so $\mathcal{H}$ must also be complete with respect to the induced norm.

3.12 The results of Exercise 3.9 (and Example 3.28) show that $S^\perp$ is spanned by the vector $v_3 = \frac{1}{3}(2, -1, 2)$.

3.13 Since a is the only element of A, it follows from the definition of $A^\perp$ that $x = (x_1, \ldots, x_k) \in A^\perp$ if and only if

$$0 = (x, a) = a_1 x_1 + \ldots + a_k x_k.$$

3.14 Let $S = \mathrm{Sp}\,\{e_{p+1}, \ldots, e_k\}$. If $x \in S$ then $x = \sum_{n=p+1}^k \lambda_n e_n$, for some numbers λ_n, $n = p+1, \ldots, k$. Any $a \in A$ has the form $a = \sum_{m=1}^p \alpha_m e_m$. By orthonormality,

$$(x, a) = \sum_{n=p+1}^{k} \sum_{m=1}^{p} \lambda_n \overline{\alpha}_m (e_n, e_m) = 0,$$

for all $a \in A$. Thus $x \in A^\perp$, and so $S \subset A^\perp$. Now suppose that $x \in A^\perp$ and write $x = \sum_{n=1}^k \lambda_n e_n$. Choosing $a = \sum_{m=1}^p \lambda_m e_m \in A$, we have

$$0 = (x, a) = \sum_{m=1}^{p} |\lambda_m|^2,$$

and so $\lambda_m = 0$, $m = 1, \ldots, p$. This implies that $x = \sum_{n=p+1}^k \lambda_n e_n \in S$, and so $A^\perp \subset S$. Therefore, $S = A^\perp$.

3.15 Let $S = \{(x_n) \in \ell^2 : x_{2n-1} = 0 \text{ for all } n \in \mathbb{N}\}$. If $x \in S$ and $y \in A$ then $(x, y) = \sum_{n=1}^\infty x_n \overline{y}_n = 0$. Thus $x \in A^\perp$, and so $S \subset A^\perp$. Conversely, let $x \in A^\perp$ and suppose $x_{2m-1} \neq 0$ for some $m \in \mathbb{N}$. The vector $\widetilde{e}_{2m-1}$ in the standard orthonormal basis in ℓ^2 belongs to A, so $0 = (x, \widetilde{e}_{2m-1}) = x_{2m-1}$, which is a contradiction. Thus $x_{2m-1} = 0$ for all $m \in \mathbb{N}$, so $x \in S$. Hence $A^\perp \subset S$, and so $A^\perp = S$.

3.16 Since $A \subset \overline{A}$ we have $\overline{A}^\perp \subset A^\perp$, by part (e) of Lemma 3.29. Let $y \in A^\perp$. Then $(x, y) = 0$ for all $x \in A$. Now suppose that $x \in \overline{A}$, and $\{x_n\}$ is a sequence of elements of A such that $\lim_{n\to\infty} x_n = x$. Then $(x, y) = \lim_{n\to\infty}(x_n, y) = 0$. Thus $(x, y) = 0$ for any $x \in \overline{A}$, and so $y \in \overline{A}^\perp$. Hence $A^\perp \subset \overline{A}^\perp$, and so $A^\perp = \overline{A}^\perp$.

3.17 $X \subset X+Y$ and $Y \subset X+Y$ so $(X+Y)^\perp \subset X^\perp$ and $(X+Y)^\perp \subset Y^\perp$. Thus $(X+Y)^\perp \subset X^\perp \cap Y^\perp$. Conversely let $u \in X^\perp \cap Y^\perp$ and let $v \in X+Y$. Then $u \in X^\perp$, $u \in Y^\perp$ and $v = x+y$, where $x \in X$ and $y \in Y$. Hence

$$(v,u) = (x+y,u) = (x,u)+(y,u) = 0+0 = 0,$$

so $u \in (X+Y)^\perp$, and hence $X^\perp \cap Y^\perp \subset (X+Y)^\perp$. Thus $X^\perp \cap Y^\perp = (X+Y)^\perp$.

3.18 Let $E = \{x \in \mathcal{H} : (x,y) = 0\}$. For any $x \in \mathcal{H}$,

$$(x,y) = 0 \iff (x,\alpha y) = 0,\ \forall \alpha \in \mathbb{F} \iff x \in S^\perp,$$

since any vector in S has the form αy, so $E = S^\perp$. Now, since S is 1-dimensional it is closed, so by Corollary 3.36, $E^\perp = S^{\perp\perp} = S$.

3.19 Suppose that $U^\perp = \{0\}$. Then by Corollary 3.35 and part (c) of Lemma 3.29, $U = U^{\perp\perp} = \{0\}^\perp = \mathcal{H}$, which contradicts the assumption that $U \neq \mathcal{H}$, so we must have $U^\perp \neq \{0\}$.

The result need not be true if U is not closed. To see this let U be a dense linear subspace with $U \neq \mathcal{H}$. Then by Exercise 3.16, $U^\perp = \overline{U}^\perp = \mathcal{H}^\perp = \{0\}$. To see that such subspaces exist in general we consider the subspace S in Exercise 2.10. It is shown there that this subspace is not closed. To see that it is dense, let $y = \{y_n\}$ be an arbitrary element of ℓ^2 and let $\epsilon > 0$. Choose $N \in \mathbb{N}$ such that $\sum_{n=N}^\infty |y_n|^2 < \epsilon^2$, and define an element $x = \{x_n\} \in \ell^2$ by $x_n = y_n$ if $n < N$, and $x_n = 0$ otherwise. Clearly, $\|x-y\|^2 = \sum_{n=N}^\infty |y_n|^2 < \epsilon^2$, which proves the density of this subspace.

3.20 (a) By parts (f) and (g) of Lemma 3.29, $A^{\perp\perp}$ is a closed linear subspace containing A. Hence $\overline{\mathrm{Sp}}\, A \subset A^{\perp\perp}$ by Definition 2.23. Now suppose that Y is any closed linear subspace containing A. Then by part (e) of Lemma 3.29 (twice) and Corollary 3.35, $A^{\perp\perp} \subset Y^{\perp\perp} = Y$, and hence $A^{\perp\perp} \subset \overline{\mathrm{Sp}}\, A$, by Definition 2.23. Therefore $A^{\perp\perp} = \overline{\mathrm{Sp}}\, A$.

(b) By part (f) of Lemma 3.29, $A^\perp$ is a closed linear subspace. By Corollary 3.35, $(A^\perp)^{\perp\perp} = A^\perp$.

3.21 With the notation of Example 3.46 we have, putting $x = e_1$,

$$\sum_{n=1}^\infty |(x, e_{2n})|^2 = 0 < 1 = \|x\|^2.$$

3.22 Using Theorem 3.42 we have:

(a) $\sum_{n=1}^\infty n^{-2} < \infty$, so $\sum_{n=1}^\infty n^{-1} e_n$ converges;

(b) $\sum_{n=1}^{\infty} n^{-1} = \infty$, so $\sum_{n=1}^{\infty} n^{-1/2} e_n$ does not converge.

3.23 (a) By the given condition and Theorem 3.42 we have

$$\begin{aligned} \sum_{n=1}^{\infty} (x, e_{\rho(n)}) e_n \text{ converges} &\iff \sum_{n=1}^{\infty} |(x, e_{\rho(n)})|^2 < \infty \\ &\iff \sum_{n=1}^{\infty} |(x, e_n)|^2 < \infty \\ &\iff \sum_{n=1}^{\infty} (x, e_n) e_n \text{ converges.} \end{aligned}$$

However, since $\{e_n\}$ is an orthonormal sequence, $\sum_{n=1}^{\infty}(x, e_n)e_n$ converges (by Corollary 3.44). Hence $\sum_{n=1}^{\infty}(x, e_{\rho(n)})e_n$ converges.

(b) Since $\{e_n\}$ is an orthonormal basis, by Theorem 3.47

$$\|\sum_{n=1}^{\infty}(x, e_{\rho(n)})e_n\|^2 = \sum_{n=1}^{\infty} |(x, e_{\rho(n)})|^2 = \sum_{n=1}^{\infty} |(x, e_n)|^2 = \|x\|^2.$$

3.24 From Theorem 3.47, $x = \sum_{n=1}^{\infty}(x, e_n)e_n$, so by the continuity of the inner product,

$$\begin{aligned} (x, y) &= \Big(\lim_{k \to \infty} \sum_{n=1}^{k} (x, e_n) e_n, y \Big) = \lim_{k \to \infty} \sum_{n=1}^{k} (x, e_n)(e_n, y) \\ &= \sum_{n=1}^{\infty} (x, e_n)(e_n, y). \end{aligned}$$

3.25 The first part of the question follows immediately from the characterization of density given in part (f) of Theorem 1.25 (for any $\epsilon > 0$ there exists $k \in \mathbb{N}$ such that $k > 1/\epsilon$ and for any $k \in \mathbb{N}$ there exists $\epsilon > 0$ such that $k > 1/\epsilon$).

Next, suppose that M is separable and $N \subset M$. Then there is a countable dense set $U = \{u_n : n \in \mathbb{N}\}$ in M. Now consider an arbitrary pair $(n, k) \in \mathbb{N}^2$. If there exists a point $y \in N$ with $d(y, u_n) < 1/k$ then we put $b_{(n,k)} = y$; otherwise we ignore the pair (n, k). Let $B \subset N$ be the complete collection of points $b_{n,k}$ obtained by this process. The set B is countable (since the set $\mathbb{N}^2$ is countable – see [7] if this is unclear). Also, for any $k \geq 1$ and any $y \in N$, there exists $u_n \in U$ such that $d(y, u_n) < 1/2k$, so by the construction of B there is a point $b_{(n,2k)} \in B$ with $d(b_{(n,2k)}, u_n) < 1/2k$. It follows that $d(y, b_{(n,2k)}) < 1/k$, and so B must be dense in N. This proves that N is separable.

3.26 By Lemma 3.25 and Exercise 3.25 each of U and $U^\perp$ are separable Hilbert spaces, so by Theorem 3.52 there exist orthonormal bases $\{e_i\}_{i=1}^m$ and $(f_j)_{j=1}^n$ for U and $U^\perp$ respectively (where m, n are the dimensions of U, $U^\perp$, and may be finite or ∞). We will show that the union $B = \{e_i\}_{i=1}^m \cup \{f_j\}_{j=1}^n$ is an orthonormal basis for $\mathcal{H}$. Firstly, it is clear that B is orthonormal (since $(e_i, f_j) = 0$ for any i, j). Next, by Theorem 3.34, $x = u + v$ with $u \in U$, $v \in U^\perp$, and by Theorem 3.47,

$$x = u + v = \sum_{i=1}^m (u, e_i)e_i + \sum_{j=1}^n (v, f_j)f_j$$
$$= \sum_{i=1}^m (x, e_i)e_i + \sum_{j=1}^n (x, f_j)f_j,$$

and hence, by Theorem 3.47 again, B is an orthonormal basis for $\mathcal{H}$.

3.27 (a) Consider the space $C_{\mathbb{R}}[a,b]$, with norm $\|\cdot\|$, and suppose that $f \in C_{\mathbb{R}}[a,b]$ and $\epsilon > 0$ is arbitrary. By Theorem 1.39 there exists a real polynomial $p_1(x) = \sum_{k=0}^n \alpha_k x^k$ such that $\|f - p_1\| < \epsilon/2$. Now, for each $k = 0, \ldots, n$, we choose rational coefficients β_k, such that $|\beta_n - \alpha_n| < \epsilon/(2n\gamma^k)$ (where $\gamma = \max\{|a|, |b|\}$), and let $p_2(x) = \sum_{k=0}^n \beta_k x^k$. Then

$$\|p_1 - p_2\| \le \sum_{k=0}^n |\beta_n - \alpha_n|\gamma^k < \epsilon/2,$$

and hence $\|f - p_2\| < \epsilon$, which proves the result (by part (f) of Theorem 1.25). For the complex case we apply this result to the real and imaginary parts of $f \in C_{\mathbb{C}}[a,b]$.

(b) Next, consider the space $L^2_{\mathbb{R}}[a,b]$, with norm $\|\cdot\|$, and suppose that $f \in L^2_{\mathbb{R}}[a,b]$ and $\epsilon > 0$ is arbitrary. By Theorem 1.61 there exists a function $g \in C_{\mathbb{R}}[a,b]$ such that $\|f - g\| < \epsilon/2$. Now, by part (a) there exists a polynomial p with rational coefficients such that $\|g - p\|_C < \epsilon/(2\sqrt{|b-a|})$ (where $\|\cdot\|_C$ denotes the norm in $C_{\mathbb{R}}[a,b]$), so the $L^2_{\mathbb{R}}[a,b]$ norm $\|g - p\| < \epsilon/2$ and hence $\|f - p\| < \epsilon$, which proves the result in the real case. Again, for the complex case we apply this result to the real and imaginary parts of f.

Now, the set of polynomials with rational coefficients is countable, and we have just shown that it is dense in $C[a,b]$, so this space is separable.

3.28 (a) The $2n$th order term in the polynomial u_n is clearly x^{2n} and the $2n$th derivative of this is $(2n)!$.

(b) It suffices to consider the case $0 \le m < n$. Now, if $0 \le k < n$, then by k integrations by parts,

$$\begin{aligned}\int_{-1}^{1} x^k \frac{d^n u_n}{dx^n}\,dx &= x^k \frac{d^{n-1}u_n}{dx^{n-1}}\Big|_{-1}^{1} - k\int_{-1}^{1} x^{(k-1)}\frac{d^{n-1}u_n}{dx^{n-1}}\,dx\\ &\ \vdots\\ &= (-1)^k k! \int_{-1}^{1} \frac{d^{n-k}u_n}{dx^{n-k}}\,dx\\ &= (-1)^k k! \frac{d^{n-k-1}u_n}{dx^{n-k-1}}\Big|_{-1}^{1} = 0.\end{aligned}$$

Since P_m has order $m < n$, it follows that $(P_m, P_n) = 0$.

(c) Using (a) and a standard reduction formula we obtain

$$\begin{aligned}\int_{-1}^{1} \frac{d^n u_n}{dx^n}\frac{d^n u_n}{dx^n}\,dx &= (-1)^n \int_{-1}^{1} \frac{d^{2n}u_n}{dx^{2n}} u_n\,dx\\ &= (-1)^n (2n)!\,2\int_0^{\pi/2} (\sin\theta)^{2n+1}\,d\theta\\ &= 2(2n)!\frac{(2n)(2n-2)\dots 2}{(2n+1)(2n-1)\dots 1}\\ &= (2^n n!)^2 \frac{2}{2n+1}.\end{aligned}$$

(d) Parts (b) and (c) show that the given set is orthonormal. Thus, by Theorem 3.47 it suffices to show that $\mathrm{Sp}\,\{e_n : n \ge 0\}$ is dense in $\mathcal{H}$. Now, a simple induction argument shows that any polynomial of order k can be expressed as a linear combination of the functions $e_1, \dots, e_k$ (since, for each $n \ge 0$, the polynomial e_n is of order n). Thus, the set $\mathrm{Sp}\,\{e_n : n \in \mathbb{N}\}$ coincides with the set of polynomials on $[-1, 1]$, which is dense in $\mathcal{H}$ by Exercise 3.27. This proves the result.

Chapter 4

4.1 As $|f(x)| \le \sup\{|f(y)| : y \in [0, 1]\} = \|f\|$ for all $x \in X$,

$$|T(f)| = \left|\int_0^1 f(x)\,dx\right| \le \int_0^1 |f(x)|\,dx \le \int_0^1 \|f\|\,dx = \|f\|.$$

Hence T is continuous.

4.2 (a) As $|g(x)| \leq \|g\|_\infty$ a.e.,

$$|f(x)g(x)|^2 \leq |f(x)|^2 \|g\|_\infty^2 \text{ a.e.}$$

Therefore

$$\int_X |fg|^2 \, d\mu \leq \|g\|_\infty^2 \int_X |f|^2 \, d\mu < \infty,$$

since $f \in L^2$. Thus $fg \in L^2$. Moreover

$$\|fg\|_2^2 = \int_X |fg|^2 \, d\mu \leq \|g\|_\infty^2 \int_X |f|^2 \, d\mu = \|f\|_2^2 \|g\|_\infty^2.$$

(b) $\|T(f)\|_2^2 = \|fh\|_2^2 \leq \|f\|_2^2 \|h\|_\infty^2$ so T is continuous.

4.3 By the Cauchy–Schwarz inequality, for any $x \in \mathcal{H}$,

$$|f(x)|^2 = |(x, y)|^2 \leq \|x\|^2 \|y\|^2$$

so f is continuous.

4.4 (a) We have to show that $\|(0, 4x_1, x_2, 4x_3, x_4, \ldots)\|_2^2 < \infty$. Now

$$\begin{aligned} &\|(0, 4x_1, x_2, 4x_3, x_4, \ldots)\|_2^2 \\ &\qquad = 16|x_1|^2 + |x_2|^2 + 16|x_3|^2 + |x_4|^2 + \ldots \\ &\qquad \leq 16 \sum_{n=1}^{\infty} |x_n|^2 \\ &\qquad < \infty, \end{aligned}$$

as $\{x_n\} \in \ell^2$. Hence $(0, 4x_1, x_2, 4x_3, x_4, \ldots) \in \ell^2$.

(b) T is continuous since

$$\|T(\{x_n\})\|_2^2 = \|(0, 4x_1, x_2, 4x_3, x_4, \ldots)\|_2^2 \leq 16 \|\{x_n\}\|_2^2.$$

4.5 Let $p_n \in \mathcal{P}$ be defined by $p_n(t) = t^n$. Then

$$\|p_n\| = \sup\{|p_n(t)| : t \in [0, 1]\} = 1,$$

while $\|T(p_n)\| = \|p_n'(1)\| = n$. Therefore there does not exist $k \geq 0$ such that $\|T(p)\| \leq k\|p\|$ for all $p \in \mathcal{P}$ and so T is not continuous.

4.6 (a) In the solution of Exercise 4.1 we showed that

$$|T(f)| = \left| \int_0^1 f(x) \, dx \right| \leq \int_0^1 |f(x)| \, dx \leq \int_0^1 \|f\| \, dx = \|f\|.$$

Hence $\|T\| \leq 1$.

(b) $|T(g)| = \left| \int_0^1 g(x)\,dx \right| = 1$. As $\|g\| = \sup\{g(x) : x \in [0,1]\} = 1$,

$$1 = |T(g)| \leq \|T\|\|g\| = \|T\|.$$

Combining this with the result in part (a) it follows that $\|T\| = 1$.

4.7 In the solution of Exercise 4.2 we showed that

$$\|T(f)\|_2^2 = \|fh\|_2^2 \leq \|f\|_2^2 \|h\|_\infty^2.$$

Hence $\|T(f)\|_2 \leq \|f\|_2 \|h\|_\infty$ and so $\|T\| \leq \|h\|_\infty$.

4.8 In the solution of Exercise 4.4 we showed that

$$\|T\{x_n\}\|_2^2 = \|(0, 4x_1, x_2, 4x_3, x_4, \ldots)\|_2^2 \leq 16\|\{x_n\}\|_2^2.$$

Therefore $\|T\{x_n\}\|_2 \leq 4\|\{x_n\}\|_2$ and so

$$\|T\| \leq 4.$$

Moreover $\|(1, 0, 0, \ldots)\|_2 = 1$ and

$$\|T(1, 0, 0, \ldots)\|_2 = \|(0, 4, 0, \ldots)\|_2 = 4.$$

Thus $\|T\| \geq 4$ and so $\|T\| = 4$.

4.9 Since T is an isometry, $\|T(x)\| = \|x\|$ for all $x \in X$, so T is continuous and $\|T\| \leq 1$. Also, if $\|x\| = 1$ then

$$1 = \|x\| = \|T(x)\| \leq \|T\|\|x\| \leq \|T\|$$

and so $\|T\| = 1$.

4.10 By the Cauchy–Schwarz inequality, for all $x \in \mathcal{H}$,

$$|f(x)|^2 = |(x, y)|^2 \leq \|x\|^2 \|y\|^2.$$

Thus $\|f\| \leq \|y\|$. However,

$$f(y) = (y, y) = \|y\|^2$$

and so $\|f\| \geq \|y\|$. Therefore $\|f\| = \|y\|$.

4.11 By the Cauchy–Schwarz inequality,

$$\|Tx\| = \|(x, y)z\| = |(x, y)|\|z\| \leq \|x\|\|y\|\|z\|.$$

Hence T is bounded and $\|T\| \leq \|y\|\|z\|$.

4.12 As $\mathcal{M}$ is a closed linear subspace of a Hilbert space, $\mathcal{M}$ is also a Hilbert space. Since $f \in \mathcal{M}'$ there exists $y \in \mathcal{M}$ such that $f(x) = (x, y)$ for all $x \in \mathcal{M}$ and $\|f\| = \|y\|$ by Theorem 4.31. If we define $g : \mathcal{H} \to \mathbb{C}$ by $g(x) = (x, y)$ for all $x \in \mathcal{H}$ then $g \in \mathcal{H}'$ and $\|g\| = \|y\|$ by Example 4.30 and Exercise 4.10. Hence $g(x) = (x, y) = f(x)$ for all $x \in \mathcal{M}$ and $\|f\| = \|y\| = \|g\|$.

4.13 (a) As $\{x_n\} \in c_0$ and $c_0 \subseteq \ell^\infty$, the result follows from Lemma 4.3.

(b) First note that if b and c are distinct elements of ℓ^1 then $f_b \neq f_c$ so there is at most one element a of ℓ^1 such that $f = f_a$. Also $\widetilde{e}_n \in c_0$ for all $n \in \mathbb{N}$ since $c_0 \subseteq \ell^\infty$. Let $a_n = f(\widetilde{e}_n)$ for all $n \in \mathbb{N}$. If $\mathcal{S}$ is the linear subspace of c_0 consisting of sequences with only finitely many non-zero terms then $\mathcal{S}$ is dense in c_0. Let x be an element of $\mathcal{S}$ where $x = (x_1, x_2, x_3, \ldots, x_n, 0, 0, \ldots)$. Then

$$f(x) = f(\sum_{j=1}^{n} x_j \widetilde{e}_j) = \sum_{j=1}^{n} x_j f(\widetilde{e}_j) = \sum_{j=1}^{n} x_j a_j.$$

Let $m \in \mathbb{N}$ and $\{b_n\}$ be a sequence in $\mathcal{S}$ such that $b_n = 0$ if $n > m$ and $a_n b_n = |a_n|$ for $n \leq m$. Then

$$\sum_{j=1}^{m} |a_j| = \left| \sum_{j=1}^{m} a_j b_j \right| \leq \|f\| \|\{b_n\}\|_\infty = \|f\|.$$

Therefore $\sum_{j=1}^{\infty} |a_j| \leq \|f\|$ so $\{a_n\} \in \ell^1$ and $\|a_n\|_1 \leq \|f\|$. Finally as the continuous functions f and f_a agree on the dense subset $\mathcal{S}$ it follows that $f = f_a$ by Corollary 1.29.

(c) The map $T : \ell^1 \to (c_0)'$ defined by $T(a) = f_a$ is a linear transformation which maps ℓ^1 onto $(c_0)'$ by part (a). From the inequalities $\|f_a\| \leq \|a\|_1$ and $\|a\|_1 \leq \|f\| = \|f_a\|$ it follows that $\|T(a)\| = \|f_a\| = \|a\|_\infty$. Hence T is an isometry.

4.14 By Lemma 4.33,

$$\|T(R)\| = \|PRQ\| \leq \|P\| \|R\| \|Q\|.$$

Thus T is bounded and $\|T\| \leq \|P\| \|Q\|$.

4.15 (a)

$$\begin{aligned} T^2(x_1, x_2, x_3, x_4, \ldots) &= T(T(x_1, x_2, x_3, x_4, \ldots)) \\ &= T(0, 4x_1, x_2, 4x_3, x_4, \ldots) \\ &= (0, 0, 4x_1, 4x_2, 4x_3, 4x_4, \ldots). \end{aligned}$$

(b) From the result in part (a)

$$\|T^2(\{x_n\})\|_2^2 = \|(0,0,4x_1,4x_2,4x_3,4x_4,\ldots)\|_2^2 = 16\|\{x_n\}\|_2^2,$$

and hence $\|T^2\| \leq 4$. Moreover $\|(1,0,0,\ldots)\|_2 = 1$ and

$$\|T^2(1,0,0,\ldots)\|_2 = \|(0,0,4,0,\ldots)\|_2 = 4.$$

Thus $\|T^2\| \geq 4$ and so $\|T^2\| = 4$.

As $\|T\| = 4$ by Exercise 4.8, it follows that $\|T\|^2 \neq \|T^2\|$.

4.16 As S and T are isometries,

$$\|(S \circ T)(x)\| = \|S(T(x))\| = \|T(x)\| = \|x\|,$$

for all $x \in X$. Hence $S \circ T$ is an isometry.

4.17 (a) As

$$T_1^{-1}T_1 = I = T_1T_1^{-1},$$

T_1^{-1} is invertible with inverse T_1.

(b) As

$$(T_1T_2)(T_2^{-1}T_1^{-1}) = T_1(T_2T_2^{-1})T_1^{-1} = T_1T_1^{-1} = I$$

and

$$(T_2^{-1}T_1^{-1})(T_1T_2) = T_2^{-1}(T_1^{-1}T_1)T_2 = T_2^{-1}T_2 = I,$$

T_1T_2 is invertible with inverse $T_2^{-1}T_1^{-1}$.

4.18 The hypothesis in the question means that the identity map $I : X \to X$ from the Banach space $(X, \|\cdot\|_2)$ to the Banach space $(X, \|\cdot\|_1)$ is bounded. Since I is one-to-one and maps X onto X it is invertible by Theorem 4.43. Since the inverse of I is I itself, we deduce that I is a bounded linear operator from $(X, \|\cdot\|_1)$ to $(X, \|\cdot\|_2)$. Therefore, there exists $r > 0$ such that $\|x\|_2 \leq r\|x\|_1$ for all $x \in X$, and so $\|x\|_1$ and $\|x\|_2$ are equivalent.

4.19 (a) Let $\alpha = \inf\{|c_n| : n \in \mathbb{N}\}$. Then $|c_n| \geq \alpha$ for all $n \in \mathbb{N}$ and so $d_n = \dfrac{1}{c_n} \leq \dfrac{1}{\alpha}$ for all $n \in \mathbb{N}$. Thus $d = \{d_n\} \in \ell^\infty$. Now

$$T_cT_d\{x_n\} = T_c\{d_nx_n\} = \{c_nd_nx_n\} = \{x_n\}$$

and

$$T_dT_c\{x_n\} = T_d\{c_nx_n\} = \{d_nc_nx_n\} = \{x_n\}.$$

Hence $T_cT_d = T_dT_c = I$.

(b) As $\lambda \notin \{c_n : n \in \mathbb{N}\}^-$ then $\inf\{|c_n - \lambda| : n \in \mathbb{N}\} > 0$. Hence if $b_n = c_n - \lambda$ then $\{b_n\} \in \ell^\infty$ and as $\inf\{|b_n| : n \in \mathbb{N}\} > 0$ then T_b is invertible by part (a). But $T_b\{x_n\} = \{(c_n - \lambda)x_n\} = \{c_n x_n\} - \{\lambda x_n\} = (T_c - \lambda I)\{x_n\}$ so $T_b = T_c - \lambda I$. Thus $T_c - \lambda I$ is invertible.

4.20 Since $\|\widetilde{e}_n\| = 1$ for all $n \in \mathbb{N}$ and

$$T_c(\widetilde{e}_n) = (0, 0, \dots, 0, \frac{1}{n}, 0, \dots)$$

so that

$$\lim_{n\to\infty} \|T_c(\widetilde{e}_n)\| = \lim_{n\to\infty} \frac{1}{n} = 0,$$

it follows that T_c is not invertible by Corollary 4.49.

4.21 As $\lim_{x\to 0} T_n = T$ there exists $N \in \mathbb{N}$ such that

$$\|T_n - T\| < 1$$

when $n \geq N$. Now

$$\|I - T_n^{-1}T\| = \|T_n^{-1}(T_n - T)\| \leq \|T_n^{-1}\|\|T_n - T\| < 1$$

when $n \geq N$. Thus $T_N^{-1}T$ is invertible by Theorem 4.40 and so $T = T_N T_N^{-1} T$ is invertible.

4.22 (a) f^y is a function from S to X and $\{\|f^y(s)\| : s \in S\} = \{\|T_s(y)\| : s \in S\}$ and hence is bounded. Therefore $f^y \in F_b(S, X)$.

(b) Let $\{(y_n, \phi(y_n))\}$ be a sequence in $\mathcal{G}(\phi)$ which converges to (y, g) where $y \in Z$ and $g \in F_b(S, X)$. Then

$$\lim_{n\to\infty} \|g(s) - \phi(y_n)(s)\| \leq \lim_{n\to\infty} \|g - \phi(y_n)\|_b = 0$$

and so, as T_s is continuous,

$$g(s) = \lim_{n\to\infty} \phi(y_n)(s) = \lim_{n\to\infty} f^{y_n}(s) = \lim_{n\to\infty} T_s(y_n) = T_s(y).$$

Thus $g(s) = f^y(s) = \phi(y)(s)$ and therefore $g = \phi(y)$. Hence $\mathcal{G}(\phi)$ is closed.

(c) Since ϕ is bounded by the closed graph theorem (Corollary 4.44) we have

$$\|T_s(y)\| = \|f^y(s)\| \leq \|f^y\|_b = \|\phi(y)\|_b \leq \|\phi\|\|y\|,$$

for all $s \in S$ and all $y \in Y$. Therefore $\|T_s\| \leq \|\phi\|$ for all $s \in S$ and so $\{\|T_s\| : s \in S\}$ is bounded by $\|\phi\|$.

Chapter 5

5.1 Let $\{x_n\}, \{y_n\} \in \ell^2$ and let $\{z_n\} = T_c^*(\{y_n\})$. From

$$(\{c_n x_n\}, \{y_n\}) = (T_c\{x_n\}, \{y_n\}) = (\{x_n\}, \{z_n\}),$$

we obtain $\sum_{n=1}^{\infty} c_n x_n \overline{y_n} = \sum_{n=1}^{\infty} x_n \overline{z_n}$. This is satisfied for all $\{x_n\} \in \ell^2$ if $\overline{z_n} = c_n \overline{y_n}$ (or $z_n = \overline{c_n} y_n$) for all $n \in \mathbb{N}$. Hence, writing $\overline{c} = \{\overline{c_n}\}$ we have $(T_c)^* = T_{\overline{c}}$ by the uniqueness of the adjoint.

5.2 Let $x = \{x_n\}, y = \{y_n\} \in \ell^2$ and let $z = \{z_n\} = (T)^*(\{y_n\})$. Since $(Tx, y) = (x, z)$, we have

$$\begin{aligned}((0, 4x_1, x_2, 4x_3, x_4, \ldots)&, (y_1, y_2, y_3, y_4, \ldots)) \\ &= ((x_1, x_2, x_3, x_4, \ldots), (z_1, z_2, z_3, z_4, \ldots)).\end{aligned}$$

Therefore

$$4x_1\overline{y_2} + x_2\overline{y_3} + 4x_3\overline{y_4} + \ldots = x_1\overline{z_1} + x_2\overline{z_2} + x_3\overline{z_3} + \ldots .$$

This is satisfied if $z_1 = 4y_2, z_2 = y_3, z_3 = 4y_4, \ldots$. Hence it follows that $T^*(y) = (4y_2, y_3, 4y_4, \ldots)$ by the uniqueness of the adjoint.

5.3 $(Tx, w) = ((x, y)z, w) = (x, y)(z, w) = \overline{(w, z)}(x, y) = (x, (w, z)y)$. Thus $T^*(w) = (w, z)y$ by the uniqueness of the adjoint.

5.4 (a) For all $x \in \mathcal{H}$ and $y \in \mathcal{K}$ we have

$$\begin{aligned}(x, (\mu R + \lambda S)^* y) &= ((\mu R + \lambda S)x, y) \\ &= (\mu Rx + \lambda Sx, y) \\ &= \mu(Rx, y) + \lambda(Sx, y) \\ &= \mu(x, R^*y) + \lambda(x, S^*y) \\ &= (x, \overline{\mu}R^*y) + (x, \overline{\lambda}S^*y) \\ &= (x, (\overline{\mu}R^* + \overline{\lambda}S^*)y).\end{aligned}$$

Thus $(\lambda S + \mu R)^* = \overline{\lambda}S^* + \overline{\mu}R^*$ by the uniqueness of the adjoint.

(b) For all $x \in \mathcal{H}$ and $z \in \mathcal{L}$ we have

$$(TRx, z) = (Rx, T^*z) = (x, R^*T^*z).$$

Hence $(TR)^* = R^*T^*$ by the uniqueness of the adjoint.

5.5 (a) If $x \in \text{Ker } T$ then $Tx = 0$ so $T^*Tx = 0$. Hence $x \in \text{Ker } (T^*T)$.

Conversely if $x \in \text{Ker } (T^*T)$, we have $T^*Tx = 0$. Hence

$$(x, T^*Tx) = 0,$$

and so $(Tx, Tx) = 0$. Therefore $Tx = 0$ and thus $x \in \text{Ker } T$.

Combining these results it follows that $\text{Ker } T = \text{Ker } (T^*T)$.

(b)

$$\begin{aligned} \overline{\text{Im } T^*} &= ((\text{Im } T^*)^\perp)^\perp && \text{by Corollary 3.36} \\ &= (\text{Ker } T)^\perp && \text{by Lemma 5.11} \\ &= (\text{Ker } T^*T)^\perp && \text{by part (a)} \\ &= ((\text{Im } (T^*T)^*)^\perp)^\perp && \text{by Lemma 5.11} \\ &= ((\text{Im } T^*T)^\perp)^\perp && \text{as } (T^*T)^* = T^*T \\ &= \overline{\text{Im } T^*T} && \text{by Corollary 3.36.} \end{aligned}$$

5.6 $A^* = \begin{bmatrix} 1 & 0 \\ 1 & 1 \end{bmatrix}$, so $AA^* = \begin{bmatrix} 2 & 1 \\ 1 & 1 \end{bmatrix}$, while $A^*A = \begin{bmatrix} 1 & 1 \\ 1 & 2 \end{bmatrix}$. Therefore A is not normal.

5.7 (a) $(T_c)^* = T_{\overline{c}}$ so

$$\begin{aligned} (T_c)^*T_c(\{x_n\}) &= T_{\overline{c}}T_c(\{x_n\}) \\ &= T_{\overline{c}}(\{c_n x_n\}) \\ &= \{\overline{c_n} c_n x_n\} \\ &= \{|c_n|^2 x_n\} \\ &= T_{|c|^2}\{x_n\}. \end{aligned}$$

Therefore $(T_c)^*T_c = T_{|c|^2}$.

$$\begin{aligned} T_c(T_c)^*(\{x_n\}) &= T_c T_{\overline{c}}(\{x_n\}) \\ &= T_c(\{\overline{c_n} x_n\}) \\ &= \{c_n \overline{c_n} x_n\} \\ &= \{|c_n|^2 x_n\} \\ &= T_{|c|^2}\{x_n\}. \end{aligned}$$

Thus $T_c(T_c)^* = T_{|c|^2}$ and so T_c is normal.

(b) $T^*T(\{x_n\}) = T^*(0, 2x_1, x_2, 2x_3, x_4, \ldots) = (4x_1, x_2, 4x_3, x_4, \ldots)$.
$TT^*(\{x_n\}) = T(2x_2, x_3, 2x_4, x_5, \ldots) = (0, 4x_2, x_3, 4x_4, \ldots)$.
Hence $T^*T \neq TT^*$ so T is not normal.

5.8 By first using the identity in Lemma 3.14(b), with $u = Tx$ and $v = Ty$, and the hypothesis in Lemma 5.29 we obtain

$$\begin{aligned} 4(Tx, y) &= (T(x+y), x+y) - (T(x-y), x-y) \\ &\quad + i(T(x+iy), x+iy) - i(T(x-iy), x-iy) \\ &= (S(x+y), x+y) - (S(x-y), x-y) \\ &\quad + i(S(x+iy), x+iy) - i(S(x-iy), x-iy) \\ &= 4(Sx, y), \end{aligned}$$

using the identity in Lemma 3.14(b) with $u = Sx$ and $v = Sy$.
Thus $(Tx, y) = (Sx, y)$ for all $x, y \in \mathcal{H}$ and so $Tx = Sx$ for all $x \in \mathcal{H}$ by Exercise 3.1. Hence $S = T$.

5.9 $(TT^*x, x) = (T^*x, T^*x) = \|T^*x\|^2 = \|Tx\|^2$ (by the assumption in the exercise) $= (Tx, Tx) = (T^*Tx, x)$. Thus, $T^*T = TT^*$ by Exercise 5.8 and so T is normal.

5.10 (a) If $c_n \in \mathbb{R}$ for all $n \in \mathbb{N}, c = \overline{c}$ and so $(T_c)^* = T_{\overline{c}} = T_c$. Hence T_c is self-adjoint.

(b) If $|c_n| = 1$ for all $n \in \mathbb{N}$, then $T_c(T_c)^* = T_{|c|^2} = I$ and similarly $(T_c)^*T_c = I$. Therefore T_c is unitary.

5.11 By Lemma 5.8 and Theorem 5.10, we have

$$(T^*ST)^* = T^*S^*T^{**} = T^*ST,$$

as S is self-adjoint. Hence T^*ST is self-adjoint

5.12 By Lemma 5.14, A^* is invertible and $(A^*)^{-1} = (A^{-1})^*$. However, as A is self-adjoint, $A^* = A$ so $A^{-1} = (A^{-1})^*$. Hence A^{-1} is self-adjoint

5.13 As S and T are self-adjoint, $(ST)^* = T^*S^* = TS$ by Lemma 5.8. Therefore $ST = (ST)^*$ if and only if $ST = TS$.

5.14 (a) $U^*U^{**} = U^*U = I$ and similarly $U^{**}U^* = I$, since U is unitary. Hence U^* is unitary. U is an isometry, by Theorem 5.30, so $\|U\| = 1$. Since U^* is also unitary, $\|U^*\| = 1$.

(b) $(U_1U_2)^* = U_2^*U_1^*$ so

$$(U_1U_2)^*U_1U_2 = U_2^*U_1^*U_1U_2 = U_2^*U_2 = I,$$

as U_1, $U_2 \in \mathcal{U}$ and similarly $U_1U_2(U_1U_2)^* = I$. Hence $U_1U_2 \in \mathcal{U}$.
As U_1 is unitary $(U_1)^* = U_1^{-1}$ and so $U_1^{-1} \in \mathcal{U}$.

(c) Let $\{U_n\}$ be a sequence in $\mathcal{U}$ which converges to $U \in B(\mathcal{H})$. Then $\{U_n^*\}$ converges to U^* so $\{U_nU_n^*\}$ converges to UU^*. However, $U_nU_n^* = I$ for all $n \in \mathbb{N}$ and so $UU^* = I$. Similarly, $U^*U = I$ so $U \in \mathcal{U}$. Hence $\mathcal{U}$ is a closed subset of $B(\mathcal{H})$.

5.15 As U is unitary,

$$\begin{aligned}\|T\| &= \|UU^*TUU^*\| \\ &\leq \|U\|\|U^*TU\|\|U^*\| \\ &= \|U\|\|f(T)\|\|U^*\| \\ &= \|U^*TU\| \\ &\leq \|U^*\|\|T\|\|U\| \\ &= \|T\|\end{aligned}$$

for all $T \in B(\mathcal{H})$. Hence $\|f(T)\| = \|T\|$ and so f is an isometry.

5.16 (a) $(1,0,0,\ldots)$ and $(0,1,0,0,\ldots)$ are in ℓ^2 and

$$T(1,0,0,\ldots) = (1,0,0,\ldots) = 1(1,0,0,\ldots)$$

so 1 is an eigenvalue of T with eigenvector $(1,0,0,\ldots)$. Also

$$T(0,1,0,0,\ldots) = (0,-1,0,0,\ldots) = (-1)(0,1,0,0,\ldots)$$

so -1 is an eigenvalue of T with eigenvector $(0,1,0,0,\ldots)$.

(b) $T^2 = I$ as

$$\begin{aligned} T^2(x_1,x_2,x_3,x_4,\ldots) &= T(x_1,-x_2,x_3,-x_4,\ldots) \\ &= (x_1,x_2,x_3,x_4,\ldots). \end{aligned}$$

Thus $\sigma(T^2) = \{1\}$ and thus since $\sigma(T^2) = (\sigma(T))^2$ it follows that $\sigma(T) \subseteq \{-1,1\}$. But from part (a), 1 and -1 are eigenvalues of T, so $\{-1,1\} \subseteq \sigma(T)$. Hence $\sigma(T) = \{-1,1\}$.

5.17 We have

$$S^*S(x_1,x_2,x_3,\ldots) = S^*(0,x_1,x_2,x_3,\ldots) = (x_1,x_2,x_3,\ldots)$$

so $S^*S = I$. On the other hand

$$SS^*(x_1,x_2,x_3,\ldots) = S(x_2,x_3,\ldots) = (0,x_2,x_3,\ldots).$$

Thus $SS^*(1,0,0,\ldots) = (0,0,0,\ldots) = 0(1,0,0,\ldots)$ so 0 is an eigenvalue of SS^* with eigenvector $(1,0,0,\ldots)$.

5.18 (a) Clearly,

$$T_c(\widetilde{e}_m) = (0,0,\ldots,0,c_m,0,\ldots) = c_m(0,0,\ldots,0,1,0,\ldots) = c_m\widetilde{e}_m.$$

Hence c_m is an eigenvalue of T_c with eigenvector $\widetilde{e}_m$.

(b) $\{c_n : n \in \mathbb{N}\} \subseteq \sigma(T_c)$ by part (a) and so, as $\sigma(T_c)$ is closed,

$$\{c_n : n \in \mathbb{N}\}^- \subseteq \sigma(T_c).$$

5.19 (a) $T^*(x_1,x_2,x_3,x_4,\ldots) = (4x_2,x_3,4x_4,\ldots)$ by Exercise 5.2. Hence

$$\begin{aligned} (T^*)^2(x_1,x_2,x_3,x_4,\ldots) &= T^*(T^*(x_1,x_2,x_3,x_4,\ldots)) \\ &= T^*(4x_2,x_3,4x_4,x_5,\ldots) \\ &= (4x_3,4x_4,4x_5,4x_6,\ldots). \end{aligned}$$

We need to find a non-zero $\{x_n\} \in \ell^2$ such that

$$(4x_3,4x_4,4x_5,\ldots) = (\mu x_1,\mu x_2,\mu x_3,\ldots),$$

that is, $4x_{n+2} = \mu x_n$ for all $n \in \mathbb{N}$. Let $x_1 = x_2 = 1$ and $x_{2n-1} = x_{2n} = \left(\frac{\mu}{4}\right)^{n-1}$ for $n \geq 2$. Then $\{x_n\}$ is non-zero and, as $|\mu| < 4$,

$$\sum_{n=1}^{\infty} |x_n|^2 = 2\sum_{n=0}^{\infty} \left(\frac{|\mu|}{4}\right)^{2(n-1)} < \infty,$$

so $\{x_n\} \in \ell^2$. Thus $(T^*)^2(\{x_n\}) = \mu\{x_n\}$ and so μ is an eigenvalue of $(T^*)^2$ with eigenvector $\{x_n\}$.

(b) $\{\lambda \in \mathbb{C} : |\lambda| < 4\} \subseteq \sigma((T^*)^2)$ by part (a) and Lemma 5.34. Thus, $\{\overline{\lambda} \in \mathbb{C} : |\lambda| < 4\} \subseteq \sigma(T^2)$ by Lemma 5.37. However, from elementary geometry

$$\{\overline{\lambda} \in \mathbb{C} : |\lambda| < 4\} = \{\lambda \in \mathbb{C} : |\lambda| < 4\},$$

so $\{\lambda \in \mathbb{C} : |\lambda| < 4\} \subseteq \sigma(T^2)$. As $\sigma(T^2)$ is closed by Theorem 5.36

$$\{\lambda \in \mathbb{C} : |\lambda| \leq 4\} \subseteq \sigma(T^2).$$

Hence if $|\lambda| \leq 2$ then $\lambda \in \sigma(T)$ otherwise $\lambda^2 \notin \sigma(T^2)$ by Theorem 5.39. This implies that $|\lambda|^2 > 4$, which is a contradiction. On the other hand, if $\lambda \in \sigma(T)$ then $\lambda^2 \in \sigma(T^2)$ by Theorem 5.39. Hence

$$|\lambda|^2 \leq r_\sigma(T^2) \leq \|T^2\| = 4$$

by Theorem 5.36 and Exercise 4.15. Hence $\sigma(T) = \{\lambda \in \mathbb{C} : |\lambda| \leq 2\}$.

5.20 (a) A is self-adjoint so its norm is the maximum of $|\lambda_1|$ and $|\lambda_2|$ where λ_1 and λ_2 are the eigenvalues of A. The characteristic equation of A is

$$\det \begin{bmatrix} 1-\lambda & 1 \\ 1 & 2-\lambda \end{bmatrix} = 0,$$

that is, $(1-\lambda)(2-\lambda)-1 = 0$. Hence the eigenvalues of A are $\frac{3 \pm \sqrt{9-4}}{2}$ or $\frac{3 \pm \sqrt{5}}{2}$. Thus the norm of A is $\frac{3+\sqrt{5}}{2}$.

(b) $\|B\|^2 = \|B^*B\| = \left\|\begin{bmatrix} 1 & 1 \\ 1 & 2 \end{bmatrix}\right\|$. Thus, $\|B\| = \sqrt{\frac{3+\sqrt{5}}{2}}$ by part (a).

5.21 $(S^n)^* = (S^*)^n = S^n$ as S is self-adjoint. Hence S^n is self-adjoint and thus by Theorem 5.43

$$\begin{aligned} \|S^n\| &= \sup\{|\mu| : \mu \in \sigma(S^n)\} \\ &= \sup\{|\lambda^n| : \lambda \in \sigma(S)\} \\ &= (\sup\{|\lambda| : \lambda \in \sigma(S)\})^n \\ &= \|S\|^n. \end{aligned}$$

5.22 Since $S-\lambda I$ is self-adjoint and $\sigma(S-\lambda I)=\{0\}$ by Theorem 5.39, it follows that
$$\|S-\lambda I\| = r_\sigma(S-\lambda I) = 0,$$
by Theorem 5.43. Hence $S-\lambda I = 0$ and so $S=\lambda I$.

5.23 Let T be the linear transformation defined on ℓ^2 by
$$T(x_1,x_2,x_3,x_4,\ldots) = (0,x_1,0,x_3,0,\ldots).$$
Then T is bounded as
$$\begin{aligned}\|T(x_1,x_2,x_3,x_4,\ldots)\|^2 &= \|(0,x_1,0,x_3,0,\ldots)\|^2\\ &\leq \|(x_1,x_2,x_3,x_4,\ldots)\|^2.\end{aligned}$$
Moreover, T is non-zero but
$$\begin{aligned}T^2(x_1,x_2,x_3,x_4,\ldots) &= T(0,x_1,0,x_3,0,\ldots)\\ &= (0,0,0,0,0,\ldots).\end{aligned}$$
Hence $T^2=0$, so if $\lambda\in\sigma(T)$ then $\lambda^2\in\sigma(T^2)=\{0\}$. On the other hand $T(0,1,0,0,\ldots)=(0,0,0,0,0,\ldots)$, so 0 is an eigenvalue for T and so $0\in\sigma(T)$. Hence $\sigma(T)=\{0\}$.

5.24 As A is self-adjoint, A^{-1} is also self-adjoint. As A is positive, $\lambda\geq 0$ for all $\lambda\in\sigma(A)$. Hence
$$\sigma(A^{-1}) = \{\lambda^{-1} : \lambda\in\sigma(A)\}\subseteq [0,\infty)$$
and so A is positive.

5.25 $Px=\sum_{n=1}^{J}(Px,e_n)e_n$ since $Px\in\mathcal{M}$, by Theorem 3.47. Hence
$$Px = \sum_{n=1}^{J}(Px,e_n)e_n = \sum_{n=1}^{J}(x,Pe_n)e_n = \sum_{n=1}^{J}(x,e_n)e_n,$$
as P is self-adjoint and $e_n\in\mathcal{M}$ so $Pe_n=e_n$ for $1\leq n\leq J$.

5.26 Since $P^2=P$,
$$S^2=(2P-I)^2=4P^2-4P+I=I.$$
If $\lambda\in\sigma(S)$ then $\lambda^2\in\sigma(S^2)=\sigma(I)=\{1\}$. Then, $\lambda=\pm 1$ and so
$$\sigma(S)\subseteq\{-1,1\}.$$
As $P=\dfrac{1}{2}(S+I)$ it follows that $\sigma(P)=\dfrac{1}{2}\sigma(S)+1\subseteq\dfrac{1}{2}\{0,2\}=\{0,1\}$ by Theorem 5.39.

5.27 (a) PQ is self-adjoint as $PQ = QP$. Also

$$(PQ)^2 = PQPQ = PPQQ = PQ,$$

as $PQ = QP$ and P and Q are orthogonal projections.

(b) As P is self-adjoint, $(PQx, y) = (Qx, Py)$. Hence

$$\begin{aligned} PQ = 0 &\iff (PQx, y) = 0 \text{ for all } x \text{ and } y \in \mathcal{H} \\ &\iff (Qx, Py) = 0 \text{ for all } x \text{ and } y \in \mathcal{H} \\ &\iff \text{Im } Q \text{ is orthogonal to Im } P, \end{aligned}$$

as every element of Im Q is of the form Qx for some $x \in \mathcal{H}$ and every element of Im P is of the form Py for some $y \in \mathcal{H}$.

5.28 (a) $\Rightarrow$ (b). If $x \in \mathcal{H}$ then

$$Px \in \text{Im } P \subseteq \text{Im } Q.$$

Hence $QPx = Px$ so $QP = P$.
(b) $\Rightarrow$ (c). As P and Q are orthogonal projections $P^* = P$ and $Q^* = Q$ so $P = P^* = (QP)^* = P^*Q^* = PQ$.
(c) $\Rightarrow$ (d). $\|Px\| = \|PQx\| \leq \|P\|\|Qx\| = \|Qx\|$ as $\|P\| = 1$.
(d) $\Rightarrow$ (e). As P is an orthogonal projection,

$$(Px, x) = (P^2x, x) = (Px, P^*x) = (Px, Px) = \|Px\|^2,$$

and similarly $(Qx, x) = \|Qx\|^2$ so

$$(Px, x) = \|Px\|^2 \leq \|Qx\|^2 = (Qx, x).$$

Hence $((Q - P)x, x) \geq 0$ for all $x \in \mathcal{H}$ and so, as $Q - P$ is self-adjoint, $P \leq Q$.
(e) $\Rightarrow$ (a). If $y \in \text{Im } P$, let $y = Qy + z$ where $Qy \in \text{Im } Q$ and $z \in (\text{Im } Q)^\perp$ be the orthogonal decomposition. Since $y \in \text{Im } P$ and $P \leq Q$,

$$\|y\|^2 = (y, y) = (Py, y) \leq (Qy, y) = (Q^2y, y) = (Qy, Qy) = \|Qy\|^2,$$

so $\|z\|^2 = 0$ as $\|y\|^2 = \|Qy\|^2 + \|z\|^2$. Hence $y = Qy \in \text{Im } Q$ so Im $P \subseteq$ Im Q.

5.29 Let $h, k \in C_{\mathbb{R}}(\sigma(S))$ be defined by $h(x) = 1$ and $k(x) = x$ for all $x \in \sigma(S)$. For $1 \leq j \leq n$, let $f_j : \sigma(S) \to \mathbb{R}$ be the function defined by

$$f_j(x) = \begin{cases} 1 & \text{if } x = \lambda_j, \\ 0 & \text{if } x \neq \lambda_j. \end{cases}$$

As $\sigma(S)$ is finite, $f_j \in C_{\mathbb{R}}(\sigma(S))$ with $f_j^2 = f_j$ and $f_j f_m = 0$ if $j \neq m$. In addition $\sum_{j=1}^{n} f_j = h$ and $\sum_{j=1}^{n} \lambda_j f_j = k$. Let $P_j = f_j(S)$. Then P_j is self-adjoint and

$$P_j^2 = f_j^2(S) = f_j(S) = P_j,$$

by Lemma 5.57, so P_j is an orthogonal projection. Similarly,

$$P_j P_k = (f_j f_k)(S) = 0$$

if $j \neq k$. Moreover, by Lemma 5.57 again,

$$I = h(S) = \sum_{j=1}^{n} f_j(S) = \sum_{j=1}^{n} P_j,$$

and

$$S = k(S) = \sum_{j=1}^{n} \lambda_j f_j(S) = \sum_{j=1}^{n} \lambda_j P_j.$$

5.30 (a) As S is self-adjoint so is S^2 and $I - S^2$. Also $\sigma(S) \subseteq [-1, 1]$ as S is self-adjoint and $\|S\| \leq 1$. Hence

$$\sigma(S^2) = (\sigma(S))^2 \subseteq [0, 1]$$

and so $\sigma(I - S^2) \subseteq [0, 1]$.

(b) As $I - S^2$ is positive, $I - S^2$ has a square root. If p is any polynomial, $Sp(I - S^2) = p(I - S^2)S$ so $S(I - S^2)^{1/2} = (I - S^2)^{1/2}S$, by Theorem 5.58. Let $U_1 = S + i(I - S^2)^{1/2}$ and $U_2 = S - i(I - S^2)^{1/2}$. As $(I - S^2)^{1/2}$ is self-adjoint $U_1^* = U_2$ and so

$$U_1^* U_1 = (S + i(I - S^2)^{1/2})(S - i(I - S^2)^{1/2}) = S^2 + (I - S^2) = I.$$

Similarly $U_1 U_1^* = I$. Hence U_1 is unitary, and U_2 is also unitary as $U_1^* = U_2$.

5.31 (a) The characteristic equation of A is $(1-\lambda)(9-\lambda) = 0$, so the eigenvalues of A are 1 and 9. A normalized eigenvector corresponding to the eigenvalue 1 is $\frac{1}{\sqrt{2}}(1, 1)$ and a normalized eigenvector corresponding to the eigenvalue 9 is $\frac{1}{\sqrt{2}}(1, -1)$. Let $U = \frac{1}{\sqrt{2}}\begin{bmatrix} 1 & 1 \\ 1 & -1 \end{bmatrix}$. Then $U^*AU = D$ where $D = \begin{bmatrix} 1 & 0 \\ 0 & 9 \end{bmatrix}$. Let $E = \begin{bmatrix} 1 & 0 \\ 0 & 3 \end{bmatrix}$. Then the square root of A is $UEU^* = \begin{bmatrix} 2 & -1 \\ -1 & 2 \end{bmatrix}$.

(b) We follow the method in the proof of Theorem 5.59. The matrix $B^* = \frac{1}{\sqrt{2}}\begin{bmatrix} 2-i & 2+i \\ 2i-1 & -1-2i \end{bmatrix}$, so $B^*B = \begin{bmatrix} 5 & -4 \\ -4 & 5 \end{bmatrix}$. By the solution to part (a) it follows that $C = (B^*B)^{\frac{1}{2}} = \begin{bmatrix} 2 & -1 \\ -1 & 2 \end{bmatrix}$. Now $C^{-1} = \frac{1}{3}\begin{bmatrix} 2 & 1 \\ 1 & 2 \end{bmatrix}$ so $U = BC^{-1} = \frac{1}{\sqrt{2}}\begin{bmatrix} 1 & i \\ 1 & -i \end{bmatrix}$. Hence the polar decomposition of B is $\left(\frac{1}{\sqrt{2}}\begin{bmatrix} 1 & i \\ 1 & -i \end{bmatrix}\right)\begin{bmatrix} 2 & -1 \\ -1 & 2 \end{bmatrix}$.

Chapter 6

6.1 For any bounded sequence $\{x_n\}$ in X, the sequence $T_0x_n = 0$, $n = 1, 2, \ldots$, clearly converges to zero, so T_0 is compact. Alternatively, T_0 is bounded and has finite rank so is compact.

6.2 Exercise 4.11 shows that T is bounded and it clearly has finite rank ($\operatorname{Im} T \subset \operatorname{Sp}\{z\}$), so T is compact.

6.3 Suppose that T has the property that for any sequence of vectors $\{x_n\}$ in the closed unit ball $\overline{B_1(0)} \subset X$, the sequence $\{Tx_n\}$ has a convergent subsequence, and let $\{y_n\}$ be an arbitrary bounded sequence in X. Since $\{y_n\}$ is bounded there is a number $M > 0$ such that $\|y_n\| \le M$ for all $n \in \mathbb{N}$. Thus the sequence $\{M^{-1}y_n\}$ lies in the closed unit ball $\overline{B_1(0)}$, so by the above assumption the sequence $\{TM^{-1}y_n\} = \{M^{-1}Ty_n\}$ has a convergent subsequence, and hence $\{Ty_n\}$ must have a convergent subsequence. This shows that T is compact. The reverse implication is trivial.

6.4 For any $m, n \in \mathbb{N}$, $m \ne n$, we have $\|e_m - e_n\|^2 = (e_m - e_n, e_m - e_n) = 2$. Thus the members of any subsequence of $\{e_n\}$ are all a distance $\sqrt{2}$ apart, and hence no subsequence can be a Cauchy sequence, so no subsequence converges.

6.5 The space $B(X,Y)$ is a Banach space (see Theorem 4.27), and Theorems 6.3 and 6.9 show that $K(X,Y)$ is a closed linear subspace of $B(X,Y)$. Therefore, by Theorem 2.28(e), $K(X,Y)$ is a Banach space.

6.6 It follows from Exercise 5.5(b) and Lemma 6.13 that either the numbers $r(T)$ and $r(T^*T)$ are both infinite, or they are finite and equal.

6.7 Suppose that the sequence $\{\|Te_n\|\}$ does not converge to 0. Then by compactness of T there exists a subsequence $\{e_{n(r)}\}$ and a vector $y \ne 0$ such that

$$\|Te_{n(r)} - y\| < r^{-1}, \quad r = 1, 2, \ldots.$$

Now, for each $k \in \mathbb{N}$, let $x_k = \sum_{r=1}^{k} r^{-1} e_{n(r)}$. Since $\sum_{r=1}^{\infty} r^{-2} < \infty$ and the sequence $\{e_{n(r)}\}$ is orthonormal, Theorem 3.42 shows that the sequence $\{x_k\}$ converges, so is bounded. However,

$$\|Tx_k\| = \Big\| \sum_{r=1}^{k} \frac{1}{r} T e_{n(r)} \Big\| \geq \sum_{r=1}^{k} \frac{1}{r} \|y\| - \sum_{r=1}^{k} \frac{1}{r^2} \to \infty,$$

which contradicts the boundedness of T.

6.8 (a) Using part (c) of Theorem 3.47 we have

$$\begin{aligned} \sum_{n=1}^{\infty} \|Te_n\|^2 &= \sum_{n=1}^{\infty} \Big(\sum_{m=1}^{\infty} |(Te_n, f_m)|^2 \Big) \\ &= \sum_{m=1}^{\infty} \Big(\sum_{n=1}^{\infty} |(e_n, T^* f_m)|^2 \Big) = \sum_{m=1}^{\infty} \|T^* f_m\|^2 \end{aligned}$$

(the change of order of the infinite sums is valid since all the terms in the summation are positive). A similar calculation also shows that

$$\sum_{n=1}^{\infty} \|Tf_n\|^2 = \sum_{m=1}^{\infty} \|T^* f_m\|^2,$$

which proves the formulae in part (a) of Theorem 6.16. It follows from these formulae that $\sum_{n=1}^{\infty} \|Te_n\|^2 < \infty$ if and only if $\sum_{n=1}^{\infty} \|Tf_n\|^2 < \infty$, so the Hilbert–Schmidt condition does not depend on the basis.

(b) It follows from the results in part (a), by putting the basis $\{f_n\}$ equal to $\{e_n\}$, that $\sum_{n=1}^{\infty} \|Te_n\|^2 < \infty$ if and only if $\sum_{n=1}^{\infty} \|T^* e_n\|^2 < \infty$, which proves part (b) of the theorem.

(c) We use Corollary 6.10. Since $\{e_n\}$ is an orthonormal basis, any $x \in \mathcal{H}$ can be written as $x = \sum_{n=1}^{\infty} (x, e_n) e_n$, by Theorem 3.47. For each $k \in \mathbb{N}$ we now define an operator $T_k \in B(\mathcal{H})$ by

$$T_k x = T\Big(\sum_{n=1}^{k} (x, e_n) e_n \Big) = \sum_{n=1}^{k} (x, e_n) T e_n.$$

Clearly, $r(T_k) \le k$. Also, for any $x \in \mathcal{H}$,

$$\begin{aligned}\|(T_k - T)x\| &= \Big\| \sum_{n=1}^{k}(x, e_n)Te_n - \sum_{n=1}^{\infty}(x, e_n)Te_n \Big\| \\ &\le \sum_{n=k+1}^{\infty} |(x, e_n)|\, \|Te_n\| \\ &\le \Big(\sum_{n=k+1}^{\infty} |(x, e_n)|^2 \Big)^{1/2} \Big(\sum_{n=k+1}^{\infty} \|Te_n\|^2 \Big)^{1/2} \\ &\le \|x\| \Big(\sum_{n=k+1}^{\infty} \|Te_n\|^2 \Big)^{1/2}\end{aligned}$$

(using the Cauchy–Schwarz inequality for ℓ^2 and Theorem 3.47). Hence

$$\|T_k - T\| \le \Big(\sum_{n=k+1}^{\infty} \|Te_n\|^2 \Big)^{1/2},$$

and since the series on the right converges, we have $\lim_{k\to\infty} \|T_k - T\| = 0$. Thus Corollary 6.10 shows that T is compact.

(d) If S, $T \in B(\mathcal{H})$ are Hilbert–Schmidt and $\alpha \in \mathbb{F}$, then

$$\sum_{n=1}^{\infty} \|\alpha Te_n\|^2 = |\alpha|^2 \sum_{n=1}^{\infty} \|Te_n\|^2 < \infty,$$

$$\begin{aligned}\sum_{n=1}^{\infty} \|(S+T)e_n\|^2 &\le \sum_{n=1}^{\infty}(\|Se_n\| + \|Te_n\|)^2 \\ &\le 2\sum_{n=1}^{\infty}(\|Se_n\|^2 + \|Te_n\|^2) < \infty,\end{aligned}$$

so αT and $S + T$ are Hilbert–Schmidt and hence the set of Hilbert–Schmidt operators is a linear subspace.

6.9 (a) Suppose that T is Hilbert–Schmidt. Then

$$\sum_{n=1}^{\infty} \|STe_n\|^2 \le \sum_{n=1}^{\infty} \|S\|^2 \|Te_n\|^2 < \infty,$$

so ST is Hilbert–Schmidt. The proof is not quite so straightforward when S is Hilbert–Schmidt because the set of vectors $\{Te_n\}$ need not be an orthonormal basis. However, we can can turn this case into the first case by taking the adjoint; thus, $(ST)^* = T^*S^*$, and by part (b) of Theorem 6.16, S^* is Hilbert–Schmidt, so $(ST)^*$ is Hilbert–Schmidt, and hence ST is Hilbert–Schmidt.

(b) If T has finite rank then by Exercise 3.26 there exists an orthonormal basis $\{e_n\}$ such that $e_n \in \operatorname{Im} T$ if $1 \leq n \leq r(T) < \infty$, and $e_n \in (\operatorname{Im} T)^\perp$ if $n > r(T)$. Now, by Lemma 5.11, $(\operatorname{Im} T)^\perp = \operatorname{Ker} T^*$ so

$$\sum_{n=1}^{\infty} \|T^* e_n\|^2 = \sum_{n=1}^{r(T)} \|T^* e_n\|^2 < \infty.$$

Thus T^* is Hilbert–Schmidt and so, by part (b) of Theorem 6.16, T is also Hilbert–Schmidt.

6.10 Let $\{e_n\}$ be the orthonormal basis of $L^2[-\pi, \pi]$ with $e_n(t) = (2\pi)^{-1/2} e^{int}$, $n \in \mathbb{Z}$, discussed in Corollary 3.57 (again, we could relabel this basis so that it is indexed by $n \in \mathbb{N}$, but this is a minor point). Since k is a non-zero continuous function on $[-\pi, \pi]$,

$$\|k e_n\|^2 = \frac{1}{2\pi} \int_{-\pi}^{\pi} |k(t)|^2 \, dt \not\to 0,$$

as $n \to \infty$, which, by Exercise 6.7, shows that T_k cannot be compact.

6.11 (a) Suppose that the sequence $\{\alpha_n\}$ is bounded, that is, there is a number M such that $|\alpha_n| \leq M$ for all $n \in \mathbb{N}$. Then for any $x \in \mathcal{H}$ we have

$$\|Tx\|^2 = \sum_{n=1}^{\infty} |\alpha_n|^2 |(x, e_n)|^2 \leq M^2 \sum_{n=1}^{\infty} |(x, e_n)|^2 \leq M^2 \|x\|^2$$

(by Lemma 3.41), and so T is bounded. On the other hand, if the sequence $\{\alpha_n\}$ is not bounded then for any $M > 0$ there is an integer $k(M)$ such that $|\alpha_{k(M)}| \geq M$. Then the element $e_{k(M)}$ is a unit vector and, by definition, $T e_{k(M)} = \alpha_{k(M)} f_{k(M)}$, so $\|T e_{k(M)}\| = |\alpha_{k(M)}| \geq M$. Thus T cannot be bounded.

(b) Now suppose that $\lim_{n\to\infty} \alpha_n = 0$. For each $k = 1, 2, \ldots$, define the operator $T_k : \mathcal{H} \to \mathcal{H}$ by

$$T_k x = \sum_{n=1}^{k} \alpha_n (x, e_n) f_n.$$

The operators T_k are bounded, linear and have finite rank, so are compact. By a similar argument to that in the proof of part (c) of Theorem 6.16, given in Exercise 6.8, we can show that $\|T_k - T\| \to 0$, so T is compact by Corollary 6.10.

Now suppose that the sequence $\{\alpha_n\}$ is bounded but does not tend to zero (if $\{\alpha_n\}$ is unbounded then by the first part of the exercise T is not bounded and so cannot be compact, by Theorem 6.2). Then there is a number $\epsilon > 0$ and a sequence $n(r)$, $r = 1, 2, \ldots$, such that $|\alpha_{n(r)}| \geq \epsilon$ for all r. Now, for any r, $s \in \mathbb{N}$, with $r \neq s$,

$$\begin{aligned}\|Te_{n(r)} - Te_{n(s)}\|^2 &= \|\alpha_{n(r)}f_{n(r)} - \alpha_{n(s)}f_{n(s)}\|^2 \\ &= |\alpha_{n(r)}|^2 + |\alpha_{n(s)}|^2 \geq 2\epsilon^2.\end{aligned}$$

Thus no subsequence of the sequence $\{Te_{n(r)}\}$ can be Cauchy, so no subsequence can converge. Hence T cannot be compact.

(c) By the definition of T, we have $Te_m = \alpha_m f_m$, for any $m \in \mathbb{N}$, so by Theorem 3.47

$$\|Te_m\|^2 = \sum_{n=1}^{\infty} |(Te_m, f_n)|^2 = \sum_{n=1}^{\infty} |\alpha_n(f_m, f_n)|^2 = |\alpha_m|^2,$$

so

$$\sum_{m=1}^{\infty} \|Te_m\|^2 = \sum_{m=1}^{\infty} |\alpha_m|^2,$$

and the required result follows immediately.

(d) Clearly $f_n \in \operatorname{Im} T$ for any integer n with $\alpha_n \neq 0$, so the result follows from the linear independence of the set $\{f_n\}$.

6.12 (a) If $y \in \operatorname{Im} T$ then there exists x such that $y = Tx$, so y has the given form with $\xi_n = (x, e_n)$, for $n \in \mathbb{N}$, and by Lemma 3.41, $\{\xi_n\} \in \ell^2$. On the other hand, if y has the given form with $\{\xi_n\} \in \ell^2$ then by Theorem 3.42, we can define a vector x by $x = \sum_{n=1}^{\infty} \xi_n e_n$, and it is clear from the definition of T that $y = Tx$.

Next, suppose that infinitely many of the numbers α_n are non-zero and $\lim_{n\to\infty} \alpha_n = 0$. Choose a sequence $\{\xi_n\}$ such that $\{\alpha_n \xi_n\} \in \ell^2$, but $\{\xi_n\} \notin \ell^2$ (see below), and define the vector y by

$$y = \sum_{n=1}^{\infty} \alpha_n \xi_n e_n = \lim_{k\to\infty} \sum_{n=1}^{k} \alpha_n \xi_n e_n.$$

Since the finite sums in this formula belong to $\operatorname{Im} T$ we see that $y \in \overline{\operatorname{Im} T}$. However, by the first part of the question we see that $y \notin \operatorname{Im} T$, hence $\operatorname{Im} T$ is not closed.

[You might worry about whether it is possible to choose a sequence $\{\xi_n\}$ with the above properties. One way of constructing such a sequence is as follows. For each integer $r \geq 1$, choose $n(r)$ ($> n(r-1)$ when $r > 1$) such that $\alpha_{n(r)} \leq r^{-1/2}$ (this is possible since $\alpha_n \to 0$) and let $\xi_{n(r)} = r^{-1/2}$. For any n not in the sequence $\{n(r)\}$ let $\xi_n = 0$. Then $\sum_{n=1}^{\infty} |\alpha_n \xi_n|^2 = \sum_{r=1}^{\infty} |\alpha_{n(r)} \xi_{n(r)}|^2 \leq \sum_{r=1}^{\infty} r^{-2} < \infty$, while $\sum_{n=1}^{\infty} |\xi_n|^2 = \sum_{r=1}^{\infty} r^{-1} = \infty$.]

(b) For any x, $y \in \mathcal{H}$,

$$(x, T^*y) = (Tx, y) = \sum_{n=1}^{\infty} \alpha_n (x, e_n)(f_n, y) = \sum_{n=1}^{\infty} (x, \overline{\alpha}_n (y, f_n) e_n),$$

(by the definition of the adjoint) which shows that T^* has the form given in the question. Next, from Corollaries 3.35 and 3.36, and Lemma 5.11 we have,

$$\overline{\operatorname{Im} T} = (\operatorname{Ker} T^*)^{\perp} = \mathcal{H} \iff \operatorname{Ker} T^* = \{0\},$$

since $\operatorname{Ker} T^*$ is closed, which proves the second result. It follows from this that to prove the third result it is sufficient to find sequences $\{\alpha_n\}$, $\{e_n\}$, $\{f_n\}$, such that T is compact and $\operatorname{Ker} T^* = \{0\}$. Since we are now supposing that $\mathcal{H}$ is separable we may let $\{e_n\}$ be an orthonormal basis for $\mathcal{H}$ and let $\{f_n\} = \{e_n\}$, and choose $\{\alpha_n\}$ such that $\alpha_n \neq 0$ for all n and $\lim_{n \to \infty} \alpha_n = 0$. Then by part (b) of Exercise 6.11, T is compact. Now suppose that $T^*x = 0$. Then $(T^*x, e_k) = 0$, for each $k \in \mathbb{N}$, and

$$0 = (T^*x, e_k) = \sum_{n=1}^{\infty} \overline{\alpha}_n (x, e_n)(e_n, e_k) = \overline{\alpha}_k (x, e_k).$$

Since, $\alpha_k \neq 0$, for all k, this shows that $(x, e_k) = 0$. Hence, since $\{e_n\}$ is a basis, this shows that $x = 0$, and so $\operatorname{Ker} T^* = \{0\}$, which proves the result.

(c) This follows immediately from the formulae for Tx and T^*x, since $\alpha_n = \overline{\alpha}_n$, for $n \in \mathbb{N}$.

6.13 If $\mathcal{H}$ is not separable then it is not possible to find a compact operator on $\mathcal{H}$ with dense range since, by Theorem 6.8, $\overline{\operatorname{Im} T}$ is separable.

6.14 (a) Suppose that A is relatively compact and $\{a_n\} \subset A$. Then $\{a_n\} \subset \overline{A}$ and $\overline{A}$ is compact, so $\{a_n\}$ must have a convergent subsequence (converging to an element of $\overline{A} \subset M$).

Now suppose that any sequence $\{a_n\}$ in A has a subsequence which converges (in M). We must show that $\overline{A}$ is compact, so suppose that $\{x_n\}$ is a sequence in $\overline{A}$. Then for each $n \in \mathbb{N}$ there exists $a_n \in A$ such that $d(a_n, x_n) < n^{-1}$. By assumption, the sequence $\{a_n\}$ has a subsequence $\{a_{n(r)}\}$ which converges to a point $y \in M$ say. It can easily be shown that the corresponding subsequence $\{x_{n(r)}\}$ must also converge to y, and since $\overline{A}$ is closed we have $y \in \overline{A}$. This shows that $\overline{A}$ is compact.

(b) Let $x \in \overline{A}$ and $\epsilon > 0$. By the definition of closure, there exists $a \in A$ such that $d(x,a) < \epsilon/2$, and by the definition of density, there exists $b \in B$ such that $d(a,b) < \epsilon/2$. Hence, $d(x,b) < \epsilon$, so B is dense in $\overline{A}$.

(c) We may suppose that $A \neq \emptyset$, since otherwise the result is trivial. Suppose that the property does not hold for some integer r_0. Choose a point $a_1 \in A$. Since the set $\{a_1\}$ is finite it follows from our supposition about r_0 that there is a point $a_2 \in A$ such that $d(a_2,a_1) \geq r_0^{-1}$. Similarly, by induction, we can construct a sequence $\{a_n\}$ in A with the property that for any integers $m,\, n \geq 1$, $d(a_m,a_n) \geq r_0^{-1}$. However, this property implies that no subsequence of the sequence $\{a_n\}$ can be a Cauchy sequence, and so no subsequence can converge. But this contradicts the compactness of A, so the above supposition must be false and a suitable set B_r must exist for each r.

(d) Let $B = \bigcup_{r=1}^{\infty} B_r$, where the sets B_r are those constructed in part (c). The set B is countable and dense in A (for any $a \in A$ and any $\epsilon > 0$ there exists $r > \epsilon^{-1}$ and $b \in B_r$ with $d(a,b) < r^{-1} < \epsilon$), so A is separable.

6.15 We follow the proofs of Theorems 3.40 and 3.52. Let $\{w_n\}$ be a countable, dense sequence in X. Clearly, $\overline{\mathrm{Sp}}\,\{w_n\} = X$. Let $\{y_n\}$ be the subsequence obtained by omitting every member of the sequence $\{w_n\}$ which is a linear combination of the preceding members. By construction, the sequence $\{y_n\}$ is linearly independent. Also, any finite linear combination of elements of $\{w_n\}$ is a finite linear combination of elements of $\{y_n\}$, so $\mathrm{Sp}\,\{w_n\} = \mathrm{Sp}\,\{y_n\}$, and hence $\overline{\mathrm{Sp}}\,\{y_n\} = X$ (by Lemma 2.24). For each $k \in \mathbb{N}$, let $U_k = \mathrm{Sp}\,\{y_1,\ldots,y_k\}$. We will construct a sequence of vectors $\{x_n\}$ inductively. Let $x_1 = y_1/\|y_1\|$. Now suppose that for some $k \in \mathbb{N}$ we have constructed a set of unit vectors $\{x_1,\ldots,x_k\}$, with k elements, which has the properties: $\mathrm{Sp}\,\{x_1,\ldots,x_k\} = U_k$, and if $m \neq n$ then $\|x_m - x_n\| \geq \alpha$. By applying Theorem 2.25 to the spaces U_k and U_{k+1} we see that there exists a unit vector $x_{k+1} \in U_{k+1}$ such that

$$\|x_{k+1} - y\| \geq \alpha, \quad y \in U_k.$$

In particular, this holds for each vector x_m, $m = 1,\ldots,k$. Thus the set $\{x_1,\ldots,x_k,x_{k+1}\}$, with $k+1$ elements, also has the above properties. Since X is infinite-dimensional this inductive process can be continued indefinitely to yield a sequence $\{x_n\}$ with the above properties. Thus we have $\overline{\mathrm{Sp}}\,\{x_n\} = \overline{\mathrm{Sp}}\,\{y_n\} = \overline{\mathrm{Sp}}\,\{w_n\} = X$ (since the sequence $\{w_n\}$ is dense in X), which completes the proof.

6.16 Since $\mathcal{H}$ is not separable it follows from Theorem 6.8 that $\overline{\mathrm{Im}\,T} \neq \mathcal{H}$, so $\mathrm{Ker}\,T = \overline{\mathrm{Im}\,T}^{\perp} \neq \{0\}$ (see Exercise 3.19). Thus there exists $e \neq 0$ such that $Te = 0$, that is, e is an eigenvector of T with eigenvalue 0.

6.17 The operator S has the form discussed in Exercise 6.11 (with $\{\widetilde{e}_n\}$ the standard orthonormal basis in ℓ^2, and for each $n \geq 1$, $f_n = \widetilde{e}_{n+1}$, $\alpha_n = 1/n$), so it follows from the results there that S is compact. The proof that T is compact is similar.

The proof that $\sigma_{\mathrm{p}}(S) = \emptyset$ is similar to the solution of Example 5.35. By Theorems 6.18 and 6.25, $\sigma(S) = \sigma_{\mathrm{p}}(S) \cup \{0\}$, so $\sigma(S) = \{0\}$.

Next, $0 \in \sigma_{\mathrm{p}}(T)$ since $T\widetilde{e}_1 = 0$. Now suppose that $\lambda \neq 0$ is an eigenvalue of T with corresponding eigenvector $a \in \ell^2$. Then from the equation $Ta = \lambda a$ we can easily show that $a_n = (n-1)!\lambda^n a_1$, $n \geq 1$, which does not give an element of ℓ^2 unless $a_1 = 0$, in which case we have $a = 0$, which is a contradiction. Thus $\lambda \neq 0$ is an not eigenvalue of T, and so, again by Theorems 6.18 and 6.25, we have $\sigma(T) = \sigma_{\mathrm{p}}(T) = \{0\}$.

To show that $\operatorname{Im} S$ is not dense in ℓ^2 we note that any element $y \in \operatorname{Im} S$ has $y_1 = 0$, so $\|\widetilde{e}_1 - y\| \geq 1$, and hence $\operatorname{Im} S$ cannot be dense in ℓ^2. To show that $\operatorname{Im} T$ is dense in ℓ^2 consider a general element $y \in \ell^2$ and an arbitrary $\epsilon > 0$. Letting $y^k = (y_1, \ldots, y_k, 0, 0, \ldots)$, there exists $k \geq 1$ such that $\|y - y^k\| < \epsilon$. Now, defining

$$x^k = (0, y_1, 2y_2, \ldots, ky_k, 0, 0, \ldots),$$

we see that $y^k = Tx^k$, so $y^k \in \operatorname{Im} T$, and hence $\operatorname{Im} T$ is dense in ℓ^2.

6.18 Suppose that (u, v), $(x, y) \in X \times Y$. Then

$$\begin{aligned} M[(u,v) + (x,y)] &= M(u+x, v+y) \\ &= (A(u+x) + B(v+y), C(u+x) + D(v+y)) \\ &= (Au + Bv, Cu + Dv) + (Ax + By, Cx + Dy) \\ &= M(u,v) + M(x,y), \end{aligned}$$

using the linearity of A, B, C, D. Similarly, $M(\alpha(x,y)) = \alpha M(x,y)$, for $\alpha \in \mathbb{C}$, so $M \in L(X \times Y)$. Now, by the definition of the norm on $X \times Y$,

$$\begin{aligned} \|M(x,y)\| &= \|Ax + By\| + \|Cx + Dy\| \\ &\leq \|Ax\| + \|By\| + \|Cx\| + \|Dy\| \\ &\leq (\|A\| + \|C\|)\|x\| + (\|B\| + \|D\|)\|y\| \\ &= K\|(x,y)\|, \end{aligned}$$

where $K = \max\{\|A\| + \|C\|, \|B\| + \|D\|\}$, which shows that $M \in B(X \times Y)$.

Next, if C^{-1} exists then by matrix multiplication we see that

$$M_1 \begin{bmatrix} C^{-1} & -C^{-1}D \\ 0 & I_V \end{bmatrix} = \begin{bmatrix} I_U & 0 \\ 0 & I_V \end{bmatrix} = I, \quad \begin{bmatrix} C^{-1} & -C^{-1}D \\ 0 & I_V \end{bmatrix} M_1 = I$$

(these matrix multiplications are valid so long as we keep the operator compositions ('multiplications') in the correct order). This, together with a similar calculation for M_2, shows that M_1 and M_2 are invertible, with inverses

$$M_1^{-1} = \begin{bmatrix} C^{-1} & -C^{-1}D \\ 0 & I_V \end{bmatrix}, \quad M_2^{-1} = \begin{bmatrix} C^{-1} & 0 \\ -EC^{-1} & I_V \end{bmatrix}.$$

The second result follows from

$$M_1 \begin{bmatrix} x \\ y \end{bmatrix} = \begin{bmatrix} 0 \\ 0 \end{bmatrix} \iff Cx + Dy = 0 \text{ and } y = 0 \iff x \in \operatorname{Ker} C \text{ and } y = 0.$$

Similarly,

$$M_2 \begin{bmatrix} x \\ y \end{bmatrix} = \begin{bmatrix} 0 \\ 0 \end{bmatrix} \iff Cx = 0 \text{ and } Ex + y = 0,$$

so $\operatorname{Ker} M_2 = \{(x, y) : x \in \operatorname{Ker} C \text{ and } y = -Ex\}$.

6.19 By the definition of the inner product on $\mathcal{M} \times \mathcal{N}$,

$$\begin{aligned} (M(u, v), (x, y)) &= ((Au + Bv, Cu + Dv), (x, y)) \\ &= (Au + Bv, x) + (Cu + Dv, y) \\ &= (u, A^*x) + (v, B^*x) + (u, C^*y) + (v, D^*y) \\ &= ((u, v), (A^*x + C^*y, B^*x + D^*y)) \end{aligned}$$

from which the result follows.

6.20 For all u, $w \in \mathcal{M}$,

$$\begin{aligned} (Cu, w) &= ((T - \lambda I)u, w) = (u, (T^* - \overline{\lambda} I)w) = (P_{\mathcal{M}}u, (T^* - \overline{\lambda} I)w) \\ &= (u, P_{\mathcal{M}}(T^* - \overline{\lambda} I)w), \end{aligned}$$

which, by the definition of the adjoint, proves that $C^* = P_{\mathcal{M}}(T^* - \overline{\lambda} I)_{\mathcal{M}}$. Note that $(T^* - \overline{\lambda} I)w$ may not belong to $\mathcal{M}$, so the projection $P_{\mathcal{M}}$ is needed in the construction of C^* to obtain an operator from $\mathcal{M}$ into $\mathcal{M}$.

6.21 (a) If S is invertible then $B^{-1}S^{-1}$ is a bounded inverse for T, so T is invertible. Also, $S = TB^{-1}$, so a similar argument shows that if T is invertible then S is invertible.

(b) $x \in \operatorname{Ker} T \iff Bx \in \operatorname{Ker} S$, so the result follows from the invertibility of B.

6.22 If $\sigma(T)$ is not a finite set then by Theorem 6.25 the operator T has infinitely many distinct, non-zero eigenvalues λ_n, $n = 1, 2, \dots$, and for each n there is a corresponding non-zero eigenvector e_n. By Lemma 1.14 the set $E = \{e_n : n \in \mathbb{N}\}$ is linearly independent, and since $e_n = \lambda_n^{-1} T e_n \in \operatorname{Im} T$, we have $E \subset \operatorname{Im} T$. Thus $r(T) = \infty$.

6.23 Suppose that $v \in (\text{Ker}\,(T - \lambda I))^\perp$ is another solution of (6.5). Then by subtraction we obtain $(T - \lambda I)(u_0 - v) = 0$ which implies that

$$u_0 - v \in \text{Ker}\,(T - \lambda I) \cap (\text{Ker}\,(T - \lambda I))^\perp,$$

and so $u_0 - v = 0$. Thus u_0 is the unique solution in the subspace $(\text{Ker}\,(T - \lambda I))^\perp$. Hence the function $S_\lambda : \text{Im}\,(T - \lambda I) \to \text{Ker}\,(T - \lambda I)^\perp$ constructed in Theorem 6.29 is well defined, and we may now use this notation.

Next, multiplying (6.27) by $\alpha \in \mathbb{C}$ we obtain $(T - \lambda I)(\alpha S_\lambda(p)) = \alpha p$, so that $\alpha S_\lambda(p)$ is the unique solution of (6.27) in $(\text{Ker}\,(T - \lambda I))^\perp$ when the right hand side is αp, so we must have $\alpha S_\lambda(p) = S_\lambda(\alpha p)$. Similarly, by considering (6.27), and the same equation but with $q \in \text{Im}\,(T - \lambda I)$ on the right hand side, we see that $S_\lambda(p+q) = S_\lambda(p) + S_\lambda(q)$. This shows that S_λ is linear.

6.24 Since $\mathcal{N}$ is invariant under T we have $T_\mathcal{N} \in L(\mathcal{N})$. For any bounded sequence $\{x_n\}$ in $\mathcal{N}$ we have $T_\mathcal{N} x_n = T x_n$ for all $n \in \mathbb{N}$, so the sequence $\{T_\mathcal{N} x_n\}$ has a convergent subsequence, since T is compact. For any $x, y \in \mathcal{N}$ we have $(T_\mathcal{N} x, y) = (Tx, y) = (x, Ty) = (x, T_\mathcal{N} y)$, so $T_\mathcal{N}$ is self-adjoint.

6.25 Since $\mathcal{H}$ is infinite-dimensional there exists an orthonormal sequence $\{e_n\}$ in $\mathcal{H}$. Define the operator $T : \mathcal{H} \to \mathcal{H}$ by

$$Tx = \sum_{n=1}^{\infty} \lambda_n (x, e_n) e_n.$$

By Exercises 6.11 and 6.12, this operator is compact and self-adjoint (since the numbers λ_n are real). Also, $Te_n = \lambda_n e_n$, for each $n \in \mathbb{N}$, so the set of non-zero eigenvalues of T contains the set $\{\lambda_n\}$. Now suppose that $\lambda \neq 0$ is an eigenvalue of T, with eigenvector $e \neq 0$, but $\lambda \notin \{\lambda_n\}$. Then by Theorem 6.33, $(e, e_n) = 0$, $n \in \mathbb{N}$, so by part (a) of Theorem 3.47, $e = 0$. But this is a contradiction, so the set of non-zero eigenvalues of T must be exactly $\{\lambda_n\}$.

6.26 Since $\lambda_n > 0$, for each $1 \le n \le r(S)$, the numbers $\sqrt{\lambda_n}$ are real and strictly positive. Thus, by Exercises 6.11 and 6.12, $\widetilde{R}$ is compact and self-adjoint. Also,

$$(\widetilde{R}x, x) = \sum_{n=1}^{r(S)} \sqrt{\lambda_n} (x, e_n)(e_n, x) = \sum_{n=1}^{r(S)} \sqrt{\lambda_n} |(x, e_n)|^2 \ge 0,$$

so $\widetilde{R}$ is positive. Finally,

$$\begin{aligned}\widetilde{R}^2 x = \widetilde{R}\Big(\sum_{n=1}^{r(S)} \sqrt{\lambda_n}(x, e_n)e_n\Big) &= \sum_{n=1}^{r(S)} \sqrt{\lambda_n}(x, e_n)\widetilde{R}e_n \\ &= \sum_{n=1}^{r(S)} (\sqrt{\lambda_n})^2 (x, e_n)e_n = Sx,\end{aligned}$$

using the representation of Sx in Theorem 6.34, and the formula $\widetilde{R}e_n = \sqrt{\lambda_n}e_n$ (which follows immediately from the definition of $\widetilde{R}$).

Clearly, we can define other square root operators by changing $\sqrt{\lambda_n}$ to $-\sqrt{\lambda_n}$ when n belongs to various subsets of the set $\{j : 1 \le j \le r(S)\}$. These square root operators will not be positive, so this result does not conflict with the uniqueness result in Theorem 5.58.

6.27 By definition, for each $n = 1, \ldots, r(S)$,

$$S^* S e_n = \mu_n^2 e_n.$$

Also, to satisfy the first equation in (6.17), we define

$$f_n = \frac{1}{\mu_n} S e_n.$$

Combining these formulae we obtain

$$S^* f_n = \mu_n e_n,$$

which is the second equation in (6.17). To see that the set $\{f_n\}_{n=1}^{r(S)}$ is orthonormal we observe that if $1 \le m,\ n \le r(S)$ then

$$\begin{aligned}(f_m, f_n) &= \frac{1}{\mu_m \mu_n}(Se_m, Se_n) = \frac{1}{\mu_m \mu_n}(S^* S e_m, e_n) \\ &= \frac{\mu_m^2}{\mu_m \mu_n}(e_m, e_n),\end{aligned}$$

so the orthonormality of the set $\{f_n\}_{n=1}^{r(S)}$ follows from that of the set $\{e_n\}_{n=1}^{r(S)}$. Now, any $x \in \mathcal{H}$ has an orthogonal decomposition $x = u + v$, with $u \in \overline{\operatorname{Im} S^* S}$, $v \in \operatorname{Ker} S^* S = \operatorname{Ker} S$, and, by Theorem 6.34, $\{e_n\}_{n=1}^{r(S)}$ is an orthonormal basis for $\overline{\operatorname{Im} S^* S}$, so

$$Sx = Su = S\Big(\sum_{n=1}^{r(S)} (u, e_n)e_n\Big) = \sum_{n=1}^{r(S)} \mu_n (x, e_n) f_n$$

(since $(u, e_n) = (x, e_n)$, for each $1 \le n \le r(S)$), which proves (6.18).

Finally, let $\mu > 0$ be a singular value of S, that is $\nu = \mu^2$ is an eigenvalue of $S^*S = S^2$. By Theorem 5.39, $\nu = \lambda^2$ for some $\lambda \in \sigma(S)$ so $\lambda \neq 0$ is an eigenvalue of S and $\mu = |\lambda|$.

6.28 No. If, for each integer $n \geq 1$, we write $\alpha_n = \mu_n e^{i\theta_n}$, with μ_n real and non-negative and define $g_n = e^{i\theta_n} f_n$, then the sequence $\{g_n\}$ is orthonormal and it can easily be seen that we obtain the same operator S if we repeat the constructions in Exercise 6.11 using the sequences $\{\mu_n\}$, $\{e_n\}$, $\{g_n\}$.

6.29 Following the constructions in Exercise 6.27,

$$S = \begin{bmatrix} 0 & 1 \\ 0 & 0 \end{bmatrix}, \quad S^* = \begin{bmatrix} 0 & 0 \\ 1 & 0 \end{bmatrix}, \quad S^*S = \begin{bmatrix} 0 & 0 \\ 0 & 1 \end{bmatrix},$$

so the only non-zero eigenvalue of S^*S is 1, with corresponding eigenvector $e = (0,1)$. Hence, $f = Se = (1,0)$, and for any $x = (x_1, x_2)$,

$$Sx = (x,e)f = x_2 \begin{bmatrix} 1 \\ 0 \end{bmatrix}.$$

6.30 Let $\{e_n\}_{n=1}^{r(S)}$ and $\{f_n\}_{n=1}^{r(S)}$ be the orthonormal sets found in Exercise 6.27. By (6.18) and Theorem 3.22 or Theorem 3.42,

$$\begin{aligned} \sum_{n=1}^{\infty} \|Sg_n\|^2 &= \sum_{n=1}^{\infty} \Big(\sum_{m=1}^{r(S)} |\mu_m|^2 |(g_n, e_m)|^2 \Big) \\ &= \sum_{m=1}^{r(S)} |\mu_m|^2 \Big(\sum_{n=1}^{\infty} |(e_m, g_n)|^2 \Big) = \sum_{m=1}^{r(S)} |\mu_m|^2, \end{aligned}$$

where the reordering of the summations is permissible because all the terms are real and non-negative, and $\sum_{n=1}^{\infty} |(e_m, g_n)|^2 = \|e_m\|^2 = 1$ (by Theorem 3.47).

6.31 Combining Exercises 6.12 and 6.27 we see that $\operatorname{Im} S$ is not closed if there are infinitely many non-zero numbers μ_n, and this is equivalent to $r(S) = \infty$.

Chapter 7

7.1 Taking the complex conjugate of equation (7.6) yields

$$(I - \mu K)\overline{u} = f$$

(using the assumption that k, μ and f are real-valued). Since the solution of (7.6) is unique, we have $u = \overline{u}$, that is, u is real-valued.

7.2 Rearranging (7.6) and using the given form of k yields

$$u = f + \mu K u = f + \mu \sum_{j=1}^{n} \alpha_j p_j,$$

where the (unknown) coefficients α_j, $j = 1, \ldots, n$, have the form

$$\alpha_j = \int_a^b q_j(t) u(t)\, dt.$$

This shows that if (7.6) has a solution u then it must have the form (7.11). Now, substituting the formula (7.11) into (7.6), and using the given definitions of the coefficients w_{ij} and β_j, yields

$$\begin{aligned} 0 &= \mu \sum_{j=1}^{n} \alpha_j p_j - \mu K \Big(f + \mu \sum_{j=1}^{n} \alpha_j p_j \Big) \\ &= \mu \sum_{j=1}^{n} (\alpha_j - \beta_j) p_j - \mu^2 \sum_{j=1}^{n} \sum_{i=1}^{n} \alpha_j w_{ij} p_i \\ &= \mu \sum_{i=1}^{n} \Big(\alpha_i - \beta_i - \mu \sum_{j=1}^{n} w_{ij} \alpha_i \Big) p_i. \end{aligned}$$

Thus (7.6) is equivalent to this equation, and since the set $\{p_1, \ldots, p_n\}$ is linearly independent, this equation is equivalent to the matrix equation (7.12). From this it follows that equation (7.6), with $f = 0$, has a non-zero solution u if and only if equation (7.12), with $\beta = 0$, has a non-zero solution α, so the sets of characteristic values are equal. The remaining results follow immediately.

7.3 For the given equation we have, in the notation of Exercise 7.2,

$$p_1(s) = s, \quad p_2(s) = 1, \quad q_1(t) = 1, \quad q_2(t) = t,$$

and so

$$W = \begin{bmatrix} \frac{1}{2} & 1 \\ \frac{1}{3} & \frac{1}{2} \end{bmatrix}, \quad \beta = \begin{bmatrix} \int_0^1 f(t)\, dt \\ \int_0^1 t f(t)\, dt \end{bmatrix}.$$

Hence the characteristic values are the roots of the equation

$$0 = \det(I - \mu W) = (1 - \tfrac{1}{2}\mu)^2 - \tfrac{1}{3}\mu^2,$$

which gives the first result. Now, putting $\mu = 2$, equation (7.12) becomes

$$\begin{bmatrix} 0 & -2 \\ -\frac{2}{3} & 0 \end{bmatrix} \begin{bmatrix} \alpha_1 \\ \alpha_2 \end{bmatrix} = \begin{bmatrix} \beta_1 \\ \beta_2 \end{bmatrix},$$

which, with (7.11), yields the solution

$$u(s) = f(s) - 3s \int_0^1 t f(t)\, dt - \int_0^1 f(t)\, dt.$$

7.4 Rearranging equation (7.6) yields

$$u = f + \mu K u. \tag{8.1}$$

By assumption, f is continuous, and by Lemma 7.1 the term Ku is continuous, so it follows immediately that u is continuous, that is, $u \in X$, which proves (a). Next, μ is a characteristic value of the operator $K_{\mathcal{H}}$ if and only if there is $0 \neq u \in \mathcal{H}$ such that $(I - \mu K_{\mathcal{H}})u = 0$. But by the result just proved, $u \in X$, so μ is a characteristic value of the operator K_X. Since $X \subset \mathcal{H}$, the converse assertion is trivial. This proves (b). Finally, from (7.2), (8.1) and the inequality mentioned in the hint (writing $\gamma = M(b-a)^{1/2}$),

$$\begin{aligned}\|u\|_X &\le \|f\|_X + |\mu|\|Ku\|_X \le \|f\|_X + |\mu|\gamma\|u\|_{\mathcal{H}} \\ &\le \|f\|_X + |\mu|\gamma C\|f\|_{\mathcal{H}} \le (1 + |\mu|\gamma C(b-a)^{1/2})\|f\|_X,\end{aligned}$$

which proves (c).

7.5 The integral equation has the form

$$\int_a^s u(t)\, dt = f(s), \quad s \in [a,b]. \tag{8.2}$$

If this equation has a solution $u \in C[a,b]$, then the left-hand side is differentiable, and differentiating with respect to s yields

$$u(s) = f'(s), \quad s \in [a,b]. \tag{8.3}$$

Since $u \in C[a,b]$, this implies that $f' \in C[a,b]$, so f must be in $C^1[a,b]$. Conversely, if $f \in C^1[a,b]$ then the formula (8.3) yields a solution of (8.2) which belongs to $C[a,b]$.

Next, for each integer $n \ge 1$ let $f_n(s) = \sin ns$, and let $u_n = n \cos ns$ be the corresponding solution of (8.2). Then $\|u_n\|_X = n = n\|f_n\|_X$ (when $n(b-a) > \pi$), so the solution does not depend continuously on f.

7.6 The proof follows the proof of Theorem 4.40. The main change required is to notice that here the series $\sum_{n=0}^{\infty} \|T^n\|_Y$ converges due to the quasi-nilpotency condition and the standard root test for convergence of real series (see Theorem 5.11 in [2]) while the proof of Theorem 4.40 used the convergence of the series $\sum_{n=0}^{\infty} \|T\|_Y^n$, which was due to the condition $\|T\|_Y < 1$ imposed there and the ratio test for real series.

Next, suppose that $0 \neq \lambda \in \mathbb{C}$. Then the operator $\lambda I - T$ can be written as $\lambda(I - \lambda^{-1}T)$, and this operator has an inverse given by $\lambda^{-1}(I - \lambda^{-1}T)^{-1}$, if the latter inverse operator exists. But this follows from the result just proved since, if T is quasi-nilpotent then αT is quasi-nilpotent for any $\alpha \in \mathbb{C}$

(since $\|(\alpha T)^n\|_Y^{1/n} = |\alpha|\|T^n\|_Y^{1/n} \to 0$ as $n \to \infty$). Thus, by definition, λ is not in the spectrum of T.

The proof that the Volterra integral operator K is quasi-nilpotent on the space X follows the proof of Lemma 7.13, using the norm $\|\cdot\|_X$ rather than the norm $\|\cdot\|_{\mathcal{H}}$. For instance, the first inequality in the proof of Lemma 7.13 now takes the form

$$|(Ku)(s)| \le \int_a^s |k(s,t)||u(t)|\,dt \le M(s-a)\|u\|_X.$$

The rest of the proof is similar. The final result follows immediately from the results just proved.

7.7 (a) Suppose that $u \in Y_i$ and $w \in X$ satisfy the relation $w = T_i u$. Then (7.16), (7.17) and $w = u''$ hold. Substituting these into equation (7.14) tranforms it into (7.18), and reversing this process tranforms (7.18) into (7.14). Thus (7.14) holds with this u if and only if (7.18) holds with this w.

(b) Suppose that $u \in Y_b$ and $w \in X$ satisfy the relation $w = T_b u$. Then (7.21) holds. Now, if u satisfies (7.20) then $w = f - qu$, and by substituting this into (7.21) we obtain (7.22). Conversely, if u satisfies (7.22) then we can rewrite this equation as $u = G_0(f - qu)$, which, by comparison with (7.21), shows that $u'' = w = f - qu$, and hence u satisfies (7.20).

7.8 We first consider the case $\lambda < 0$ and write $\nu = \sqrt{-\lambda} > 0$. The general solution of the differential equation is then $A \sin \nu s + B \cos \nu s$. Substituting this into the boundary values yields

$$B = 0, \qquad A \sin \nu\pi = 0.$$

Clearly, $A = 0$ will not yield a normalized eigenfunction, so the second condition becomes $\sin(\nu\pi) = 0$, and hence the negative eigenvalues are given by $\lambda_n = -n^2$, $n \in \mathbb{N}$ (negative integers n yield the same eigenvalues and eigenfunctions, so we need not include them). The corresponding normalized eigenfunctions are $e_n = (2/\pi)^{1/2} \sin ns$ (putting $A = (2/\pi)^{1/2}$).

Now suppose that $\lambda > 0$ and write $\nu = \sqrt{\lambda} > 0$. The general solution of the differential equation is now $Ae^{\nu s} + Be^{-\nu s}$. Substituting this into the boundary values and solving the resulting pair of equations in this case leads to the solution $A = B = 0$, for any $\nu > 0$, which is not compatible with a non-zero eigenfunction. Thus there are no eigenvalues in this case. It can be shown in a similar manner that $\lambda = 0$ is not an eigenvalue.

7.9 We first consider the case $\lambda > 0$ and write $\nu = \sqrt{\lambda} > 0$. The general solution of the homogeneous differential equation is then $Ae^{\nu s} + Be^{-\nu s}$. We now find the functions u_l, u_r, used in the construction of the Green's function in Theorem 7.24. From the initial conditions for u_l we obtain

$$A + B = 0, \ \nu A - \nu B = 1.$$

Hence,

$$A = \frac{1}{2\nu}, \ B = -\frac{1}{2\nu},$$

and a similar calculation for u_r leads to the functions

$$\begin{aligned} u_l(s) &= \frac{1}{2\nu}(e^{\nu s} - e^{-\nu s}) = \frac{1}{\nu}\sinh \nu s, \\ u_r(s) &= -\frac{1}{2\nu}(e^{\nu(\pi - s)} - e^{-\nu(\pi - s)}) = -\frac{1}{\nu}\sinh \nu(\pi - s). \end{aligned}$$

The constant ξ_0 is given by

$$\xi_0 = -u_r(0) = \frac{1}{\nu}\sinh \nu\pi.$$

Hence, by Theorem 7.24, the Green's function in this case is

$$g(\lambda, s, t) = \begin{cases} -\dfrac{\sinh \nu s \sinh \nu(\pi - t)}{\nu \sinh \nu\pi}, & \text{if } 0 \le s \le t \le \pi, \\ -\dfrac{\sinh \nu(\pi - s) \sinh \nu t}{\nu \sinh \nu\pi}, & \text{if } 0 \le t \le s \le \pi. \end{cases}$$

Similar calculations in the case $\lambda < 0$ (putting $\nu = \sqrt{-\lambda} > 0$) leads to the Green's function

$$g(\lambda, s, t) = \begin{cases} -\dfrac{\sin \nu s \sin \nu(\pi - t)}{\nu \sin \nu\pi}, & \text{if } \ 0 \le s \le t \le \pi, \\ -\dfrac{\sin \nu(\pi - s) \sin \nu t}{\nu \sin \nu\pi}, & \text{if } \ 0 \le t \le s \le \pi. \end{cases}$$

Clearly, the function $g(\lambda, s, t)$ is singular at the eigenvalues of the boundary value problem (that is, when $\sin \nu\pi = 0$ in the second case).

7.10 Using the hint,

$$\begin{aligned} \lambda(u, u) = (u'' + qu, u) &= \int_a^b u''(s)u(s)\, ds + (qu, u) \\ &= [u''(s)u(s)]_a^b - \int_a^b u'(s)u'(s)\, ds + (qu, u) \\ &= -(u', u') + (qu, u) \le \|q\|_X(u, u) \end{aligned}$$

(using the boundary conditions $u(a) = u(b) = 0$ in (7.27)). Hence the result follows, since $(u, u) \neq 0$.

7.11 (a) The proof of Theorem 3.54 constructs a suitable function as a trigonometric polynomial on the interval $[0, \pi]$. An even simpler proof would construct a suitable function here as an ordinary polynomial. The argument can easily be extended to an arbitrary interval $[a, b]$.

(b) Write p_δ as $p_\delta(s) = p_{1,\delta}(s-a) + p_{2,\delta}(s-a)^2 + p_{3,\delta}(s-a)^3$, for some constants $p_{1,\delta}$, $p_{2,\delta}$, $p_{3,\delta}$ (this cubic polynomial satisfies the required condition at a). To ensure that v_δ is C^2 at the point $a+\delta$ we require that the derivatives $p_\delta^{(i)}(a+\delta) = w^{(i)}(a+\delta)$, $i = 0, 1, 2$. These conditions comprise a set of three linear equations for the coefficients $p_{i,\delta}$ which can be solved (do it) to yield

$$\begin{aligned} p_{1,\delta} &= \frac{3}{\delta} w_0 - 2w_1 + \frac{1}{2}\delta w_2, \\ p_{2,\delta} &= -\frac{3}{\delta^2} w_0 + \frac{3}{\delta} w_1 - w_2, \\ p_{3,\delta} &= \frac{1}{\delta^3} w_0 - \frac{1}{\delta^2} w_1 + \frac{1}{2\delta} w_2 \end{aligned}$$

(writing $w_i = w^{(i)}(a+\delta)$, $i = 0, 1, 2$). Having found p_δ, we then define v_δ as described in the question.

(c) From the values of the coefficients found in part (b) we see that

$$|p_{1,\delta}| \le C_1\delta^{-1}, \quad |p_{2,\delta}| \le C_1\delta^{-2}, \quad |p_{3,\delta}| \le C_1\delta^{-3},$$

where $C_1 > 0$ is a constant which depends on w, but not on δ; similarly for C_2, C_3 below. Thus, by the construction of v_δ, we see that

$$\begin{aligned} \|w - v_\delta\|_{\mathcal{H}}^2 &\le \int_0^\delta |w|^2\,ds + \int_0^\delta |p_\delta|^2\,ds \\ &\le C_2\Big\{\,\delta + \int_0^\delta \big((\delta^{-1}s)^2 + (\delta^{-2}s^2)^2 + (\delta^{-3}s^3)^2\big)\,ds\,\Big\} \\ &\le C_3\delta, \end{aligned}$$

which proves part (c).

(d) Finally, consider arbitrary $z \in \mathcal{H}$ and $\epsilon > 0$. By part (a) there exists $w \in C^2[a,b]$ with $\|z - w\|_{\mathcal{H}} < \epsilon/3$; by parts (b) and (c) there exists $v \in C^2[a,b]$ with $v(a) = 0$ and $\|w - v\|_{\mathcal{H}} < \epsilon/3$; by a similar method near b we can construct $u \in C^2[a,b]$ with $u(a) = 0$, $u(b) = 0$ (that is, $u \in Y_b$), and $\|v - u\|_{\mathcal{H}} < \epsilon/3$. Combining these results proves that $\|z - u\|_{\mathcal{H}} < \epsilon$, and so proves that Y_b is dense in $\mathcal{H}$.

7.12 (a) It is clear from the definition of the Liouville transform that the transformed function $\widetilde{u}$ satisfies the boundary conditions in (7.33). Now, applying the chain rule to the formula $u(s) = p(s)^{-1/4}\widetilde{u}(t(s))$, we obtain

$$\begin{aligned}\frac{du}{ds} &= -\frac{1}{4}p^{-5/4}\frac{dp}{ds}\widetilde{u} + p^{-1/4}\frac{d\widetilde{u}}{dt}p^{-1/2}\\ &= -\frac{1}{4}p^{-5/4}\frac{dp}{ds}\widetilde{u} + p^{-3/4}\frac{d\widetilde{u}}{dt},\\ \frac{d}{ds}\left(p\frac{du}{ds}\right) &= \frac{d}{ds}\left(-\frac{1}{4}p^{-1/4}\frac{dp}{ds}\right)\widetilde{u} - \frac{1}{4}p^{-1/4}\frac{dp}{ds}\frac{d\widetilde{u}}{dt}p^{-1/2}\\ &\quad + \frac{1}{4}p^{-3/4}\frac{dp}{ds}\frac{d\widetilde{u}}{dt} + p^{1/4}\frac{d^2\widetilde{u}}{ds^2}p^{-1/2}\\ &= -\frac{1}{4}p^{-1/4}\left(-\frac{1}{4}p^{-1}\left(\frac{dp}{ds}\right)^2 + \frac{d^2p}{ds^2}\right)\widetilde{u} + p^{-1/4}\frac{d^2\widetilde{u}}{ds^2}\end{aligned}$$

(note that, to improve the readability, we have omitted the arguments s and t in these calculations, but it is important to keep track of them: throughout, u and p have argument s, while $\widetilde{u}$ has argument t). Substituting these formulae into (7.31) gives (7.33).

(b) By the definition of the change of variables (7.32) and the standard formula for changing variables in an integral we see that

$$\int_0^c \widetilde{u}(t)\overline{\widetilde{v}(t)}\,dt = \int_a^b u(s)\overline{v(s)}p(s)^{1/2}p(s)^{-1/2}\,ds = \int_a^b u(s)\overline{v(s)}\,ds.$$

(c) Using integration by parts and the boundary conditions we see that, for $u,\ v \in Y_b$,

$$\begin{aligned}(Tu, v) &= \int_a^b ((pu')'\overline{v} + qu\overline{v})\,ds = [pu'\overline{v}]_a^b + \int_a^b (-pu'\overline{v}\,' + qu\overline{v})\,ds\\ &= [-up\overline{v}\,']_a^b + \int_a^b (u(p\overline{v}\,')' + qu\overline{v})\,ds = (u, Tv)\end{aligned}$$

(again for readability we have omitted the argument s in these calculations).

Further Reading

In this book we have not discussed many applications of functional analysis. This is not because of a lack of such applications, but, conversely, because there are so many that their inclusion would have made this text far too long. Nevertheless, applications to other areas can provide a stimulus for the study of further developments of functional analysis. Mathematical areas in which functional analysis plays a major role include ordinary and partial differential equations, integral equations, complex analysis and numerical analysis. There are also many uses of functional analysis in more applied sciences. Most notable perhaps is quantum theory in physics, where functional analysis provides the very foundation of the subject.

Often, functional analysis provides a general framework and language which allows other subjects to be developed succinctly and effectively. In particular, many applications involve a linear structure of some kind and so lead to vector spaces and linear transformations on these spaces. When these vector spaces are finite-dimensional, standard linear algebra often plays a crucial role, whereas when the spaces are infinite-dimensional functional analysis is likely to be called upon. However, although we have only mentioned the linear theory, there is more to functional analysis than this. In fact, there is an extensive theory of non-linear functional analysis, which has many applications to inherently non-linear fields such as fluid dynamics and elasticity. Indeed, non-linear functional analysis, together with its applications, is a major topic of current research. Although we have not been able to touch on this, much of this theory depends crucially on a sound knowledge of the linear functional analysis that has been developed in this book.

There are a large number of functional analysis books available, many at a very advanced level. For the reader who wishes to explore some of these areas further we now mention some books which could provide a suitable starting point. References [8], [11], [12] and [16] discuss similar material to that in this book, and are written at about the same level (some of these assume different prerequisites to those of this book, for example, a knowledge of topology or Zorn's lemma). In addition, [8] contains several applications of functional analysis, in particular to ordinary and partial differential equations, and to numerical analysis. It also studies several topics in non-linear functional analysis.

For a more advanced treatment of general functional analysis, a reasonably wide-ranging text for which knowledge of the topics in this book would be a prerequisite is [14]. An alternative is [15]. Rather more specialized and advanced textbooks which discuss particular aspects of the theory are as follows:

- for further topics in the theory of Banach spaces see [6];
- for the theory of algebras of operators defined on Hilbert spaces see [9];
- for integral equations see [10];
- for partial differential equations see [13];
- for non-linear functional analysis see [17].

References

[1] T. S. Blyth and E. F. Robertson, *Basic Linear Algebra*, Springer, SUMS, 1998.

[2] J. C. Burkill, *A First Course in Mathematical Analysis*, CUP, Cambridge, 1962.

[3] J. C. Burkill and H. Burkill, *A Second Course in Mathematical Analysis*, CUP, Cambridge, 1970.

[4] M. Capiński and E. Kopp, *Measure, Integral and Probability*, Springer, SUMS, 1999.

[5] C. W. Curtis, *Linear Algebra, an Introductory Approach*, Springer, New York, 1984.

[6] J. Diestel, *Sequences and Series in Banach Spaces*, Springer, New York, 1984.

[7] P. R. Halmos, *Naive Set Theory*, Springer, New York, 1974.

[8] V. Hutson and J. S. Pym, *Applications of Functional Analysis and Operator Theory*, Academic Press, New York, 1980.

[9] R. V. Kadison and J. R. Ringrose, *Fundamentals of the Theory of Operator Algebras, Vol. I*, Academic Press, New York, 1983.

[10] R. Kress, *Linear Integral Equations*, Springer, New York, 1989.

[11] E. Kreyszig, *Introductory Functional Analysis with Applications*, Wiley, New York, 1978.

[12] J. D. Pryce, *Basic Methods of Linear Functional Analysis*, Hutchinson, London, 1973.

[13] M. Renardy and R. C. Rogers, *An Introduction to Partial Differential Equations*, Springer, New York, 1993.

[14] W. Rudin, *Functional Analysis*, McGraw-Hill, New York, 1973.

[15] A. E. Taylor and D. C. Lay, *Introduction to Functional Analysis*, Wiley, New York, 1980.

[16] N. Young, *An Introduction to Hilbert Space*, CUP, Cambridge, 1988.

[17] E. Zeidler, *Nonlinear Functional Analysis and its Applications, Vol. I*, Springer, New York, 1986.

Notation Index

Index